# Studies in Computational Intelligence

Volume 1134

The series "Studies in Computational Intelligence" (SCI) publishes new developments and advances in the various areas of computational intelligence—quickly and with a high quality. The intent is to cover the theory, applications, and design methods of computational intelligence, as embedded in the fields of engineering, computer science, physics and life sciences, as well as the methodologies behind them. The series contains monographs, lecture notes and edited volumes in computational intelligence spanning the areas of neural networks, connectionist systems, genetic algorithms, evolutionary computation, artificial intelligence, cellular automata, self-organizing systems, soft computing, fuzzy systems, and hybrid intelligent systems. Of particular value to both the contributors and the readership are the short publication timeframe and the world-wide distribution, which enable both wide and rapid dissemination of research output.

Indexed by SCOPUS, DBLP, WTI Frankfurt eG, zbMATH, SCImago.

All books published in the series are submitted for consideration in Web of Science.

Pedro Yobanis Piñero Pérez · Janusz Kacprzyk ·
Rafael Bello Pérez · Iliana Pérez Pupo
Editors

# Computational Intelligence in Engineering and Project Management

Springer

*Editors*
Pedro Yobanis Piñero Pérez
Artificial Intelligence for a Sustainable Development Group
IADES
Havana, Cuba

Rafael Bello Pérez
Central University of Las Villas
Santa Clara, Cuba

Janusz Kacprzyk
Systems Research Institute
Polish Academy of Sciences
Warsaw, Poland

Iliana Pérez Pupo
Artificial Intelligence for a Sustainable Development Group
IADES
Havana, Cuba

ISSN 1860-949X ISSN 1860-9503 (electronic)
Studies in Computational Intelligence
ISBN 978-3-031-50494-5 ISBN 978-3-031-50495-2 (eBook)
https://doi.org/10.1007/978-3-031-50495-2



This Springer imprint is published by the registered company Springer Nature Switzerland AG
The registered company address is: Gewerbestrasse 11, 6330 Cham, Switzerland

Paper in this product is recyclable.

# Preface

Project management is being transformed through the introduction of different artificial intelligence techniques. In each of the areas of knowledge in project management, new algorithms and techniques associated with computational intelligence are gradually introduced that facilitate the treatment of information uncertainty. Progress is gradually being made in digital transformation with an agile approach, which impacts the various areas of human knowledge that are managed by projects.

For example, in the construction sector, BIM technologies have become widespread and are gradually imposing themselves. In this way, the simulation in virtual environments of the complete execution of projects is facilitated. In this way, it is about mitigating the difficulties of traditional project management, where the following statistics are manifested. Around 34% of the resources invested during a project's execution are misused and wasted. Approximately, 95% of the information that is generated in the project management process is later underused in new projects. This situation not only occurs in construction projects; in the case of software projects, several sources identify that, in medium- and large-sized projects, approximately 50% need to be renegotiated and 19% are canceled.

In this context, the main difficulties are associated with deficiencies in the decision-making processes. In addition, deficiencies persist in the control and monitoring processes of the projects. In general, efforts to integrate project management technologies with computational intelligence and other intelligent data analysis techniques are still incipient. In this sense, many opportunities are identified for the introduction of techniques that facilitate the treatment of investment and agility in decision-making.

In this context, we describe Artificial Intelligence (AI) as the set of software and hardware techniques that simulate different human reasoning methods and the collective intelligence of natural ecosystems to solve complex problems where the algorithm for its solution has non-polynomial complexity, or there are no algorithms for its solution. Many authors use the terms computational intelligence and soft computing interchangeably to refer to the same set of techniques as a branch of artificial intelligence. In this sense, Computational Intelligence is presented as the branch of AI that encompasses various techniques aimed at simulating human tolerance in decision-making processes in environments with uncertainty and imprecision. Among the precursor techniques of computational intelligence are: evolutionary algorithms, artificial neural networks, fuzzy set theory and fuzzy systems. However, other areas such as the rough set, linguistic data summary, natural language processing, the conversational systems, fuzzy cognitive maps, collective intelligence, the neutrosophic theory and other fuzzy logic extensions are contributing to the application and extension of computational intelligence. This book presents a selection of papers with different experiences in the application of computational intelligence techniques to help decision-making in Project Management.

The book is organized into four parts according to the nature of the works presented.

Havana, Cuba — Pedro Yobanis Piñero Pérez
Warsaw, Poland — Janusz Kacprzyk
Santa Clara, Cuba — Rafael Bello Pérez
Havana, Cuba — Iliana Pérez Pupo

# Acknowledgments

Project management is being transformed through the introduction of different artificial intelligence techniques. In each of the areas of knowledge in project management, new algorithms and techniques associated with computational intelligence are gradually introduced that facilitate the treatment of information uncertainty. Progress is gradually being made in digital transformation with an agile approach, which impacts the various areas of human knowledge that are managed by projects. This book collects a group of experiences in the application of Computational Intelligence techniques in Project Management. The majority of papers propose new algorithms and techniques to help decision-making in different project management processes. In addition, different reviews and critical analysis works are presented that facilitate the analysis of trends in the development of computational intelligence and its evolution.

From the project management point of view, most of the work is presented with a strong focus on agile development. New algorithms are presented that combine different techniques to solve specific problems.

We would like to thank all the engineers, professors and researchers without whose efforts this book could not have been written. We are indebted to Dr. Tom Ditzinger and Mr. Holger Schaepe for their comments that enabled us to correct errors and advance satisfactorily in the development of this book. Thanks to them for their dedication and help to implement and finish this important publication project on time while maintaining the highest publication standards. We would also like to thank Saranya Sakkarapani, from Springer, for her help and support.

Pedro Yobanis Piñero Pérez
Janusz Kacprzyk
Rafael Bello Pérez
Iliana Pérez Pupo

# Contents

## Computational Intelligence in Project Planning and Monitoring



## Decision-Making in Project-Oriented Organizations Supported by Computational Intelligence's Techniques

# Critical Review of Computational Intelligence in Project Management

# Conversational Systems and Computational Intelligence, A Critical Analysis

**Yuniesky Orlando Vasconcelo Mir, Pedro Yobanis Piñero Pérez, Iliana Pérez Pupo, Luis Alvarado Acuña, and Rafael Bello Pérez**

**Abstract** The increase in research on conversational systems and their applications, constitutes the main motivation of this work. In this research, a critical analysis of the growth of smart chatbots and their combination with computational intelligence techniques is carried out. The analysis is oriented on three fundamental fronts. First, a review protocol is applied to identify the main schools and centers of knowledge. Then the conversational systems are characterized based on the level of inclusion of artificial intelligence techniques. Finally, the integration of conversational systems with different computational intelligence techniques is reviewed. The analysis identifies that there are many opportunities and lines open to research. In particular, the need to strengthen the application of neutrosophic theory and sets for the evaluation of conversations is identified. Also, the need to combine linguistic data summarization techniques and reinforcement learning is identified to improve training methods and reduce the computational cost of conversational systems responses.

**Keywords** Chatbots · Computational intelligence · Conversational systems · Linguistic data summarization · Neutrosophic sets

---

Y. O. V. Mir · P. Y. Piñero Pérez · I. P. Pupo (✉)
Artificial Intelligence for a Sustainable Development Group, IADES, Havana, Cuba
e-mail: iliperezpupo@gmail.com; iades@businessredmine.com

Y. O. V. Mir
e-mail: iades@businessredmine.com

P. Y. Piñero Pérez
e-mail: pppyob@gmail.com; iades@businessredmine.com

L. A. Acuña
Departamento de Ingeniería de La Construcción, Universidad Católica del Norte, Antofagasta, Chile
e-mail: lualvar@ucn.cl

R. B. Pérez
Centro de Investigación en Informática, Universidad Central Marta Abreu de Las Villas, Santa Clara, Cuba
e-mail: rbellop@uclv.edu.cu

 
P. Y. Piñero Pérez et al. (eds.), *Computational Intelligence in Engineering and Project Management*, Studies in Computational Intelligence 1134,
https://doi.org/10.1007/978-3-031-50495-2_1

## 1 Introduction

Conversational agents or chatbots are revolutionizing the way humans and computers interact. Recent advances of the artificial intelligence, the massive adoption of mobile devices, IoT, robotics, smart homes and Industry 4.0, have favored the resurgence of conversational agents as an alternative to graphical user interfaces (Graphic User Interfaces, GUI) that for decades have dominated Human-Computer Interactions (HCI) [1].

According to [1] a chatbot (conversational agent, dialogue system) is a computer system that operates as an interface between human users and a software application, using spoken or written natural language as the primary means of communication. In general, conversational systems have their origin in typical chatbots, initially created to simulate a conversation with human users based on questions and answers in a given domain.

There are several names that identify this type of system: natural language interfaces, conversational interfaces, dialogue systems, chatbots, conversational artificial agents, intelligent virtual agents, virtual assistants, among others. In order to facilitate the understanding of this paper, the terms Chatbot and conversational systems will be used interchangeably to refer to the same concept.

In the evolution of these technologies, there are several elements that clearly differentiate typical chatbot architectures from current conversational systems [2, 3]. The main differences in the architecture between dialog systems and chatbots are the natural language understanding component and the dialog manager [1, 4]. Some elements that make it possible to differentiate conversational systems lie in the services to which they are integrated and represented (databases, task or control modules), the domain of knowledge, the modalities (text, voice, avatars) [5] and the channels in which they are deployed.

There are, different chatbot experiences and works in education, government, business, e-commerce, healthcare, entertainment, and many more [6]. In the business field, chatbots have become very common because they reduce service costs and can manage many customers simultaneously. In addition, these systems offer convenient support, are more user-friendly and engaging than traditional user's services models, for example, searching for static content in frequently asked questions (FAQ) lists.

Several authors refer that most of the conversational systems do not maintain a fluent conversation as humans wish [1]. This can be influenced by the following elements:

- Conversational systems learn from the knowledge to which they have access. If the knowledge supplied to the conversational system is biased or incomplete, the system will respond incompletely.
- Learning models for different languages that guarantee high levels of acculturation is a line open to research.

- Conversational systems generally incorporate complex learning processes, and incorporate facilities for learning from interactions with human users. This interaction and learning processes can last several months of information exchange until the answers are refined.
- There is a human predisposition for interaction with conversational systems. 70% prefer to talk to a human rather than an AI system or chatbot [7], while 53% believe that intelligent systems make biased decisions.

In 2019, 2.8 trillion dollars were generated worldwide from commercial operations where conversational systems intervened. It is also estimated that in less than five years a growth of 50 times this figure is expected. In this context, the authors of the paper identify that there are many opportunities in the development of conversational systems.

In this work, the authors carry out a systematic review focused on the combination of conversational systems with computational intelligence techniques. In this way, it is intended to create platforms as services, for the rapid development of conversational systems. This research has the following specific objectives:

- Identify the main schools of knowledge in the development of conversational systems research.
- Characterize the architectural elements of conversational systems and their limitations.
- Identify the different approaches in the integration between conversational systems with computational intelligence techniques.
- Identify different strategies for evaluating the responses of conversational systems.
- Systematize knowledge and the evolution of conversational systems from an ethical perspective.

The work is structured in the following sections. The second section presents the protocol for the systematic review and the methodology to be applied. The third section shows the results of the descriptive analysis of the systematic review. In the fourth section, the results of the study are analyzed for each of the elements proposed as objectives of the systematic review. Finally, the conclusions and future work are presented.

## 2 Methodology Used for Trend Analysis

In this section, a brief analysis of the evolution and trends in the combinations between conversational systems and computational intelligence techniques is carried out.

## 2.1 Review Protocol Used in the Search

Systematic review techniques are applied [8] and the following protocol is applied:

Step 1. Design of a systematic review protocol that allows the construction of a referential theoretical framework associated with the subject.

a. Definition of the object of study, field of action and objectives.

   - Research object: Conversational systems
   - General objective: Identify opportunities for improvement in the processes of learning and exploitation of conversational systems.
   - Research field: learning conversational systems with other soft computing techniques.

b. Definition of a bibliographic manager: zotero.
c. Definition of academic information sources for the development of the review: "*Semantic Scholar*", "*Google Scholar*", "*Scopus*" and other academic metasearch engines based on open science.
d. Definition of key phrases for conducting searches:

   - "Conversational Systems" and "Linguistic Data Summarization"
   - "Conversational Systems" and "Linguistic Summaries of Data".
   - (chatbots OR "conversational system" OR "dialogue system") AND ("Linguistic summaries" OR "Linguistic Data Summarization" OR "Linguistic Summaries of Data")
   - (chatbots OR "conversational system" OR "dialogue system") AND ("Linguistic summaries" OR "Linguistic Data Summarization" OR "Linguistic Summaries of Data") AND evaluations
   - (chatbots OR "conversational system" OR "dialogue system") AND ("Linguistic summaries" OR "Linguistic Data Summarization" OR "Linguistic Summaries of Data") AND evaluations AND neutrosophic
   - (chatbots OR "conversational system" OR "dialogue system") AND evaluations AND neutrosophic
   - "Conversational Systems" and "Neutrosophic Sets"
   - "Conversational Systems" and "Computational intelligence"
   - "Conversational Systems" and "Reinforcement Learning"

Step 2. Definition of the goals of the general bibliometric analysis in the form of research questions and inclusion-exclusion criteria:

- How has the trend of publications per year been?
- Who are the main authors?
- What are the affiliations and countries of the main authors?
- How are the publications distributed considering the types of documents in: articles, books, theses and conference or congress reports?

Step 3. Definition of goals in the form of research questions associated with computational techniques:

- What are the trends in the integration of computational intelligence with conversational systems?
- What are the main characteristics of architectural models, tendencies in the construction of conversational systems?
- What are the trends in the learning mechanisms of conversational systems?
- What are the trends in the use of reinforcement learning techniques in conversational systems?
- How has the linking of linguistic data summarization techniques behaved?
- How do the different methods of evaluating the responses of conversational systems behave?

Step 4. Sort and filter posts into the following set of categories:

- Classics: refers to pioneering publications in conversational systems, where the fundamental principles that set guidelines in the theory are exposed.
- Extensions to theories: refers to publications that extend the theory proposed in the publications understood as classic, they do not set guidelines that significantly change the methods previously proposed, although they do develop contributions to knowledge.
- Application results: publications that focus on the use of existing theory in concrete, practical settings.
- Tertiary reviews: refers to articles reviewing trends and developments in the subject in question.

Step 5. Synthesize the main trends.

Step 6. In-depth analysis of the consulted bibliography of the established categories and characterization regarding the following elements:

- General descriptive analysis of trends and main areas of application.
- Predominant strategies in soft computing techniques and algorithms, such as linguistic data summarization, for learning conversational systems.
- Analysis of integration of learning techniques with reinforcement in the interactive learning of conversational systems.
- Characterization of architectural models of conversational systems.

Step 7. Application of statistical techniques for the analysis of the revision.

Step 8. Identify open lines of research.

## 2.2 *Resultados Preliminares Del Análisis De Las Tendencias*

Search is performed on the terms "Conversational systems" and "computational intelligence" [9]. A growing trend is identified in the publications as shown in Fig. 1.

Among the authors who have published the most in the last ten years on these topics, the following stand out: Augello, et al. [7]. While the institutions most prominent are European and Japanese universities see Fig. 2.

However, by country, it is identified that the United States reported the largest number of publications in the study. Although regions such as Europe and the Asia Pacific Region also group a considerable number of publications in the last 10 years, see Figs. 3 and 4.

67% of the articles have been published in conferences, while 19% have been published in journals with different levels of impact.

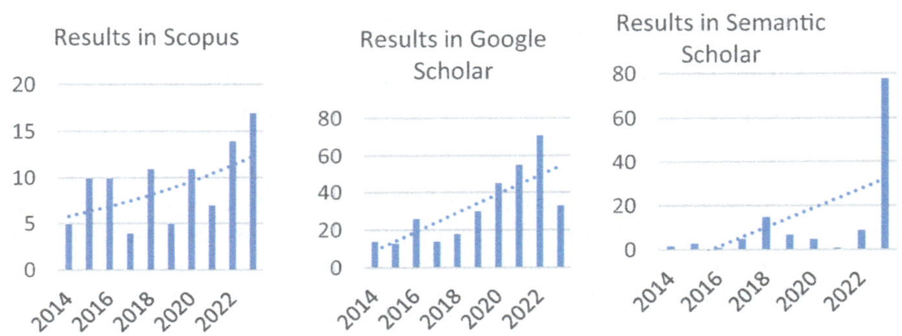


**Fig. 1** Trends in conversational systems and computational intelligence hybridity

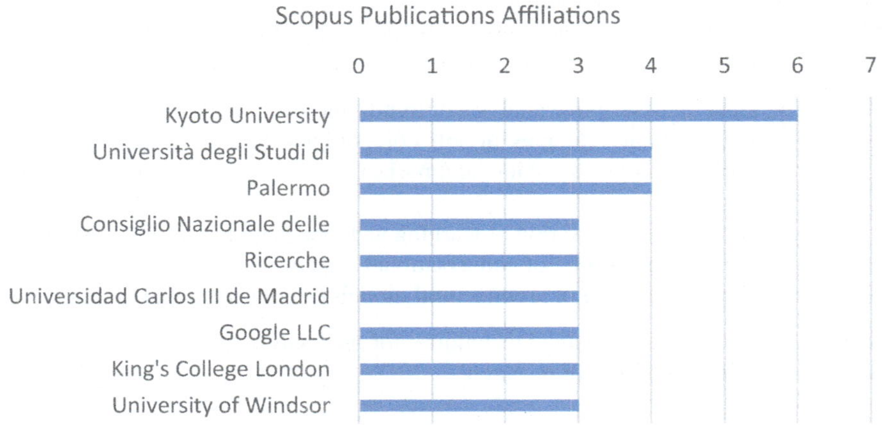


**Fig. 2** Affiliations with best results in conversational systems and computational intelligence hybridity, SCOPUS

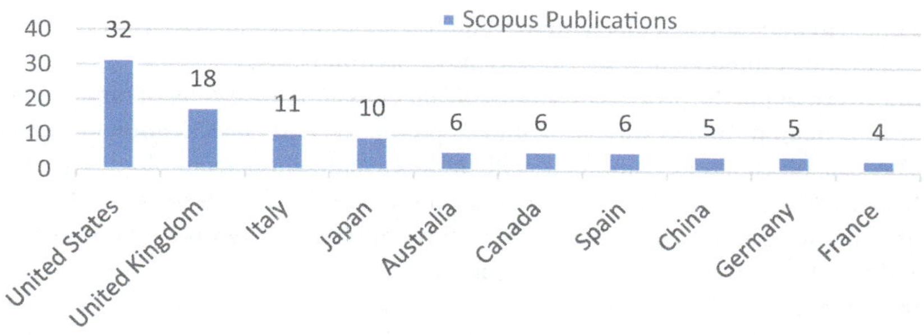


**Fig. 3** Results of Documents by country/territory using in 1993–2023, SCOPUS

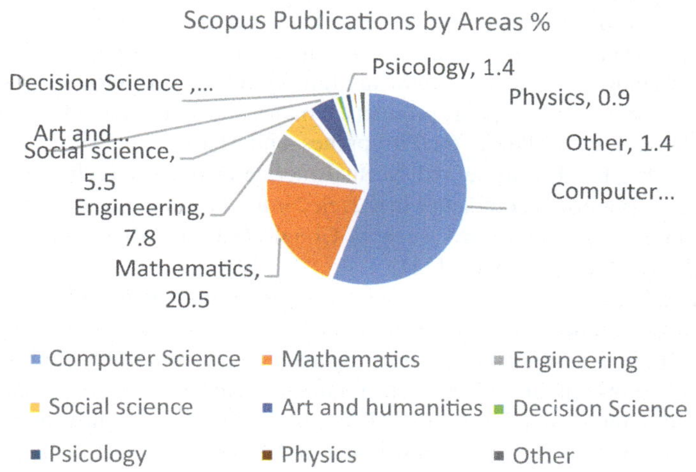


**Fig. 4** Results by area, SCOPUS

### *2.3 Results of the Analysis of the Applications of Conversational Systems*

Among the benefits and challenges in the use of conversational systems, the following stand out:

- Improved efficiency and humanization of work in scenarios associated with Sentiment analysis, Customer service, Fraud detection, Prediction and classification.
- Customer satisfaction and enhance availability: Users can be made to receive systems availability to answer questions 24 h a day, 7 days a week. It is said that 66% of customers expect companies to understand their needs. This element constitutes part of the remains of chatbots in the generation of empathic and specific responses [7].

- Time saving: a particularly important element in scenarios associated with electronic commerce and service-oriented environments.
- Improvement in the process of sentiment analysis and processing of large volumes of information. In this way, it facilitates prediction and classification processes.
- The use of chatbots combined with new virtualization technologies (Docker, Kubernates, etc.) allows scaling based on the real need for resources. Better use of computing resources can be achieved by mitigating high power consumption and providing a better customer experience.
- In general, conversational systems are desired to be capable of generating responses to customers in no more than 30 s [10]. This element constitutes a quality criterion.

Regarding applications, the Asia-Pacific region is the one with the largest number of applications of conversational systems. Different countries in the Asia-Pacific region use conversational systems as innovative solutions to stay in the competitive market landscape. These systems are rapidly introduced into the retail market for electronic commerce support products. Countries featured include China, India, Indonesia, Vietnam, Malaysia, the Philippines, and Thailand.

On the other hand, Europe and North America also show rapid growth, especially because important companies in the telecommunications sector are developing new technologies in this regard, such as Google (Alexa), Microsoft and OpenAI (ChatGPT), IBM, among others [11].

In general, the main areas of application are commercialization, marketing, banking, telecommunications, education, medical services [12]. However, there are sectors such as the architecture engineering and construction (AEC) industry where the use of chatbots remains lagging. The findings reveal that Conversational AI applications hold immense benefits for the AEC industry, but it is currently underexplored [13]. The authors of this paper identify that in the case of project management and engineering, the use of conversational systems will have a significant impact on the introduction of agile work methods, changing the way projects are managed. Another sector where these technologies are not yet widely exploited is in sports. The authors of this paper believe that the conversational style will have a significant impact in many sectors. In the case of sports, these technologies will change the way in which communications between coaches and athletes are established. They will also contribute to the introduction of good practices in physical culture and systematicity in sport.

Some works report surveys that show an increase in the introduction of conversational systems. It is identified that only 19% of the participating entities did not express interest in incorporating these techniques, while 62% already have inclusion plans and the rest have already incorporated chatbots into their management, see Fig. 5.

Some analyzes indicate that in scenarios where Smart Chatbots are installed, they intervene in the following way [14]:

- Approximately 37% of inquiries may be related to simple questions or emergency treatment.

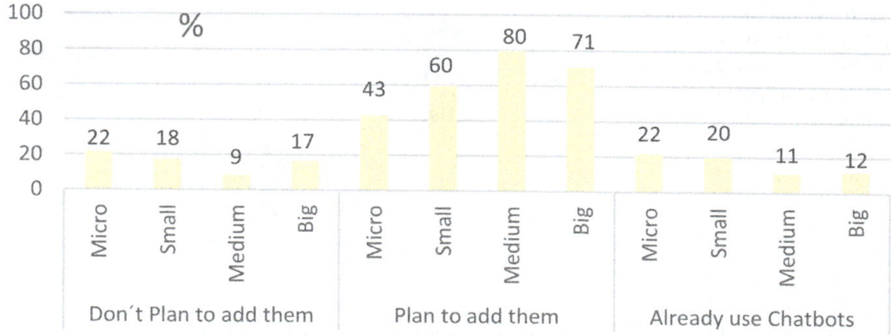


**Fig. 5** Intension de uso de los chatbots en diferentes tipos de empresas

- 34% of them are associated with the search for services or assistance.
- 35% of detailed search services for answers.
- Approximately 27% of the purchases of simple items and 13% of purchases in expensive items.
- 27% of the payment of the invoices and 22% of the inspirations in sales actions.
- They participate in 33% of reservations in hotels and restaurants.
- They are increasingly used in the computerization of call centers.
- Some 53% of respondents say that waiting too long for answers is the most frustrating part of interacting with businesses.
- If the alternative were to wait 15 min for a response, 62% of consumers would rather talk to a chatbot than a human agent. While 38% of users prefer to wait for the service representative.
- Approximately 4% of business owners who have introduced Smart Chatbots are dissatisfied. While 74% of the owners show satisfaction and 22% behave in a neutral way.

## 3 Results of the Analysis of the Architectural Models of Conversational Systems

The architectural models of the simplest conversational systems are based on FAQ models and rule-based inference models. These are frequently called Dumb Chatbots. These chatbots work with simple logic, generally supported by prescriptive insights and rules. When asked by the user, the system identifies the intention and, using the rules, issues a response.

While more complex models generally combine machine learning techniques and natural language processing [7]. Among the elements that characterize conversational systems are identified:

- Application domains: in this aspect they are classified as closed domain chatbots [15] and open domain chatbots such as the models proposed by ChatGPT.

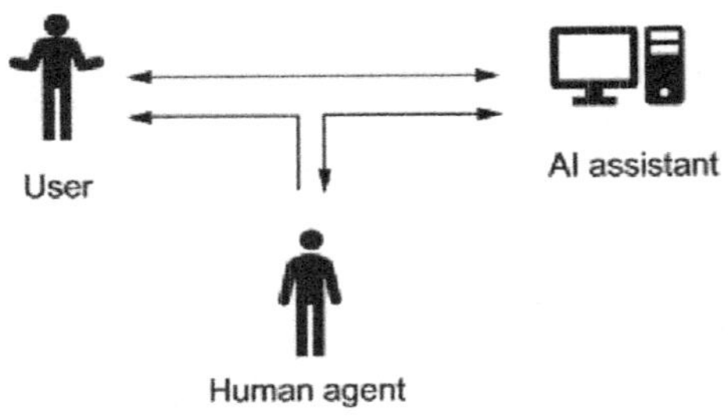


**Fig. 6** Model for assistance conversational system, combining IA assistant and human-agent specialized responses

- Concepts and forms of representation and storage of knowledge.
- Regarding the level of incorporation of artificial intelligence techniques in their management, they are classified as: Dumb Chatbots that do not incorporate artificial intelligence techniques and Smart Chatbots (AI chatbots) that combine different artificial intelligence techniques at different levels [16].
- Characterization regarding the level of intervention of human agents and artificial intelligence. In this sense, several levels are identified [17]: a basic level where there are conversational systems actively assisted by humans, a medium level and an advanced level where conversational systems with greater autonomy are located. In addition, hybrid approaches can be found in the use of conversational systems.

## *3.1 Basic Smart Conversational Characterization*

On the other hand, AI chatbots can be classified into three groups depending on the level of incorporation of artificial intelligence techniques [1]. The basic level is represented by systems, where users communicate with the conversational system directly, but the chatbot is assisted by a human agent, see Fig. 6.

Two fundamental processes in this smart chatbot type are explained in detail below. Knowledge representation and active learning are supported by the intervention of human agents.

## *3.2 Representation of Knowledge in Conversational Systems*

Regarding the structures and forms of knowledge representation, the following concepts are generally managed:

- Intensiones: representan de forma clara procesos o acciones que el usuario transmite.
- Intentions: they clearly represent processes or actions that the user transmits.

- Expressions: linguistic expressions that imply or identify the statement of an intention
- Responses: linguistic expressions that show responses associated with intentions
- Entities: they are nouns that represent the objects or entities that can participate in a dialogue.
- Synonyms: element that identifies different terms with the same meaning and that are important for the learning process of conversational systems. Although, they can also represent expressions that may have the same meaning or refer to the same entity.
- Protoforms or lookups: refers to prototypes of expressions that can help to systematize the way in which user questions are presented.
- Rules that relate intentions to possible responses.
- Scripts or Stories: elements that identify frequent dialogue sequences that can be used to predict dialogues and speed up response speed.
- Knowledge bases that allow the organization of information and indexing to guarantee reliable answers as quickly as possible.

All these elements constitute knowledge entities that must be stored or represented in some way. From the computational point of view, in this sense, the different forms of traditional knowledge representation of artificial intelligence stand out, which include connectionist models that represent neural networks. In addition, the different storage models of structured databases and non-sql databases are highlighted, the latter with greater influence in open domains with a large amount of information.

On the other hand, it is important to index the information in the different mechanisms and forms of presentation, to achieve agility in the search for information. The study identifies that despite the results achieved in the form of knowledge representation, there are needs and opportunities for the development of basic and applied research in this direction.

## *3.3 Active Learning Supported by Human Agents*

In this model, learning techniques are combined with human assistants that intervene in both learning processes and response processes. Human agents generally intervene under an active learning approach, supplying knowledge in a prescriptive way to the knowledge base of the intelligent system [18, 19]. Frequently, this knowledge is built using the concepts explained in the previous section. From time to time, with an offline approach, a learning process occurs that improves the response model of the intelligent agent.

In these models, human agents also intervene when the system does not have an answer to a question. In this context, when the conversational system does not know the answer to a question, it sends an alert to the human expert to issue the final answer. In this way, human intervention is alleviated.

The result is a continuous cycle of learning and improvement, ensuring that AI chatbots and human experts gradually improve their performance. This model is generally used in environments with closed domains. Some of the limitations of this model are listed below:

- Unfriendly training interfaces and slow learning process.
- These chatbots trained on a certain ontology generally supported by rules. Once you encounter an unknown domain or unexpected types of conversation, Rule-based Chatbots tend to refuse to answer questions and do not have a high level of interaction [1].
- These models generally incorporate deep neural networks for the response and interaction process with users, which implies a high consumption of computational resources. Generally supported by high human intervention.

## *3.4 Large Languaje Model (LLM) Chatbots Characterization*

Another architectural model of conversational systems are those supported by the Large Language Model (LLM). These models are often supported by deep neural networks trained to perform natural language recognition and processing tasks. They are capable of learning from large volumes of data as well as recognizing, translating, predicting, or generating text or other content.

The most common LLM architecture is based on the existence of a transformation model made up of an encoder and a decoder. The input texts are scanned to identify the tokens present. Then, using comparison functions, relationships between the tokens are discovered.

LLM models are often made up of multiple layers of neural networks: recurrent layers, feedforward layers, embedding layers, and attention layers. These different layers cooperate in processing the input text and generating the response.

Three types of LLM stand out in the architectures analyzed:

- Generic language models that have the ability to predict the next word based on the language in the training data. These models are used in information retrieval tasks.
- Instruction-tuned language models that are trained to give answers to questions or instructions given as input.
- These models generally incorporate sentiment analysis elements and are capable of generating text or code. Dialogue-tuned language models are trained to have a dialogue by predicting the next response.

The construction and exploitation of LLMs can be organized at different times, as described below.

- Training: These models are trained using large sets of textual data, obtained from public data sources such as Wikipedia, GitHub, or others. This step is characterized by the use of unsupervised learning algorithms aimed at discovering relationships between words. In addition, in this step, relationships between n-grams and contexts are discovered. It is important to identify that the quality of the data supplied for this learning process and its semantics significantly influence the future responses of the conversational system. This element can constitute a limitation of the LLM if it is not carefully analyzed.
- Fine-Tuning: This step occurs when you want the conversational system to have a specific function or task. For example, if you want the conversational system to perform translation activities, it must receive a setting to perform this activity.
- Prompt-tuning: Performs a function similar to fine-tuning in that it trains a model to perform a specific task through precise interactions and concrete examples. For example:

    Customer review: This is a correct way!
    Customer sentiment: positive
    Customer review: we hate people
    Customer sentiment: negative

In this example, the Prompt-based training environment helps train the system by pointing out items that are not correct. A new approach in training based on prompt-tuning is the techniques of Conversational Driven Design.

Several examples of use of LLM systems are listed below:

- Information retrieval: oriented to the search for information, they are capable of producing information in response to a query. In addition, they can summarize and communicate the answer using conversational styles. Examples of these systems are Bing and Google (Alexa) [20]. Another example of this type of system is BERT (Bidirectional Encoder Representations from Transformers) also developed by Google with the ability to understand natural language and answer questions.
- Sentiment analysis: applications used by LLMs for the analysis of sentiment and emotions. They allow for avoiding expressions of hate in conversational spaces. An example of this platform is REVE Chat [21].
- Text generation: applications that can generate new texts or expressions from the knowledge learned. For example, the ChatGPT family developed by OpenAI, and the different generations, such as GPT-3 and GPT-4. From these, other specific task models have been developed, such as EinsteinGPT [22], developed by Salesforce, which constitutes a CRM system, and Bloomberg's BloombergGPT [23] which specializes in finance.
- Code generation: works similarly to text generation. In this approach, LLMs learn patterns and are able to generate code for certain functionalities.
- Chatbots oriented to user services: these chatbots provide informational, or reservation services aimed at users in specific domains. They are the most widely used in trade activity and are aimed at creating good communication with users, interpreting queries entered by users, and providing answers.

LLM models have certain limitations; here are some of them:

- These models present risks associated with digital ethics when they are not managed or developed correctly. They can learn from data sets without the prior consent of the people involved, violating the privacy of the information. Mismanagement of these systems can lead to the dissemination of erroneous or biased information about certain phenomena [1].
- They can provide biased information: the data used to train language models significantly influences the results and responses to queries.
- Current models do not properly handle copyright licenses, frequently reusing written content without obtaining permission from the owners. They may expose users who use them to situations of copyright infringement.
- Scaling: generally, the training and exploitation of these systems are expensive in terms of computational resources. This situation means that the use of these systems implies high energy consumption. For example, Megatron-Turing uses hundreds of NVIDIA DGX A100 multi-GPU servers, consuming 6.5 kW per use. Along with a lot of power to cool, these models need a lot of power and leave behind large carbon footprints. For its part, the energy consumed in a BERT training is equivalent to a trans-American flight.
- Request misunderstanding: Chatbots often misinterpret requests because they cannot understand the correct intent of the customer. Factors such as:

  - Training set with low concept separability and unwanted results after training.
  - More than 50% of all searches will be done by voice, for the convenience that this represents. However, difficulties persist for chatbots to understand accents or cultural dialects when identifying intentions.

Authors such as the renowned Chomsky [24] carry out a critical analysis of ChatGPT, where elements that constitute limitations and opportunities for improvement of LLM systems are identified. For example, Chomsky refers that:

- The human mind has the ability to make decisions with small amounts of information; Create explanations without the need to infer raw correlations between data points. In this sense, the strengthening of interactive learning models, combining chatbots with reinforcement learning techniques, becomes an opportunity for improvement.
- The deepest flaw of the ChatGPT model is the absence of the most critical capacity of any intelligence: "to say not only what is the case, what was the case, and what will be the case (that is, description and prediction), but also what is not the case and what could and could not be the case". This situation gives the opportunity to apply models that extend the analysis of ambiguity and certainty, such as the neutrosophical theory.
- Noam Chomsky identifies as the Achilles heel of machine learning the fact that it does not posit any causal mechanism or physical laws for description and prediction. In this case, the need for the development of Explanatory Artificial Intelligence (XAI) techniques is identified and the possibility of combining these

techniques with fuzzy cognitive maps or other techniques that allow managing causal relationships.
- Chomsky also refers that intelligence does not only consist of making creative conjectures, but also creative criticism, and that human-style thinking is based on possible explanations and error correction.

From this Chomsky analysis, we conclude that chatbots:

- They must store not only the correct expressions but also the incorrect expressions.
- They must have identified incorrect answers as impossible elements.
- When faced with a question, they must apply the principle that if we eliminate the impossible, what remains is what is rationally possible as an answer.

Finally, a different alternative to LLM models is the use of vector-oriented databases, which have emerged in recent years and incorporate search mechanisms based on vector similarity algorithms, such as Pinecode, weaviate, qdrant, milvus, Elasticsearch Relevance Engine (ESRE). These databases, together with the new frameworks LangChain, LlamaIndex and others, combine to provide solutions to the limitations of LLM models. These technologies allow large volumes of data to be indexed, facilitating the search for answers to different queries. However, one of the difficulties of technologies such as ElastisSearch is related to the semantic analysis of queries. Recent works address this issue and successfully combine natural language processing models and generative artificial intelligence techniques.

## 4 Integration Analysis of Conversational Systems with Specific Computational Intelligence Techniques

This section discusses research that combines conversational systems with specific computational intelligence techniques.

### *4.1 Neutrosophic Theory and Other Extensions in Conversational System Evaluations*

Strategies for evaluating the responses of conversational systems are fundamental to their behavior [17, 25]. The analysis of the following works was carried out:

- Research [26] defines more than 63 metrics that are grouped into different perspectives: user experience, information retrieval, linguistics, technology, and business.
- In [27] 23 metrics are analyzed for automatic evaluation based on automatic generation techniques. In this other study, evaluation metrics are classified into

two groups based on the algorithmic approach used to calculate them: those based on rules and those based on machine learning models.
- The work proposed by [27], defines a set of metrics that can be automatically calculated to help detect quality problems in the knowledge base of the training set for understanding natural language.
- The studies [28] group the metrics according to the following criteria: quality in the generation of answers, associated with the understanding of the answer, metrics related to aesthetics, and general evaluation.

In general, it is concluded regarding this point that the evaluation by humans is still the most accurate [17]. The most common being the use of hybrid models that combine automated strategies and human evaluations.

As a summary, the authors of this work identify that there are different evaluation criteria that allow measuring the speed of the response, the ease of use, the degree of certainty, the degree of falsehood, the level of neutrality, as well as the levels of relevance and importance of the answers. The different evaluation methods influence the following elements:

- Identification of best practices to correct defects and raise the confidence levels of end users.
- Improve learning mechanisms by promoting interactive learning strategies.
- Measure the perceived quality of users.

Assessment measures can be organized around different aspects [4, 17];

- Regarding the moment that occurs, they can be organized into: real-time evaluations and off-line evaluations.
- Regarding the agent that issues the evaluation: evaluations issued by the user and evaluations issued by the conversational system itself.

The evaluations issued by the users are focused on receiving notification from the users about the responses received by the system. An example of this approach is the issuance of signals, emojis, or others that indicate user satisfaction or dissatisfaction.

The main disadvantage of this approach is that a constant issuance of questions addressed to the user to find out about their level of satisfaction, can be invasive. The search for new methods aimed at reducing the invasion of the user's space continues to be an open line of research.

On the other hand, the evaluations issued by the system itself are generally focused on:

- The analysis of the user's observed behavior considering historical records and the analysis of the outliers in the system responses.
- The analysis of the user's profile together with contextual information and the most probable responses of the system.
- Validation of the measures used, that is, an evaluation of how well they measure what they purport to measure.
- Measurement of responses based on the principles and measures used in information retrieval systems.

### 4.1.1 Proposals for Measures for the Evaluation of Conversational Systems Inspired by Measures of Quality of Prediction and Information Retrieval

Examples of measures associated with information systems that are adapted for the evaluation of conversational systems are: accuracy (Eq. 1), precision (Eq. 2), recall (Eq. 3) and F1_score (Eq. 4), were:

- False Positive [FP]: the conversational system offers the user an incorrect answer to a human question on topics for which the system was pre-trained.
- False Negative [FN]: where the users introduce previously utterance, that is, knowledge for which the system was pre-trained, but the conversational system does not identify the utterance as known. In this case, the system gives the wrong answer to the user, indicating ignorance on the subject.
- True Positive [TP]: when the conversational system receives an utterance or question and provides the expected response correctly.
- True Negative [FP]: when the conversational system correctly recognizes when the user introduces an utterance about a new unknown topic and sends the correct messages.

Accuracy: A measure of the total quality; the result is better when accuracy is high, see Eq. 1.

$$\text{accuracy} = \frac{\text{TP} + \text{TN}}{\text{TP} + \text{TN} + \text{FP} + \text{FN}} \tag{1}$$

Precision: This measure is associated with the learning process. The best results are obtained when the conversational system makes no errors in solving known problems. A high value of this metric indicates good answers from the conversational system; see Eq. 2.

$$\text{precision} = \frac{\text{TP}}{\text{TP} + \text{NP}} \tag{2}$$

Recall: Measures the amount of correct answers cuando el sistema se enfrenta a sitauciones para las que fue pre-entrenado. The result is better when the recall is greater, see Eq. 3.

$$\text{recall} = \frac{\text{TP}}{\text{TP} + \text{FN}} \tag{3}$$

F1_score: Used to combine the precision and recall measures into a single value. The smaller this indicator is, the better the response quality of the conversational system; see Eq. 4.

$$\text{F1\_score} = 2 * \frac{\text{recall} + \text{precision}}{\text{recall} * \text{precision}} \tag{4}$$

The general formula derived by Van [29] or non-negative real, is:

$$F_{\beta}\_\text{score} = \frac{\left(1+\beta^2\right)\text{recall} * \text{precision}}{\beta^2(\text{recall} + \text{precision})} \tag{5}$$

Fall-out: The proportion of non-relevant answers that are retrieved out of all non-relevant answers:

$$\text{Fallout} = \frac{|\text{Relevant answers}| \cap |\text{Non} - \text{Relevant answers}|}{|\text{Non} - \text{Relevant answers}|} \tag{6}$$

In binary classification, fall-out is closely related to specificity and is equal to (1 – specificity). It can be looked at as the probability that a non-relevant answer is retrieved by the questions (utterances).

### 4.1.2 Neutrosophical Theory and Its Windows for the Evaluation of Conversational Systems

Conversational systems aim to provide a fluid conversation with a high level of certainty for users. At this point, the treatment of imprecision, uncertainty, and veracity is necessary to obtain better harmony with reality [30]. In this sense, fuzzy logic has been widely used to model the certainty of information [31]. However, the modeling of fuzzy sets does not adequately model the concepts of falsehood and neutrality [32, 33].

An extension to traditional fuzzy logic is the neutrosophical theory introduced by Smarandache in [34–36] to improve the treatment of indeterminate and inconsistent information in the real world. In this sense, the neutrosophic set is a generalization of the theory of fuzzy set intuitionistic, interval-valued fuzzy sets [37]. The neutrosophic set, proposed by Smarandache [38], allow simultaneous management of functions of truth, indeterminacy, and falsity in [0, 1].

For this work, it is particularly important the Definition 1 of neutrosophic sets defined in [32, 39].

**Definition 1** (*Neutrosophic set*) let X be a universe of discourse, a neutrosophic set A over X is an object of the form: $A = \{\langle \mu_A(x), \tau_A(x), \sigma_A(x)\, n\rangle{:}x \in X\}$, see Fig. 7 and it is fulfilled that:

- $\mu_A(x) \in [0, 1]$ membership function that represents the degree of certainty that the value $x$ belongs to the set $A$, see Eq. 7.
- $\tau_A(x) \in [0, 1]$ membership function that represents the degree of indeterminacy that the value $x$ belongs to the set $A$, see Eq. 8.
- $\sigma_A(x) \in [0, 1]$ membership function that represents the degree of degree of non-membership (or falsity) that the value $x$ belongs to the set $A$, see Eq. 9.
- $0 \leq \mu A(x) + \tau A(x) + \sigma A(x) \leq 3 \forall x \in X$.

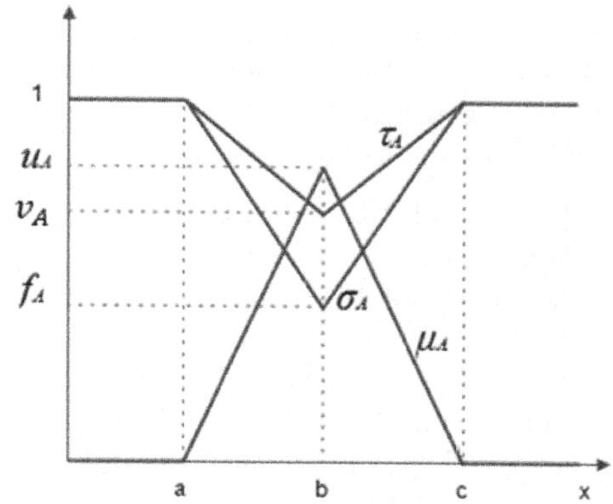


**Fig. 7** Neutrosophic set incorporating the values of truthfulness, neutrality, and falsity

$$\mu_A(x) = \begin{cases} (x-a)u_A/(b-a) & (a \le x < b) \\ u_A & (x = b) \\ (c-x)u_A/(c-b) & (b < x \le c) \\ 0 & \text{any other case} \end{cases} \tag{7}$$

$$\tau_A(x) = \begin{cases} (b-x+v_A(x-a))/(b-a) & (a \le x < b) \\ r_A & (x = b) \\ (x-b+v_A(c-x))/(c-b) & (b < x \le c) \\ 1 & \text{any other case} \end{cases} \tag{8}$$

$$\sigma_A(x) = \begin{cases} (b-x+f_A(x-a))/(b-a) & (a \le x < b) \\ f_A & (x = b) \\ (x-b+f_A(c-x))/(c-b) & (b < x \le c) \\ 1 & \text{any other case} \end{cases} \tag{9}$$

The neutrosophic theory can provide facilities for the evaluation of responses in conversations. However, in the searches carried out, no references to works that combine these techniques were found. The authors of this work consider this element an open line of research for future work.

## 4.2 Linguistic Summarization of Data in the Learning of Conversational Systems

Linguistic data summarization techniques (LDS) facilitate sentence construction in natural language to describe the information contained in databases.

These summaries express, in natural language, the relationship between concepts and have the structures "*Qy's are S*" and "*QRy's are S*" [31, 40]. The following elements make up a summary: quantifiers ($Q$), filters or qualifiers ($R$) and summarizers ($S$) [41, 42]. The components $R$ and $S$ are entities that can intervene in a conversation established by a conversational system. Below is an example of linguistic summaries and its components:

Summary 1: "A total of 41 projects are very late"
Where:
$Q$: "A total of 41" - value of "amount" entity
$R$: -
$S$: "very late" - value of "progress" entity
Summary 2: "Most high-performing specialists in the programmer role are characterized by being passive and task-oriented under normal conditions and under stress conditions".
Where:
$Q$: "Most" - value of "amount" entity
$R$: "high-performing specialists in the programmer role"
$S$: "are characterized by being passive and task-oriented under normal conditions and under stress conditions"

The main elements are formalized below:

- $Q$: quantifier, is a fuzzy set with the universe of discourse expressing a quantity, for example, "the majority", "60%" or "more than half".
- $R$: qualifier or filter, it is another attribute that determines a fuzzy subset of the object $y_i$, for example "specialists with high performance" over the object $y_i$ "Specialist or Person".
- $S$: summarizer, is an attribute with a linguistic value defined in the problem domain, for example "they are characterized by being passive" for the attribute "characterization".
- $T$: degree of truth (validity) of the summary, it is a number in the interval [0, 1] that evaluates the degree of truth of the summary; abstracts with a high $T$ value are of interest.

In general, linguistic data summary techniques are based on the construction of linguistic summaries in the form of protoforms, see Table 1. The protoforms can be used both for the construction of summaries and for modeling interactions in the form of questions and answers [43, 44].

The protoforms that have been used the most in the bibliography for the construction of linguistic summaries follow the structure proposed by Zadeh in [46] and later extended by Kacprzyk and Zadrożny in [47]. These protoforms are called classical protoforms. Classical protoforms summarize the attributes of the data set or the relationships between them [40, 48].

El The term $S^{\text{Estructure}}$ means that the variables that make it up are known in the summary; while $S^{\text{Value}}$ denotes that the value of the summarizer is unknown.

There are other variants of protoforms related to climate data [49], where the authors combine protoforms with the terms, "but" and "especially" to establish contrast and emphasis sentences, respectively. This approach makes it possible to

**Table 1** Protoforms for the linguistic summaries' construction [45]

| Type | Protoforms | Known element | Doubt (Question) |
|---|---|---|---|
| 0 | *QRy's* are *S* | All | *T* |
| 1 | *Qy's* are *S* | *S* | *Q* |
| 2 | *QRy's* are *S* | *S R* | *Q* |
| 3 | *Qy's* are *S* | $Q\ S^{\text{Estructura}}$ | $S^{\text{Value}}$ |
| 4 | *QRy's* are *S* | $Q\ S^{\text{Estructura}}, R$ | $S^{\text{Value}}$ |
| 5 | *QRy's* are *S* | Nothing | *Q R S* |

**Table 2** Combination of conversational systems with different computational intelligence techniques

| Academic search engine | 2019–2023 | 2022–2023 |
|---|---|---|
| Topic: "conversational systems" and "linguistic summaries" | | |
| Scholar google | 4 | 2 |
| Semantic scholar | 163 | 14 |

obtain more expressive linguistic summaries and constitutes an open line of research. Here are different examples:

Examples of contrast protoform:

"Many values in Ancares are normal", **but**
"Most of the values of the third week in Ancares are hot"
"Most of the values of the fourth week in Ancares are cold"

Example of emphasis protoform:

"Many values in Pontevedra-Campolongo are hot", **especially**
"Many values of the second week in Pontevedra-Campolongo are very hot"

In proportion to the growth that conversational systems have had, few publications have been identified that combine Linguistic Data Summarization (LDS) techniques with the conversational systems, see Table 2.

The few results found show that the use of linguistic data summarization techniques constitutes an unexplored line in the initial learning of other learning models of conversational systems.

It is identified in this research that the generation of linguistic summaries from data can facilitate the learning processes of conversational systems by providing precise answers to specific questions. In particular, these techniques can be employed to generate instances of intents, entities, and responses. These responses could then be enhanced using generative LLM models to add more detail or make the response more conversational.

It was also identified that conversational systems can generate multilingual responses from linguistic summaries, according to the results of Pérez Pupo [31].

**Table 3** Conversational systems with reinforcement learning

| Academic search engine | 2019–2023 | 2022–2023 |
|---|---|---|
| Topic: "conversational systems" and "reinforcement learning" | | |
| Scholar google | 937 | 353 |
| Semantic scholar | 32 | 14 |

## *4.3 Reinforcement Learning Combined with Conversational Systems*

Regarding the analysis of the combination with reinforcement learning techniques, a greater number of publications are identified [50], see Table 3.

There are different systematic reviews associated with the training of conversational systems. For example, in [50] 22 papers are reviewed between the years 2005–2020, to understand how human data is being collected for training conversational agents. These papers reveal a common use of learning from demonstration/observation, crowdsourcing methods in system training, and dataset cataloging alongside handwriting and sentence labeling.

As a summary of the review, it is identified that reinforcement learning techniques will increase the capacity of conversational systems for "in-context learning" by promoting strategies that guarantee continuous learning. As shown in Fig. 8, among the countries with the most outstanding results in this sense are the United States, China, India, and several European countries.

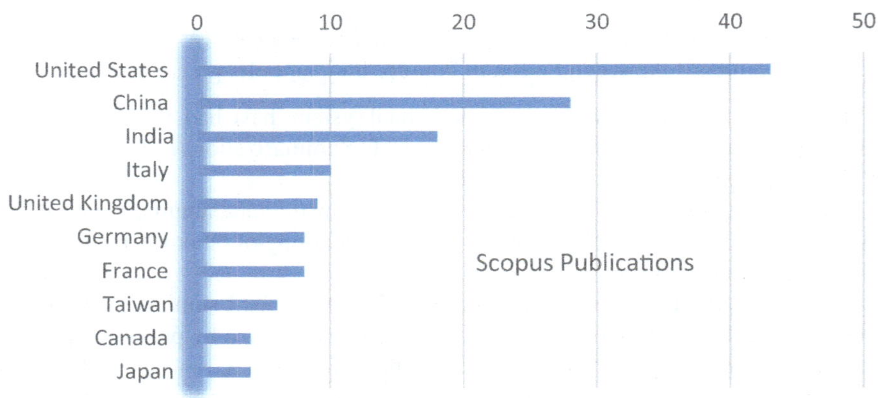


**Fig. 8** Search results of documents by country/territory using "conversational AND systems AND reinforcement AND learning" in 2002–2023, SCOPUS

## 5 Conclusions

Based on the analysis developed in this research, it is concluded tha:

- The introduction of conversational styles in dissimilar sectors will increase in different ways. These styles will significantly influence the efficiency and agility of the interactions between humans and information systems.
- The application of conversational AI in the architecture, engineering, and construction (AEC) industry is lagging. The findings reveal that conversational AI applications hold immense benefits for the AEC industry, but they are currently underexplored.
- The main challenges are focused on achieving the nuances of human language and naturalness in conversations.

Regarding the technical aspects and the architectural models, it can be concluded that:

- The hybridization of conversational systems with other computational intelligence techniques is identified as a trend, facilitating knowledge management processes and the humanization of conversations.
- There are techniques such as reinforcement learning, linguistic data summarization, and neutrosophic theories whose nature allows them to be applied immediately in conversational systems. However, not enough research has yet been carried out, which constitutes an open line of research.
- In practice, building conversational systems for each application domain is a costly process in time and effort. It is necessary to create models and platforms that facilitate the automatic or semi-automatic construction of conversational systems. In this sense, linguistic data summary techniques serve to support knowledge management processes prior to learning.
- The introduction of elements of the neutrosophic theory for the evaluation processes of conversations has not been exploited. The introduction of this theory in the evaluation of conversations would help to better model human behavior in unknown situations.
- This paper identifies that despite the results achieved in the forms of knowledge representation in conversational systems, there are needs and opportunities for the development of basic research and applications in this direction.

As future work of this research, the opportunities for improvements to artificial intelligence mentioned earlier in this article based on the criticism of Noam Chomsky are identified. In particular, the development of new architectures that combine the best practices of basic smart chatbots and LLM models is proposed. The need to design models based on neutrosophic theory for the evaluation of conversations, combining these techniques with continuous learning, is identified. As well as advancing in specific applications in scenarios such as sports, fitness, and project engineering, among other sectors that report fewer applications in the bibliographies studied.

## References


1. Galitsky B (2019) Developing enterprise chatbots: learning linguistic structures. Springer International Publishing, Cham
2. Adamopoulou E, Moussiades L (2020) An overview of chatbot technology. In: Maglogiannis I, Iliadis L, Pimenidis E (eds) Artificial intelligence applications and innovations. Springer International Publishing, Cham, pp 373–383
3. Tlili A, Shehata B, Adarkwah MA, Bozkurt A, Hickey DT, Huang R, Agyemang B (2023) What if the devil is my guardian angel: ChatGPT as a case study of using chatbots in education. Smart Learn Environ 10:15. https://doi.org/10.1186/s40561-023-00237-x
4. Lin C-C, Huang AYQ, Yang SJH (2023) A review of AI-driven conversational chatbots implementation methodologies and challenges (1999–2022). Sustainability 15:4012. https://doi.org/10.3390/su15054012
5. Li M, Wang R (2023) Chatbots in e-commerce: The effect of chatbot language style on customers' continuance usage intention and attitude toward brand. J Retail Consum Serv 71:103209. https://doi.org/10.1016/j.jretconser.2022.103209
6. Park D-M, Jeong S-S, Seo Y-S (2022) Systematic review on chatbot techniques and applications. J Inf Process Syst 18:26–47. https://doi.org/10.3745/JIPS.04.0232
7. Rojc M, Ariöz U, Šafran V, Mlakar I, Rojc M, Ariöz U, Šafran V, Mlakar I (2023) Multilingual chatbots to collect patient-reported outcomes. IntechOpen
8. Böger B, Fachi MM, Vilhena RO, Cobre AF, Tonin FS, Pontarolo R (2021) Systematic review with meta-analysis of the accuracy of diagnostic tests for COVID-19. Am J Infect Control 49:21–29. https://doi.org/10.1016/j.ajic.2020.07.011
9. Mordor Intelligence (2023) Chatbot market—Statistics. Size & Industry Trends. https://www.mordorintelligence.com/industry-reports/global-chatbot-market. Accessed 20 Aug 2023
10. Quesada JF, Griol D, Callejas Z (2019) Informe sobre sistemas conversacionales multimodales y multilingües
11. de Souza MM, de Castro C, Salgado L (2023) Conversational agents: a survey on culturally informed design practices. J Interact Syst 14:14. https://doi.org/10.5753/jis.2023.2961
12. Alsharhan A, Al-Emran M, Shaalan K (2023) Chatbot adoption: a multiperspective systematic review and future research Agenda. IEEE Trans Eng Manag 1–13. https://doi.org/10.1109/TEM.2023.3298360
13. Saka AB, Oyedele LO, Akanbi LA, Ganiyu SA, Chan DWM, Bello SA (2023) Conversational artificial intelligence in the AEC industry: a review of present status, challenges and opportunities. Adv Eng Inform 55:101869. https://doi.org/10.1016/j.aei.2022.101869
14. Liu L, Duffy VG (2023) Exploring the future development of artificial intelligence (AI) applications in chatbots: a bibliometric analysis. Int J Soc Robot 15:703–716. https://doi.org/10.1007/s12369-022-00956-0
15. Uc-Cetina V, Navarro-Guerrero N, Martin-Gonzalez A, Weber C, Wermter S (2022) Survey on reinforcement learning for language processing. Artif Intell Rev. https://doi.org/10.1007/s10462-022-10205-5
16. Aggarwal A, Tam CC, Wu D, Li X, Qiao S (2023) Artificial intelligence-based chatbots for promoting health behavioral changes: systematic review. J Med Internet Res 25:e40789. https://doi.org/10.2196/40789
17. Caldarini G, Jaf S, McGarry K (2022) A literature survey of recent advances in chatbots. Inf MDPI 13:41. https://doi.org/10.3390/info13010041
18. Kasneci E, Sessler K, Küchemann S, Bannert M, Dementieva D, Fischer F, Gasser U, Groh G, Günnemann S, Hüllermeier E, Krusche S, Kutyniok G, Michaeli T, Nerdel C, Pfeffer J, Poquet O, Sailer M, Schmidt A, Seidel T, Stadler M, Weller J, Kuhn J, Kasneci G (2023) ChatGPT for good? On opportunities and challenges of large language models for education. Learn Individ Differ 103:102274. https://doi.org/10.1016/j.lindif.2023.102274
19. Aher GV, Arriaga RI, Kalai AT (2023) Using large language models to simulate multiple humans and replicate human subject studies. In: Proceedings of the 40th international conference on machine learning. PMLR, pp 337–371

20. Land JM, Pavlik EJ, Ueland E, Ueland S, Per N, Quick K, Gorski JW, Riggs MJ, Hutchcraft ML, Llanora JD, Yun DH (2023) Evaluation of replies to voice queries in Gynecologic oncology by virtual assistants Siri, Alexa, Google, and Cortana. BioMedInform 3:553–562. https://doi.org/10.3390/biomedinformatics3030038
21. REVE Chat (2023) Customer Communication Platform by REVE Chat. In: REVE Chat. https://revechat.com/. Accessed 24 Aug 2023
22. Globant (2023) Einstein GPT: la herramienta de IA generativa de Salesforce | Globant Blog. https://stayrelevant.globant.com/es/technology/data-ai/einstein-gpt-que-impacto-podria-tener-la-herramienta-de-ia-generativa-de-salesforce-para-ti/. Accessed 24 Aug 2023
23. Wu S, Irsoy O, Lu S, Dabravolski V, Dredze M, Gehrmann S, Kambadur P, Rosenberg D, Mann G (2023) BloombergGPT: a large language model for finance
24. Chomsky N (2023) La crítica de Noam Chomsky al sistema de inteligencia artificial chat GPT—Cultura Inquieta. In: Cult Inquieta—Pensam. https://culturainquieta.com/es/pensamiento/item/20093-la-critica-de-noam-chomsky-al-sistema-de-inteligencia-artificial-chat-gpt.html. Accessed 24 Aug 2023
25. Yeh Y-T, Eskenazi M, Mehri S (2021) A Comprehensive assessment of dialog evaluation metrics
26. Peras D (2018) Chatbot evaluation metrics: review paper
27. Cañizares PC, Pérez-Soler S, Guerra E, de Lara J (2022) Automating the measurement of heterogeneous chatbot designs. In: Proceedings of the 37th ACM/SIGAPP symposium on applied computing. ACM, Virtual Event, pp 1491–1498
28. Coppola R, Ardito L (2021) Quality assessment methods for textual conversational interfaces: a multivocal literature review. Information 12:437. https://doi.org/10.3390/info12110437
29. Rijsbergen V, Joost C (1979) Information retrieval. 2nd. newton, ma. Butterworth-Heinemann, USA
30. Zadeh LA (1997) The roles of fuzzy logic and soft computing in the conception, design and deployment of intelligent systems. In: Nwana HS, Azarmi N (eds) Software agents and soft computing towards enhancing machine intelligence: concepts and applications. Springer, Berlin, Heidelberg, pp 181–190
31. Pérez Pupo I (2021) Algoritmos para la sumarización lingüística de datos para la ayuda a la toma de decisiones. Doctoral, Centro de Estudios de Gestión de Proyectos y Toma de Decisiones, Universidad de las Ciencias Informáticas
32. Pérez Pupo I, Piñero Pérez PY, García Vacacela R, Bello R, Santos Acosta O, Leyva Vázquez MY (2018) Extensions to linguistic summaries indicators based on neutrosophic theory: applications in project management decisions. Neutrosophic Sets Syst Univ N M 22:87–100
33. Smarandache F, Pramanik S (2016) New trends in neutrosophic theory and applications. Infinite Study
34. Sahin M, Abdullah Kargin FS (2018) Generalized single valued triangular neutrosophic numbers and aggregation operators for application to multi-attribute group decision making. New Trends Neutrosophic Theory Appl II:51–84
35. Smarandache F (2010) Neutrosophic set-a generalization of the intuitionistic fuzzy set. J Def Resour Manag 1:107
36. Goguen JA, Zadeh LA (1973) Fuzzy sets. Inf Control 8(1965):338–353; Zadeh LA (1971) Similarity relations and fuzzy orderings. Inf Sci 3:177–200. J Symb Log 38:656–657. https://doi.org/10.2307/2272014
37. Atanassov K, Gargov G (1989) Interval valued intuitionistic fuzzy sets. Fuzzy Sets Syst 31:343–349
38. Smarandache F (2000) Neutrosophy. Math0010099
39. Smarandache F (2016) Neutrosophic overset, neutrosophic underset, and neutrosophic offset. In: Similarly for neutrosophic over-/under-/off-logic, probability, and statistics. 160700234 Cs
40. Yager RR (1982) A new approach to the summarization of data. Inf Sci 28:69–86. https://doi.org/10.1016/0020-0255(82)90033-0

41. Pérez I, López P, Varona E, Piñero P, García R (2018) Construcción de resúmenes lingüísticos a partir rasgos de la personalidad y el desempeño en el desarrollo de software. Rev Cuba Cienc Informáticas 12:135–150
42. Kacprzyk J, Yager RR (2001) Linguistic summaries of data using fuzzy logic. Int J Gen Syst 30:133–154. https://doi.org/10.1080/03081070108960702
43. Pérez Pupo I, Piñero Pérez PY, Bello Pérez RE, García Vacacela R, Villavicencio Bermúdez N (2022) Linguistic data summarization: a systematic review. In: Piñero Pérez PY, Bello Pérez RE, Kacprzyk J (eds) Artificial intelligence in project management and making decisions. Springer International Publishing, Cham, pp 3–21
44. Pérez Pupo I, Piñero Pérez PY, Al-subhi SH, García Vacacela R, Martínez Noriega HA, Villavicencio Bermúdez N (2022) New linguistic data summarization approach for prediction problems in project management applications. In: Piñero Pérez PY, Bello Pérez RE, Kacprzyk J (eds) Artificial intelligence in project management and making decisions. Springer International Publishing, Cham, pp 23–38
45. Kacprzyk J, Zadrożny S (2009) Linguistic database summaries using fuzzy logic, towards a human-consistent data mining tool, 10
46. Zadeh LA (2002) A prototype-centered approach to adding deduction capability to search engines-the concept of protoform. In: Intelligent systems, 2002. Proceedings. 2002 first international IEEE symposium. IEEE, pp 2–3
47. Kacprzyk J, Zadrożny S (2005) Linguistic database summaries and their protoforms: towards natural language based knowledge discovery tools. Inf Sci 173:281–304. https://doi.org/10.1016/j.ins.2005.03.002
48. Hudec M, Bednárová E, Holzinger A (2018) Augmenting statistical data dissemination by short quantified sentences of natural language. J Off Stat 34:981–1010. https://doi.org/10.2478/jos-2018-0048
49. Ramos-Soto A, Martin-Rodillab P (2019) Enriching linguistic descriptions of data: a framework for composite protoforms. Fuzzy Sets Syst 26. https://doi.org/10.1016/j.fss.2019.11.013
50. Camargo J (2022) Systematic review of training methods for conversational systems: the potential of datasets validated with user experience

# Fuzzy Cognitive Maps, Extensions and Applicability as an Explanatory Artificial Intelligence Model

**Yosvany Márquez Ruiz, Pedro Yobanis Piñero Pérez, Iliana Pérez Pupo, Roberto García Vacacela, and Salah Hasan Saleh Al-Subhi**

**Abstract** The growth of the prediction capabilities of deep neural networks constitutes one of the elements that has allowed the generalization of these models to dissimilar problems. But in the particular case of decision-making in project management, in addition to the quality of the prediction, it is necessary to know the explanation of the responses. This aims to increase the confidence of the systems in decision-making and agility in project management. The explanation of the answers is also important during project cuts, a process that manifests itself as a multistage sequential decision-making problem. In this research, it is proposed to explore the potential of fuzzy cognitive maps and their extensions, considering the potential of these techniques to represent causal relationships. To carry out this work, a conceptual theoretical framework is developed based on a systematic review. It is identified that there are insufficiencies in the research reported in the bibliography consulted regarding the treatment of indeterminacy and the solution of decision-making problems in project management.

Y. M. Ruiz (✉)
Centro de Estudios de Gestión de Proyectos y Toma de Decisiones, Universidad de las Ciencias Informáticas, Havana, Cuba
e-mail: ymarquez@gmail.com; ymarquezr@uci.cu

P. Y. Piñero Pérez · I. Pérez Pupo · S. H. S. Al-Subhi
Artificial Intelligence for Sustainable Development Group, IADES, La Habana, Cuba
e-mail: iades@businessredmine.com; pppyob@gmail.com

I. Pérez Pupo
e-mail: iades@businessredmine.com; iliperezpupo@gmail.com

S. H. S. Al-Subhi
e-mail: iades@businessredmine.com

R. García Vacacela
Facultad de Especialidades Empresariales, Universidad Católica Santiago de Guayaquil, Guayaquil, Ecuador
e-mail: roberto.garcia@cu.ucsg.edu.ec; garciav.roberto@gmail.com



P. Y. Piñero Pérez et al. (eds.), *Computational Intelligence in Engineering and Project Management*, Studies in Computational Intelligence 1134,
https://doi.org/10.1007/978-3-031-50495-2_2

**Keywords** Computational intelligence · Extension cognitive maps · Fuzzy cognitive maps · Project management

## 1 Introduction

Fuzzy cognitive maps (FCMs) introduced by Kosko in 1986 [1] are among the artificial intelligence techniques that allow modeling and simulating complex and dynamic systems [2, 3]. FCMs and their extensions have been fundamentally aimed at solving prediction problems in scenarios with imprecision, uncertainty, indeterminacy, incomplete information, and vagueness [4]. There have been models that have shown utility in environments where it is necessary to guarantee the interpretability of the results [5].

Different extensions to these techniques are identified in the bibliography consulted. For example, with respect to topology, extensions can be grouped into simple FCMs [6, 7] and multiple FCMs (*m-FCMs*) [8]. Some research on FCMs that use simple topologies is listed below [9–11].

Some of the FCM extensions that are also based on simple topologies are competitive FCMs [12], case-based FCMs [4], time-dependent FCMs [13] and intuitive FCMs [14] Extensions of FCMs based on multiple topologies (m-FCMs) are used to solve more complex problems that cannot be solved with extensions based on simple topologies or an aggregation of them [15]. Examples of extensions (m-FCMs) are distributed FCMs [16], hierarchical FCMs [17], parallel FCMs [18] and multilayer FCMs [19].

Although extensions of traditional FCMs have been developed [20, 21], insufficiencies still persist, such as those listed below:

- Need to elevate the treatment of indeterminacy [22]. In this sense, different efforts are being developed associated with the use of neutrosophic theory [23–25] with FCMs.
- Increase the interpretability of the representation of causal relationships through the use of linguistic terms [26, 27]. With the aim of increasing the interpretability of the map, in the consulted bibliography, several authors propose linguistic FCMs that use computing with words (CWW) for the representation of causal relationships and inference of the map. However, the linguistic FCMs reported in the bibliography do not address indeterminacy, an element that constitutes an opportunity for their improvement and which is addressed in the present research.
- Elevate the application of cognitive maps in multi-stage decision-making environments to improve the reliability and explainability of decisions. Particularly in decision-making moments during project cuts or in the explainability of responses and predictions in conversational systems.

In this research, an exploratory study of FCMs and their extensions is carried out to identify their capacity for solving multistage sequential decision-making problems.

With potential application in decision-making processes in project cuts and conversational systems. The work is organized into four sections. The first section shows the systematic review protocol used in the exploratory study. Section 2 presents a critical analysis of FCMs and extensions of simple fuzzy cognitive maps. Section 3 provides a critical analysis of the extensions of multiple fuzzy cognitive maps. Finally, recommendations and future work are presented.

## 2 Review Protocol Used in the Exploratory Study

To develop the theoretical framework, a systematic review is carried out [28] following the protocol described below.

Systematic review protocol applied:

1. Definition of the object of study, field of action, and objective of the research.
2. Definition of a bibliographic manager.
3. Define sources of academic information for the development of the review: "*Scopus*", "*Web of Science*", "*Science Direct*" and other metasearch engines based on open science.
4. Definition of the following set of key phrases to carry out the searches: *"fuzzy cognitive maps", "fuzzy cognitive mapping", "neutrosophic cognitive maps", "fuzzy cognitive maps"* + "*cardiovascular*" + "*pregnant*".
5. Define the goals of bibliometric analysis in the form of research questions and inclusion–exclusion criteria:

    i. What has been the trend in publications per year?
    ii. Who are the main authors?
    iii. What are the affiliations and countries of the main authors?
    iv. How are the publications distributed, considering the types of documents: articles, books, theses, and conference or congress reports?
    v. Exclusion of works published in spaces with a low level of arbitration.
    vi. Inclusion of works associated with map extensions regarding forms of knowledge representation.
    vii. Exclusion of topics associated with general concept maps.

6. Sort and filter posts into the following set of categories:

    - Classic: refers to the first publications associated with fuzzy cognitive maps.
    - Extensions of theories: refers to publications that extend the theory presented in publications understood as classic, they do not set guidelines that significantly change the methods previously proposed, although they do develop contributions to knowledge.
    - Application results: publications that focus on the use of existing theory in concrete, practical settings.
    - Tertiary reviews: refer to articles reviewing trends and evolution in the topic in question.

7. Synthesize the main trends.

Within the research framework, a bibliometric analysis was developed of the publications related to FCMs during the period 1985–2023 in the academic search engines *Scopus, Web of Science*, and *Science Direct*. The following variables were analyzed: number and type of publications, most published authors, and scientific productivity by country on the topic.

Figure 1 shows the number of publications related to FCMs per year from 1985 to August 2023, from a search in the *Scopus* database under the terms "*Fuzzy Cognitive Maps*" and "*Fuzzy Cognitive Mapping*" in which a total of 2,529 publications were found. Other searches were performed on the *Web of Science* and *Science Direct* databases under the same terms, and 905 and 471 publications were recorded, respectively, as shown in Figs. 2 and 3.

The distribution by year in the three databases allowed us to observe the trend of an exponential increase in publications related to FCMs. The growing number of publications in recent years confirms the acceptance that FCMs have had by the international scientific community and shows that there is a continuous interest in this technique, which motivates research in this area of knowledge.

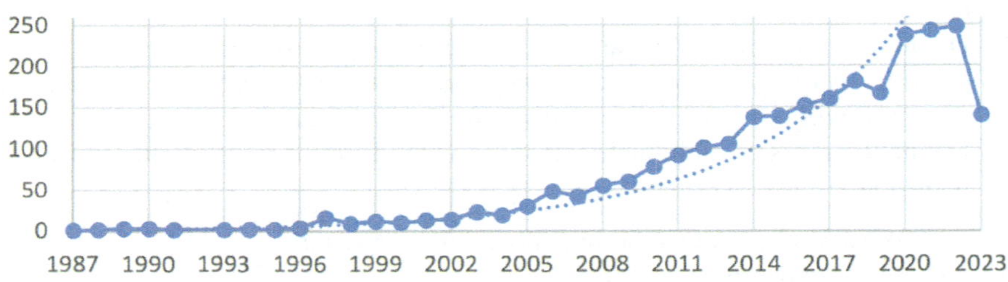


**Fig. 1** Publications about FCMs per year in scopus

**Fig. 2** Publications about FCMs per year in web of science

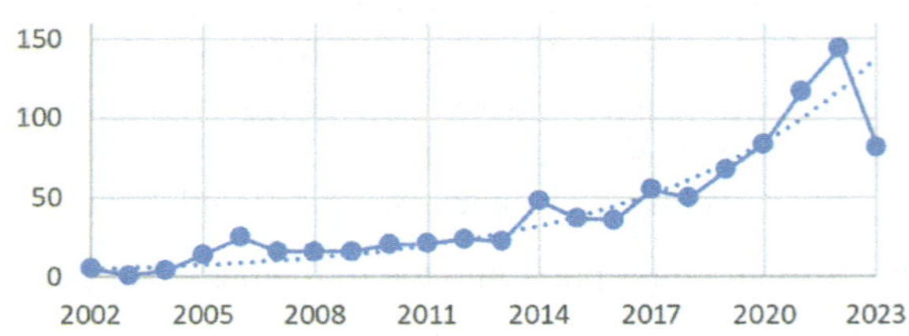


**Fig. 3** Publications about FCMs per year in science direct

**Fig. 4** Types of publications about FCMs from searches in scopus

**Fig. 5** Types of publications about FCMs from searches in web of science

**Fig. 6** Types of publications about FCMs from searches in science direct

Figures 4, 5 and 6 shows that the types of documents that predominate in the three databases consulted are magazine articles.

From the analysis of the search results in the Scopus and Web of Science databases with the term "Fuzzy Cognitive Maps", it can be seen that the most prominent countries within the scientific production on the subject during the period studied are Greece, China, and the United States, see Fig. 7.

A large part of the publications analyzed are concentrated in these three countries,, with Elpiniki I. Papageorgiou being the most published author [29, 30].

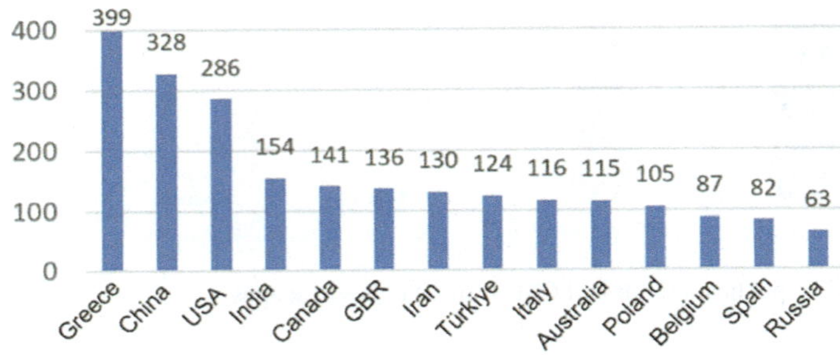


**Fig. 7** Scientific productivity about FCMs by country, scopus

# 3 Characterization and Evolution of Fuzzy Cognitive Maps

FCMs were introduced by *Kosko* in 1986 [1] as an extension of the cognitive maps developed by *Axelrod* in 1976 [31]. They are characterized by their simplicity, adaptability, and ability to deal with uncertainty, ambiguity, and the absence of information [32] providing higher levels of interpretability of the results [33]. Furthermore, its ability to represent cyclical relationships or feedback allows the modeling of dynamic systems [34], discovering patterns that are frequently hidden in these systems [12, 35].

In FCMs, generally, there are three possible types of causal relationships between concepts, see Fig. 8, which are represented by numerical values:

- Positive causality ($W_{ij}> 0$): indicates a positive causality between the concepts $C_i$ and $C_j$, , that is, the increase (decrease) in the value of $C_i$ leads to the increase (decrease) in the value of $C_j$.
- Negative causality ($W_{ij} < 0$): indicates a negative causality between the concepts $C_i$ and $C_j$, that is, the increase (decrease) in the value of $C_i$ leads to the decrease (increase) in the value of $C_j$.
- Non-existence of relationships ($W_{ij} = 0$): indicates the non-existence of a causal relationship between $C_i$ and $C_j$.

In traditional FCMs, the intensity of causal relationships are represented by numerical values [36]; that is why the treatment of the uncertainty and ambiguity of the information associated with the relationships between concepts is insufficient [37].

**Fig. 8** Fuzzy cognitive map

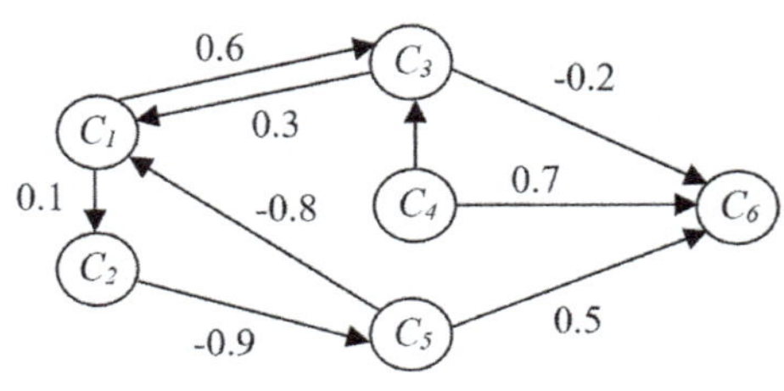

The inference process in these maps begins with the construction of an initial vector $C$ that represents the initial activation levels of all the concepts of the system that need to be analyzed. Based on an iterative process, the activation level of the concepts $C$ is calculated in each iteration, applying the activation function $f$ to the product between the vector C and the connectivity matrix w, which is composed of the weights of the causal relationships between the concepts, see Eq. 1. This process ends when the map reaches the equilibrium state, where the system converges to a set of values without reporting changes in successive iterations or because a maximum number of iterations has been reached [38, 39]. The vector $A_i$ at time $t + 1$ is calculated as follows:

$$A_i^{t+1} = f\left(A_i^t + \sum_{i=1}^{n} w_{ji} \cdot A_j^t\right) \tag{1}$$

where $f(x)$ is the activation function and $w_{ji}$ represents the weights of the causal relationships between the concepts [40]. In the bibliography consulted [41] most of the works focus on the use of the Sigmoid and Hyperbolic Tangent functions. The selection of the activation function depends on the application domain and the results obtained from its use.

In the bibliography consulted, numerous extensions of FCMs are distinguished. In the following sections, these are analyzed based on their applicability in the solution of multi-stage sequential problems, the treatment of indeterminacy, predictability, and interpretability. For analysis, we divide the extensions into two groups: simple FCMs and multiple FCMs.

## 4 Analysis of Extensions of Simple Fuzzy Cognitive Maps

Simple FCMs, as their name indicates, are made up of a single map and constitute the most commonly applied variant. For their part, multiple FCMs constitute extensions of simple FCMs; they are developed in domains where, naturally, several subsystems are identified, which in simple extensions makes knowledge engineering difficult.

### *4.1 Linguistic Fuzzy Cognitive Maps*

In traditional FCMs, both the causal relationships and the activation of concepts are described by numerical values, however, this can constitute a challenge for experts, who generally think qualitatively [27, 42]. With the aim of increasing the interpretability of the map, there are some efforts in the literature to use computing with CWW words [43] in FCMs. In the hybridization of FCMs and CWW, the most commonly used computing models with words are semantic models based on membership functions and operations based on the extension principle [44] and symbolic models [45].

### 4.1.1 A Linguistic Approach to FCMs Based on the Extension Principle

Richard et al. [46] introduce the CWW FCMs model, where the activation levels of concepts are described by linguistic terms represented by type 2 membership functions (IT2 MF). For aggregation, the weighted average operator is used, which is based on [47]. In the inference process, each output concept in each iteration is assigned the linguistic activation term of the input nodes with the highest similarity using the Jaccard similarity function. The CWW FCMs model was applied for the first time in [48] for the classification of celiac diseases.

Gupta et al. [49] propose a linguistic fuzzy cognitive map model that replaces the numerical components of the map inference process with triangular fuzzy numbers. One of the limitations of this work, pointed out by the same authors, is the lack of flexibility in the definition of triangular fuzzy numbers.

In another work [50] Frias presents a hybrid model of fuzzy cognitive maps and CWW, where the linguistic terms assigned to both the concepts of the map and the relationships between them are represented by Gaussian fuzzy numbers (GFN). A similar model was applied in [51] for a classification problem where map weights are calculated automatically from historical data using an evolutionary learning algorithm.

The main limitation of the works based on the principle of extension is the loss of information that occurs as a consequence of the linguistic approach used [50], affecting the predictive capacity of the maps.

In the opinion of the author of this investigation, this limitation could be overcome if the original aggregation values were maintained in the inference process, an element that is considered in the present investigation.

### 4.1.2 Linguistic Approach to FCMs Based on Symbolic Models

Frias and collaborators in [26] present a system based on FCMs where the inference process is carried out through a computational linguistic model based on the symbolic approach ordinal scales to represent the activation values of the concepts and causal relationships. A similar investigation was carried out in [52]. Despite its simplicity and easy interpretation of the results, this model maintains an approximation process that produces a loss of information [53].

Pérez-Teruel and collaborators in [54] propose a decision-making model based on FCMs using the 2-tuple linguistic model. The use of this model allows computing processes to be carried out with words without loss of information, based on the concept of symbolic translation [53]. In another work [55], FCMs are used with the 2-tuple linguistic model to model the consensus process in decision-making.

Gonçalves and collaborators [56] use the CWW paradigm in FCMs to model the interdependencies between risks in project portfolios. In this work, the 2-tuple linguistic model is adopted to represent the map relations.

The analyzed works use the 2-tuple linguistic model only during the construction of the map. These works do not use this linguistic model during inference, an element that constitutes a line open to research.

In general, the linguistic FCMs previously analyzed do not take into account the relationship of indeterminacy between the concepts, an element that affects the effectiveness of the map and constitutes an opportunity to improve them. Furthermore, no work on the application of these maps to the solution of multistage sequential problems is reported.

## 4.2 *Competitive Fuzzy Cognitive Maps (CFCMs)*

Competitive fuzzy cognitive maps (CFCMs) are frequently used in differential medical diagnosis and consist of two types of concepts: decisions and factors [57, 58], as show in Fig. 9.

Each decision concept represents a single diagnosis, and in order to always infer a single diagnosis, the decision concepts are mutually exclusive. This aspect of CFCMs is considered in the present investigation. CFCMs are simple structures that do not take into account the multistage sequential approach, nor do they have the capacity to represent phenomena such as indeterminacy and imprecision. On the other hand, this map has the same limitations as traditional FCMs in terms of interpretability.

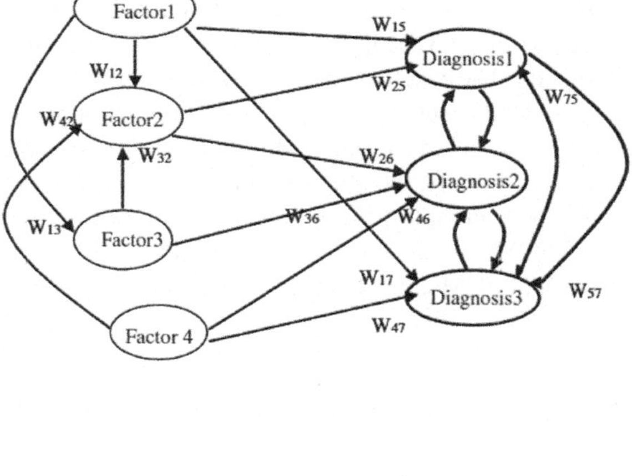


**Fig. 9** Competitive FCMs

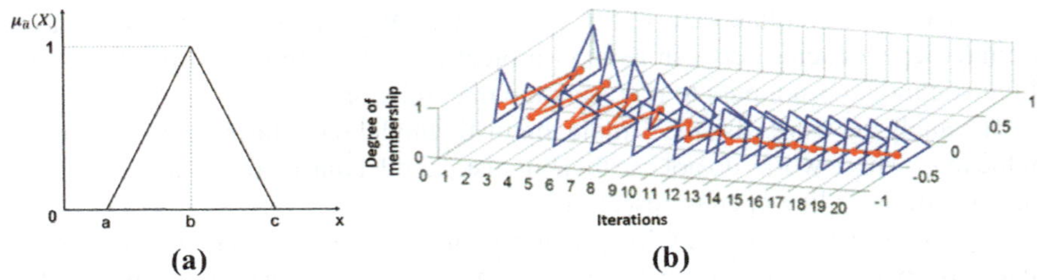


**Fig. 10** Illustration of a triangular fuzzy number (**a**) and the inference process with triangular fuzzy numbers (**b**), taken from [37]

## *4.3 Triangular Fuzzy Cognitive Maps (TrFCMs)*

In triangular fuzzy cognitive maps (TrFCMs), the numerical values of traditional FCMs are replaced by triangular fuzzy numbers [50], see Fig. 10, which improve the treatment of uncertainty and the aggregation of information from different sources [59]. Although these maps improve the treatment of uncertainty and imprecision, they do not consider the indeterminacy or falsity that exist in the relationships between concepts. Nor have they been applied to solve multistage, sequential decision-making problems.

## *4.4 Case-Based Fuzzy Cognitive Maps (CBFCMs)*

Figure 11 presents the structure of case-based FCMs (CBFCMs) [60, 61]. CBFCMs work as follows: when the decision inferred by the map is not distinctive, the case-based system (CBR) is activated [62], using the same input concepts of the map as input data. The case-based system will return the case decision with greater similarity to the stored cases. From this result, the map weights are updated by inhibiting the connections of the input concepts of the returned case with the rest of the map decisions. Finally, the map simulation process is repeated with the adjusted weights. In this way, greater differentiation is achieved between the map output decisions.

CBFCMs do not take into account the multi-stage sequential approach, and like traditional FCMs, these maps do not handle indeterminacy.

## *4.5 Fuzzy Gray Cognitive Maps (FGCMs)*

Fuzzy gray cognitive maps (FGCMs) are a combination of FCMs and gray systems' theory (GST) [63]. They were proposed to improve the treatment of uncertainty associated with the existence of different ways of representing map relationships [64]. In these maps, the strength of the relationship between two concepts is described

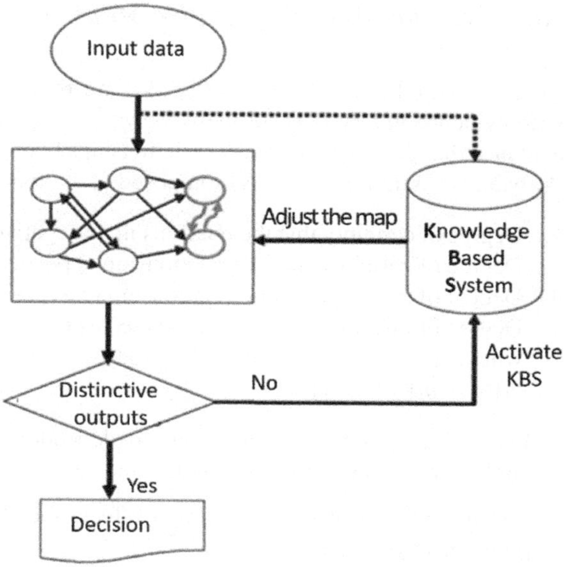


**Fig. 11** Case-based FCMs

by a gray number composed of a lower limit $\underline{W}$ and an upper limit $\overline{W}$. A gray number ($\otimes W$) is a number whose exact value is unknown, but the interval that contains it is known and is represented as $\otimes W \in \left[\underline{W}, \overline{W}\right]$ [65]:

- Black relationships: these are causal relationships with an unknown intensity, an element that can be used to represent indeterminacy. This type of relationship is assigned the interval value $[-1, 1]$.
- Gray relationships: these are causal relationships whose exact intensity is unknown. This relationship is managed by the FGCMs through an interval value of lower amplitude than the one represented in the black relationships.
- White relationships: these are causal relationships whose intensity is known exactly and therefore is represented with an exact numerical value.

A positive element of FGCMs is that they somehow try to model indeterminacy when representing black relationships. But they do so with an approach that focuses on establishing a greater or lesser degree of precision that affects everything from construction to inference [66]. As a consequence, these maps do not handle the levels of indeterminacy and falsity in a differentiated way. On the other hand, FGCMs constitute simple structures that have not been designed for scenarios with a multistage sequential approach.

## 4.6 Evidence-Based Cognitive Maps (ECMs)

Evidence-based cognitive maps (ECMs) are based on the combination of FCMs with evidence theory [67]. In these maps, the experts' preferences are considered evidence that describes the causal relationships between the concepts through Basic Probability Assignments (BPA) composed of the four elements, as shown in Fig. 12.

1. Degree of certainty that the relationship is negative and is denoted ($m\{-1\} = a$).
2. Degree of certainty that the relationship is positive and is denoted ($m\{1\} = b$).
3. Degree of certainty that no relationship exists and is denoted ($m\{0\} = c$).
4. Degree of uncertainty that is calculated as ($m\{-1, 0, 1\} = 1 - a - b - c$). where $m\{-1\}$, $m\{1\}$, $m\{0\}$, $m\{-1, 0, 1\} \in [0, 1]$, such that $m\{-1\} + m\{1\} + m\{0\} + m\{-1, 0, 1\} = 1$.

In these maps, the four components are dependent on each other. In the work [69] the authors point out the following deficiencies:

- The use of this type of restriction affects the flexibility of the expression of preferences by experts.
- The fact that experts must express multiple values for each relationship increases the complexity of the map construction process.

The authors of the paper [67] point out that these maps can generate unreasonable results due to some deficiencies in the modeling process, including the aggregation of contradictory information, which affects their predictive capacity.

A relevant element of this proposal is that it proposes a method to model indeterminacy and achieve differentiation between positive and negative relationships. But indeterminacy is represented as a calculable value based on the degrees of positive

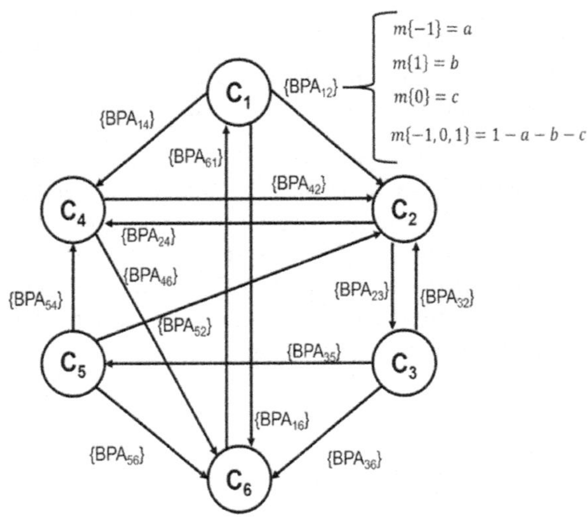


**Fig. 12** Evidence-based cognitive map, taken from [68]

and negative relationships and the non-existence of a relationship, and it cannot be expressed as a preference directly by experts. In inference, these maps combine the degrees represented in the experts' preferences to obtain an interval. As a consequence, the ability to know the degree of indeterminacy in the responses is lost. Furthermore, ECMs have not been applied to multistage, sequential decision-making problems.

### *4.7 Fuzzy Cognitive Maps Based on Distributed Degrees of Belief (BDD-FCMs)*

FCMs based on distributed degrees of belief (BDD-FCMs) were introduced in [70] and allow experts to express their preferences about the relationships between concepts through numerical values, linguistic terms, intervals or belief structures composed of more than one linguistic term [71, 72]. For example, an expert may express his/her opinion about the strength of the relationship between two concepts in the following way: he/she is 60% certain that the relationship is "negatively high," 20% is indeterminate, and 10% is "positively weak." The belief structure in this case would be {(negatively high, 0.6), (indeterminate, 0.2), (positively weak, 0.1)}.

In this map, the authors establish that the sum of the degrees of belief in a relationship cannot be greater than one. This restriction affects the flexibility of the expression of preferences by experts [73].

The inference process is carried out after transforming the belief structures into numerical values. In this process, the degree of indetermination that is not reflected in the final decisions is lost. On the other hand, the final results of the inference process are numerical, an element that affects interpretability. Another shortcoming of this extension is its inability to model multi-stage sequential decision-making problems.

### *4.8 Approximate Cognitive Networks (RCNs)*

Rough cognitive networks (RCNs) are FCMs based on rough set theory (RST) [74]. Another work that extends RCNs is fuzzy RCNs (FRCNs) proposed by Nápoles himself in [75]. As a positive element of these maps, it is noted that they take advantage of the capacity of approximate sets to handle the uncertainty caused by inconsistency.

The construction and exploitation of the RCNs include the following three stages:

- The first is associated with the process of constructing the information granules using RST, in which the positive and negative regions and borders for each decision class are obtained from the training data and a pre-constructed inseparability relationship.

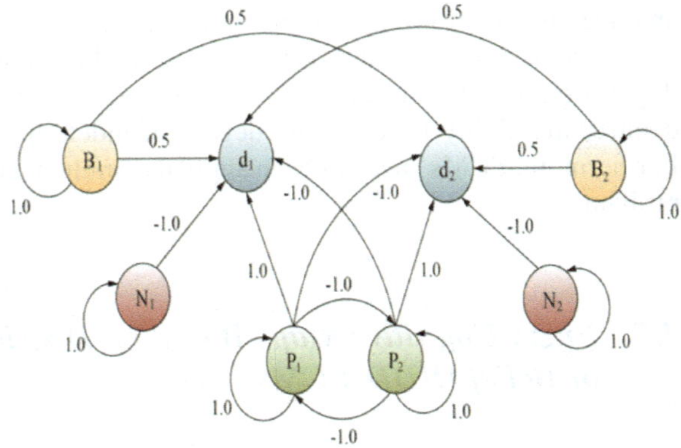


**Fig. 13** RCN for a decision-making problem with two decisions, taken from [74]

- The second stage consists of the automatic construction of the map, where the nodes correspond to the built regions, see Fig. 13.
- The third stage corresponds to the exploitation of the map through the evaluation of new instances.

Although no experts are involved in the map construction process, as the authors of the work point out, this map requires the existence of a decision system that the experts in previous stages had to clearly establish: the set of attributes that describe the objects, the set of decision attributes, and the separability relationship.

On the other hand, the map constructed in the second stage only allows calculating the level of activation associated with the decision attributes, an element that reduces the flexibility of the maps to be able to calculate the activation of the concepts involved in the problem in question. Another element to point out is that in these maps, the process of explaining the results of the inference is difficult, which makes them unintuitive for experts, affecting their interpretability. This element is pointed out by the author himself [75]. The very nature of these maps makes their application difficult for solving multistage sequential problems.

## *4.9 Rule-Based Fuzzy Cognitive Maps (RBFCMs)*

Another way of representing map relationships is through rules. This has given rise to rules-based maps, or RBFCMs. In this sense, the following approaches are presented:

- Cooperative approaches between rule-based and map-based systems:

    a. *Offline* approach based on the use of rules only in the map construction stage [76].
    b. *Online* approach where fuzzy inference systems are used to calculate the weight of the relationships between concepts during inference [77].

- Hybrid approaches with a strong connection between rule-based and map-based systems:
  a. With rules in map relations [78].
  b. Where the map is completely represented by a fuzzy inference system [79].

The *Offline* cooperative approach identified in the work [76] is based on experts expressing the relationships between the concepts of the map using rules, but these rules are only used to construct the map and identify the strength of the relationships that are finally represented numerically, and from this moment on, the inference is carried out like the traditional approaches of FCMs. A positive element of this approach is that the use of rules makes it easier for experts to express their preferences. As a negative element, it is noted that it does not sufficiently take advantage of all the knowledge expressed by the experts in the rules by following, during the inference and in the presentation of the results, a traditional approach.

In the *Online* cooperative approach that is represented in [77] the use of a Mamdani-type fuzzy inference system [80] is identified as a novel element for calculating the weight of the relationships between concepts, this is carried out dynamically during the calculation of the activation levels of each concept. As a negative element, it is noted that the experts must build the rules manually, which implies a high level of effort on their part during the construction of the map.

An example of the hybrid approach is introduced by Carvalho and collaborators [78] where the map relations are represented by rules. This approach is also inspired by Mamdani-type fuzzy inference systems. A limitation of this work is that the authors propose a set of predetermined membership functions such that the semantics of the fuzzy sets influence their amplitude, for example, the fuzzy set "*High*" has a greater area than the fuzzy set "*Low*". According to [81] this behavior limits the flexibility and applicability of the map. This same author in his work [82] points out that Carvalho's proposal has a high computational cost.

The hybrid approach, where the map is completely represented by a fuzzy inference system, is proposed in [79]. In this work, in each iteration, the fuzzy inference system is executed, and its outputs make up the set of objects that were activated and constitute input to the next iteration. In his proposal, Barriba constructs the fuzzy inference system exhaustively from the combination of all fuzzy sets. Carvalho [83] criticizes this approach due to its complexity, associated with the high number of rules needed to describe the behavior of the system. Furthermore, he criticizes the fact that the introduction of a new concept to the model exponentially increases the number of rules in the model and also forces the modification of all existing rules, affecting the versatility and applicability of the model.

In general, in all the above approaches, the treatment of indeterminacy and falsity is not identified. This element is treated in the present investigation.

In the bibliography consulted, in this type of map, the rules are constructed in a perspective manner, which increases the complexity of the construction process [81]. It is considered that this situation can be improved and constitutes a motivation for the proposals developed in this research.

It is considered that this situation can be improved and constitutes a motivation for the proposals developed in this research. A positive element of these maps is the interpretability they offer from the set of linguistic rules used to represent causal relationships, since each rule in itself provides a portion of information, and all the rules describe all the information about the system in question [81]. This element is considered in the present investigation.

On the other hand, these maps have not been applied to multistage, sequential decision-making problems. The validation of these works has been through case studies, they are algorithms that are not public, and the descriptions in the articles make comparison with these map models difficult.

### 4.9.1 Intuitionistic Fuzzy Cognitive Maps (iFCMs)

Intuitionistic fuzzy cognitive maps (iFCMs) use intuitionistic fuzzy sets (IFSs) for map construction [84]. IFSs extend fuzzy sets; just as these include a membership function, but they also incorporate a non-membership function that indicates the degree of non-membership (falsehood) of an element to a fuzzy set, see Definition 1.

**Definition 1** ([85]) Given a discourse universe X, an intuitionistic fuzzy set A in X is: $A = \{x, \mu_A(x), \gamma_A(x) | x \in X\}$, where $\mu_A : X \rightarrow [0, 1]$ and $\gamma_A : X \rightarrow [0, 1]$ they define the degrees of membership and non-membership, respectively, of an element $x \in X$ to the set$A \subset X$. For each$x \in X$, it holds that $0 \leq \mu_A \leq 1, 0 \leq \gamma_A \leq 1$ and$0 \leq \mu_A(x) + \gamma_A(x) \leq 1$. The function $\pi_A(x) = 1 - \mu_A(x) - \gamma_A(x)$ represents the degree of doubt (hesitation) of the element $x \in X$ to the set$A \subset X$, considered as the complement of the degrees of membership and non-membership [86].

The iFCMs extend the traditional FCM maps by considering the degree of doubt (*hesitation*) $\pi_A(x)$ of the experts at the time of determining the weight of the map relations. Assuming that the degree of doubt among the experts has a negative impact on the causal relationships, the activation function used in traditional FCMs is reformulated as shown in Eq. (2) [87].

$$A_i^{t+1} = f\left(A_i^t + \sum_{i=1}^{n} w_{ji}^{\mu} \cdot A_j^t - w_{ji}^{\pi} \cdot A_j^t\right) \tag{2}$$

where $w_{ji}^{\mu} \in [-1, 1]$ and $w_{ji}^{\pi} \in [-1, 1]$ represent the weight of the relationship and the degree of doubt $\pi_A(x)$ that corresponds to the connection of concept $i$ to concept $j$ [88].

Some works [14, 84] point out that iFCMs can generate confusing or unreasonable inference results as a result of operations with IFSs [85].

The inadequacy of these maps is that they are not capable of representing indeterminacy or estimating it as a concept independent of the degrees of certainty (membership) and falsity (non-membership). Furthermore, restrictions on the representation of relationships affect the flexibility of the expression of preferences by experts [73].

In the bibliography consulted, only 8% of the works use this type of map, and it is identified that these maps have been gradually replaced by the NCMs neutrosophic cognitive maps, which represent more than 30% of the bibliography consulted. On the other hand, iFCMs have not been applied to the solution of multistage sequential problems.

### 4.9.2 Neutrosophic Cognitive Maps (NCMs)

Neutrosophic theory is an area of knowledge that deals with neutrality or indeterminacy in the context of decision-making, it constitutes a generalization of fuzzy logic theory [89] and intuitionistic logic theory [85]. The neutrosophic theory offers a mathematical representation of uncertainty and indeterminacy, where each proposition has a degree of certainty, indeterminacy, or falsity [90].

In this research, the definition of indeterminacy proposed by Florentín Smarandache [91], is assumed as the degree that characterizes the elements that do not have absolute membership in a set (certainty) and cannot be assured that they do not belong. to the set itself in an absolute way (falsehood). That is, indeterminacy includes the elements that are between the opposite sets <A> and <antiA>, and is denoted by <neutA>.

In the bibliography consulted, the extension of FCMs introduced by the neutrosophic theory, known as the neutrosophic cognitive map (NCM), represents indeterminacy through the symbol $I$ [92]. However, the use of the $I$ symbol during inference does not allow the levels of indeterminacy to be quantified and causes loss of information, affecting the effectiveness of the map with respect to the prediction capacity and the interpretability of the results [93]. This situation denotes the need to develop new extensions that modify traditional inference mechanisms using mathematical models established in neutrosophic theory.

In general, the treatment of indeterminacy influences the main elements that characterize FCMs, such as topology, the way in which maps are constructed, and the inference mechanism.

A search was carried out in the period from 2003 to 2021 with the term "*Neutrosophic cognitive maps*" and 26 publications were found in the *Scopus* database and 19 publications in other databases with different levels of indexing. Of the 45 publications found, 20 works correspond to the year 2020. It is important to note that, although the number of works related to the FCMs identified in the systematic review is high (3,905), the number of publications associated with the treatment of indeterminacy is low (0.011%). A detailed analysis of these works is presented below.

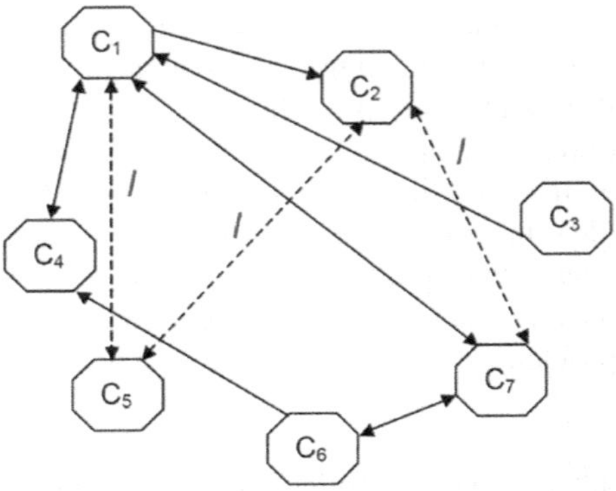


**Fig. 14** Neutrosophic cognitive map

Neutrosophic cognitive maps (NCMs) were introduced in 2003 [94] by combining FCMs with neutrosophic theory [24] and have been used in decision-making scenarios with the presence of indeterminacy [95].

An NCM is a cognitive map where the relationships between concepts are represented by values in the interval $[-1, 1]$ or by the symbol "*I*" [96] see Fig. 14. As a consequence, in this type of map, the arithmetic operations where the symbol "*I*" appears [97] are redefined. Hereinafter, this extension of the FCMs is called *NCM_Indeterminacy* to facilitate its identification and its comparison with the other extensions proposed in this research.

In inference, a normalization function is applied that transforms the activation values of the concepts in each iteration to 0, 1, or $I$ based on a threshold $k \in [0, 1]$ established as a parameter [98, 99].

Given an activation vector $A = (b_1, b_2, \ldots, b_n)$, it is established:

- $b_i = 0$, if $b_i < k$, this means that if the activation value of a concept is less than the threshold $k$, the concept is penalized by reducing its activation level to 0, causing loss of information.
- $b_i = 1$, if $b_i > k$, this means that if the activation value of a concept is greater than the threshold $k$, in this case the activation level is overestimated, which may affect the prediction ability.
- $b_i = I$, if $b_i$ is not an integer, the use of the symbol $I$ during the inference limits the ability to work differentially on the degrees of indeterminacy and the possible assignment of linguistic terms to the different levels of indeterminacy, affecting the interpretability of the map results, nor does it allow the degrees of certainty and falsity made possible by the theory of neutrosophic sets to work in an integrated manner.

## 5 Analysis of Extensions of Multiple Fuzzy Cognitive Maps

The simple extensions mentioned above are not suitable for modeling complex systems composed of multiple processes. In this way, multiple topologies of fuzzy cognitive maps (m-FCMs) arise, which are based on the decomposition of a problem into subproblems that are managed by interrelated subsystems. Each subsystem is then modeled by an individual map, but the multiple maps interact with each other to achieve an ultimate goal. In this section, these extensions are discussed with respect to the solution of multistage sequential problems and the treatment of indeterminacy.

### *5.1 Hierarchical Fuzzy Cognitive Maps (JFCMs)*

In the JFCMs hierarchical fuzzy cognitive maps, a hierarchical structure is established where there is a supervisory map at the top level and different maps at the lower levels [15], as shown in Fig. 15. The supervisory map is responsible for inferring the final outputs by evaluating all the information received from the maps at the lower levels. These maps cannot handle multistage sequential problems because, semantically, they represent content relations and not sequential relations. Furthermore, among the maps of the lower levels, relationships of sequentiality but also of competition between all are not represented.

### *5.2 Distributed Fuzzy Cognitive Maps (DFCMs)*

Distributed fuzzy cognitive maps DFCMs are based on modeling different maps in a cyclic structure where each map receives inputs from the others and delivers the outputs to the others under a bidirectional connection [16], see Fig. 16. Relationships are established between the distributed maps of cyclical competition that are not designed to represent relationships of sequentiality. As a consequence, DFCMs cannot model multistage, sequential problems.

### *5.3 Multilayer Fuzzy Cognitive Maps (MFCMs)*

The extension of multilayer fuzzy cognitive maps MFCMs is made up of small map structures organized in layers and grouped with a specific objective [100]. This structure has a main map in the upper layer linked in a hierarchical structure with the lower layers [101], as show in Fig. 17.

In multilayer maps the activation starts at the lower levels to the higher levels. Content relationships and not sequential relationships are established between the

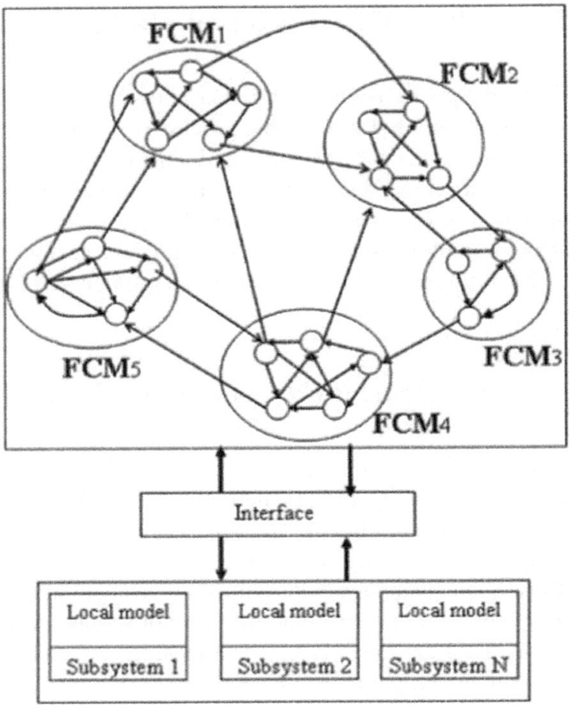


**Fig. 15** Hierarchical FCMs

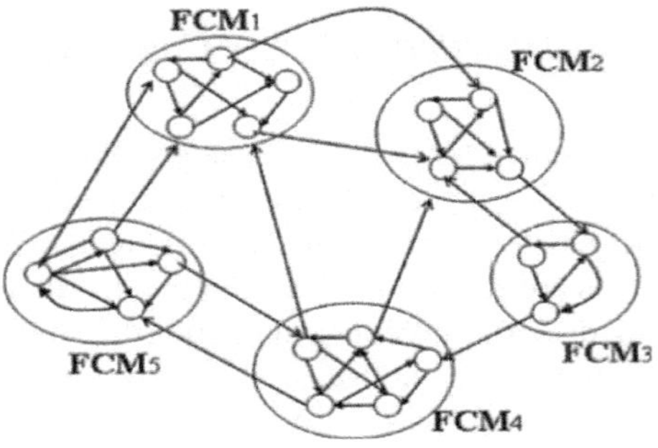


**Fig. 16** Distributed FCMs

different maps, so forcing them to represent multistage sequential problems would affect the interpretation of the results they generate.

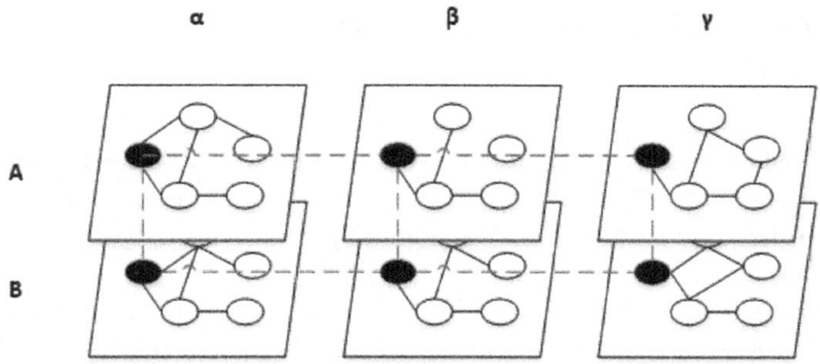


**Fig. 17** Multilayer FCMs

## 5.4 *Parallel Fuzzy Cognitive Maps (PFCMs)*

The extension of parallel fuzzy cognitive maps PFCMs establishes another different form of communication between the maps. In this case, there is a coordinator module that controls the entire simulation process under a given system of rules and restrictions [102]. This type of map, as its name indicates, allows individual execution of maps in parallel and does not establish sequential relationships between the maps, see Fig. 18.

In general, the aforementioned extensions of multiple FCMs are not sufficient for modeling some complex problems, such as multistage sequential decision-making; such as in phenomena where diagnosis, decision and prognosis are worked on as a whole. This is because these extensions establish hierarchical and non-sequential integration structures or maintain simultaneity scenarios in the activation of all concepts. On the other hand, these extensions of multiple FCMs do not address indeterminacy.

**Fig. 18** Parallel FCMs

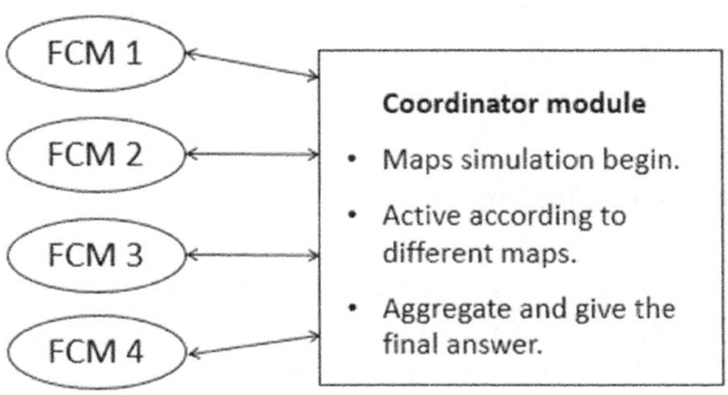

## 5.5 Analysis of Validation Methods Used in the Research Consulted

This section analyzes the main validation strategies used in 150 investigations carried out between 1999 and 2021. The following categories are used to characterize the analyzed works: case studies, data triangulation, methods triangulation and expert triangulation.

From the analysis, it can be seen that 65% of the research is validated through case studies and simple qualitative analyses. In addition, it is observed that in 53 investigations (35%) the triangulation of methods is applied, and most of them are compared with models based on traditional FCMs and their extensions proposed by other authors.

On the other hand, it can be seen that only 10 investigations (7%) combine method triangulation with data triangulation. It is also observed that most of the investigations use small databases with less than 100 cases; in a few investigations [18, 25] larger databases are used.

Regarding the comparison metrics, the precision in the final answers, the prediction error and the number of iterations to reach convergence stand out. In total, 53 works (35%) are identified that apply the triangulation of methods with different comparison metrics; 13 of them apply precision in the final answers as a metric; nine use convergence analysis; and 16 work with error in the prediction. Five publications combine precision in final responses and convergence analysis; seven papers combine convergence analysis with prediction error; and only three papers apply the three metrics. These comparison criteria are considered in the experimentation of this research.

Furthermore, in the searches carried out, a total of 45 investigations were found based on the *NCM_Indeterminacy* neutrosophic cognitive maps, of which 36 (80%) were validated using only the case study technique and simple qualitative analysis. A triangulation of methods is applied in nine investigations (20%). Only two studies combine method triangulation with data triangulation and perform statistical cross-validation tests.

# 6 Conclusions

It is identified that traditional FCMs and most of their extensions lack the representation of indeterminacy. In the bibliography consulted, extensions of FCMs are reported, such as intuitionistic fuzzy cognitive maps (iFCMs), which try to represent indeterminacy, but due to their nature, they do not achieve significant results and have been gradually replaced by neutrosophic cognitive maps (*NCMs_Indeterminacy*), which represent up to 30% of the bibliography consulted.

*NCMs_Indeterminacy* represents indeterminacy using the symbol $I$, which causes information loss in inference and affects predictability and interpretability.

Furthermore, it is concluded that extensions that combine linguistic models with FCMs improve interpretability, but these extensions do not address indeterminacy, and there are no works that show their use for multistage sequential decision-making.

In the analysis of rule-based FCMs, four approaches are identified that only handle the certainty of the information. Maps built under the cooperative approach only use the rules in the map construction stage, while maps based on the hybrid approach use the rules in the inference process. In the bibliography consulted, the validation of these works has been through case studies; they are algorithms that are not sufficiently described in the articles, which makes comparison with these maps difficult.

It is also identified that multiple cognitive maps such as JFCMs, DFCMs, MFCMs and PFCMs establish hierarchical and non-sequential structures of integration, which makes their application difficult in multistage sequential decision-making problems.

- Need for new extensions of FCMs that integrately analyze the diagnosis, decision and prognosis processes as a single decision-making problem.
- Deficiencies in FCMs and their extensions in the treatment of indeterminacy.
- There is a loss of information in the neutrosophic cognitive maps reported in the bibliography consulted, affecting their effectiveness in solving decision-making problems with respect to their prediction capacity.
- There are opportunities to improve interpretability both in the construction process and in the analysis of the inference results of the FCMs.

As future work, the need to combine fuzzy cognitive maps with other predictive models to facilitate the explainability of inference processes.

## References

1. Kosko B (1986) Fuzzy cognitive maps. Int J Man Mach Stud 24:65–75
2. Meskauskas Z, Jasinevicius R, Kazanavicius E, Petrauskas V (2020) XAI-based fuzzy SWOT maps for analysis of complex systems. In: 2020 IEEE international conference on fuzzy systems (FUZZ-IEEE). IEEE, Glasgow, UK, pp 1–8
3. Denizci A, Karadeniz S, Ulu C (2021) Fuzzy cognitive map based PI controller design. In: Kahraman C, Cevik Onar S, Oztaysi B, Sari IU, Cebi S, Tolga AC (eds) Intelligent and fuzzy techniques: smart and innovative solutions. Springer International Publishing, Cham, pp 1250–1257
4. Bakhtavar E, Valipour M, Yousefi S, Sadiq R, Hewage K (2021) Fuzzy cognitive maps in systems risk analysis: a comprehensive review. Complex Intell Syst 7:621–637. https://doi.org/10.1007/s40747-020-00228-2
5. Nápoles G, Jastrzębska A, Salgueiro Y (2021) Pattern classification with evolving long-term cognitive networks. Inf Sci 548:461–478. https://doi.org/10.1016/j.ins.2020.08.058
6. Apostolopoulos ID, Groumpos PP (2021) Advanced fuzzy cognitive maps: state-space and rule-based methodology for coronary artery disease detection. Biomed Phys Eng Express 7:045007. https://doi.org/10.1088/2057-1976/abfd83
7. Akinnuwesi BA, Adegbite BA, Adelowo F, Ima-Edomwonyi U, Fashoto G, Amumeji OT (2020) Decision support system for diagnosing Rheumatic-Musculoskeletal Disease using fuzzy cognitive map technique. Inform Med Unlock 18:100279. https://doi.org/10.1016/j.imu.2019.100279

8. Amirkhani A, Papageorgiou EI, Mohseni A, Mosavi MR (2017) A review of fuzzy cognitive maps in medicine: Taxonomy, methods, and applications. Comput Methods Programs Biomed 142:129–145. https://doi.org/10.1016/j.cmpb.2017.02.021
9. Chen H, Zhang L, Wu X (2020) Performance risk assessment in public–private partnership projects based on adaptive fuzzy cognitive map. Appl Soft Comput 93:106413. https://doi.org/10.1016/j.asoc.2020.106413
10. Nikas A, Stavrakas V, Arsenopoulos A, Doukas H, Antosiewicz M, Witajewski-Baltvilks J, Flamos A (2020) Barriers to and consequences of a solar-based energy transition in Greece. Environ Innov Soc Trans 35:383–399. https://doi.org/10.1016/j.eist.2018.12.004
11. Jahangoshai Rezaee M, Yousefi S, Baghery M, K. Chakrabortty R (2021) An intelligent strategy map to evaluate improvement projects of auto industry using fuzzy cognitive map and fuzzy slack-based efficiency model. Comput Ind Eng 151:106920. https://doi.org/10.1016/j.cie.2020.106920
12. Stylios CD, Bourgani E, Georgopoulos VC (2020) Impact and applications of fuzzy cognitive map methodologies. In: Kosheleva O, Shary SP, Xiang G, Zapatrin R (eds) Beyond traditional probabilistic data processing techniques: interval, fuzzy etc. methods and their applications. Springer International Publishing, Cham, pp 229–246
13. Groumpos PP (2023) A critical historic overview of artificial intelligence: issues, challenges, opportunities and threats. Artif Intell Appl 1. https://doi.org/10.47852/bonviewAIA3202689
14. Zhang Y, Qin J, Shi P, Kang Y (2019) High-order intuitionistic fuzzy cognitive map based on evidential reasoning theory. IEEE Trans Fuzzy Syst 27:16–30. https://doi.org/10.1109/TFUZZ.2018.2853727
15. Stylios CD, Georgopoulos VC, Malandraki GA, Chouliara S (2008) Fuzzy cognitive map architectures for medical decision support systems. Appl Soft Comput 8:1243–1251. https://doi.org/10.1016/j.asoc.2007.02.022
16. Bourgani E, Stylios C, Georgopoulos V, Manis G (2013) A study on fuzzy cognitive map structures for medical decision support systems. Atlantis Press, Milano, Italy
17. Papageorgiou EI, Salmeron JL (2012) Learning fuzzy grey cognitive maps using nonlinear Hebbian-based approach. Int J Approx Reason 53:54–65. https://doi.org/10.1016/j.ijar.2011.09.006
18. Judy MV, Soman G (2018) Parallel fuzzy cognitive map using evolutionary feature reduction for big data classification problem. In: Mandal JK, Sinha D (eds) Social transformation—Digital way. Springer Singapore, pp 226–239
19. Cogollo J, Correa A (2019) Modeling supply chain quality management using multi-layer fuzzy cognitive maps. In: 2019 IEEE international conference on fuzzy systems (FUZZ-IEEE). IEEE, New Orleans, USA, pp 1–6
20. Papageorgiou EI, Iakovidis DK (2013) Intuitionistic fuzzy cognitive maps. IEEE Trans Fuzzy Syst 21:342–354. https://doi.org/10.1109/TFUZZ.2012.2214224
21. Wang J, Guo AQ (2018) Ensemble interval-valued fuzzy cognitive maps. IEEE Access 6:38356–38366
22. Ramalingam S, Govindan K, Kandasamy WBV, Broumi S (2019) An approach for study of traffic congestion problem using fuzzy cognitive maps and neutrosophic cognitive maps-the case of indian traffic. Neutrosophic Sets Syst 30:273–283. https://doi.org/10.5281/zenodo.3569712
23. Vasantha WB, Kandasamy I, Devvrat V, Ghildiyal S (2019) Study of imaginative play in children using neutrosophic cognitive maps model. Neutrosophic Sets Syst 30. https://doi.org/10.5281/zenodo.3569702
24. Smarandache F (1999) A unifying field in logics: neutrosophic logic. In: Neutrosophy, neutrosophic set, neutrosophic probability, 2nd edn. American Research Press, Rehoboth N.M.
25. Chithra B, Nedunchezhian R (2020) Dynamic neutrosophic cognitive map with improved cuckoo search algorithm (DNCM-ICSA) and ensemble classifier for rheumatoid arthritis (RA) disease. J King Saud Univ Comput Inf Sci. https://doi.org/10.1016/j.jksuci.2020.06.011

26. Frías M, Filiberto Y, Nápoles G, Vanhoof K, Bello R (2017) Fuzzy cognitive maps reasoning with words: the ordinal case. In: 2nd international symposium on fuzzy and rough sets. Cuba
27. Frías M, Filiberto Y, Nápoles G, García-Socarrás Y, Vanhoof K, Bello R (2018) Fuzzy cognitive maps reasoning with words based on triangular fuzzy numbers. In: Castro F, Miranda-Jiménez S, González-Mendoza M (eds) Advances in soft computing. Springer International Publishing, Cham, pp 197–207
28. Gough D, Oliver S, Thomas J (2017) An introduction to systematic reviews. Los Angeles
29. Al-subhi SH, Rubio PAR, Pérez PP, Papageorgiou EI, Vacacela RG, Mahdi GSS (2020) A new neutrosophic clinical decision support model for the treatment of pregnant women with heart diseases. Reviste Investigación Operacional 41:780–790
30. Al-subhi SH, Rubio PAR, Pérez PP, Mahdi GSS, Leyva-Vázquez M (2020) Novedosa herramienta de apoyo para tomar decisiones en diagnóstico, tratamiento y pronóstico de cardiópatas embarazadas. Revista Cubana de Obstetricia y Ginecología 46
31. Axelrod R (1976) Structure of decision: the cognitive maps of political elites. Princeton University Press, Princeton, N.J.
32. Hajek P, Froelich W, Prochazka O (2020) Intuitionistic fuzzy grey cognitive maps for forecasting interval-valued time series. Neurocomputing 400:173–185. https://doi.org/10.1016/j.neucom.2020.03.013
33. Wang J, Peng Zhen, Wang X, Li C, Wu J (2020) Deep fuzzy cognitive maps for interpretable multivariate time series prediction. IEEE Trans Fuzzy Syst 1–1. https://doi.org/10.1109/TFUZZ.2020.3005293
34. Paiva BMR, Ferreira FAF, Carayannis EG, Zopounidis C, Ferreira JJM, Pereira LF, Dias PJVL (2021) Strategizing sustainability in the banking industry using fuzzy cognitive maps and system dynamics. Int J Sust Dev World 28:93–108. https://doi.org/10.1080/13504509.2020.1782284
35. Sierra LA, Yepes V, García-Segura T, Pellicer E (2018) Bayesian network method for decision-making about the social sustainability of infrastructure projects. J Clean Prod 176:521–534. https://doi.org/10.1016/j.jclepro.2017.12.140
36. Nasirzadeh F, Ghayoumian M, Khanzadi M, Cherati MR (2020) Modelling the social dimension of sustainable development using fuzzy cognitive maps. Int J Constr Manag 20:223–236. https://doi.org/10.1080/15623599.2018.1484847
37. Yesil E, Dodurka MF, Urbas L (2014) Triangular fuzzy number representation of relations in fuzzy cognitive maps. In: 2014 IEEE international conference on fuzzy systems (FUZZ-IEEE). IEEE, Beijing, China, pp 1021–1028
38. Papageorgiou EI (2014) Fuzzy cognitive maps for applied sciences and engineering: from fundamentals to extensions and learning algorithms. Springer Science & Business Media, Berlin
39. Harmati IÁ, Kóczy LT (2020) Notes on the rescaled algorithm for fuzzy cognitive maps. In: Kóczy LT, Medina-Moreno J, Ramírez-Poussa E, Šostak A (eds) Computational intelligence and mathematics for tackling complex problems. Springer International Publishing, Cham, pp 43–49
40. Aguilar J, Téran O, Sánchez H, de Mesa JG, Cordero J, Chávez D (2017) Towards a fuzzy cognitive map for opinion mining. Procedia Comput Sci 108:2522–2526. https://doi.org/10.1016/j.procs.2017.05.287
41. Bueno S, Salmeron JL (2009) Benchmarking main activation functions in fuzzy cognitive maps. Expert Syst Appl 36:5221–5229. https://doi.org/10.1016/j.eswa.2008.06.072
42. Chen T, Chiu Y-T (2021) A study of dynamic fuzzy cognitive map model with group consensus based on linguistic variables. Technol Forecast Soc Chang 171:120948. https://doi.org/10.1016/j.techfore.2021.120948
43. Mendel JM, Zadeh LA, Trillas E, Yager R, Lawry J, Hagras H, Guadarrama S (2010) What computing with words means to me [discussion forum]. IEEE Comput Intell Mag 5:20–26. https://doi.org/10.1109/MCI.2009.934561
44. Tong RM, Bonissone PP (1980) A linguistic approach to decisionmaking with fuzzy sets. IEEE Trans Syst Man Cybern 10:716–723. https://doi.org/10.1109/TSMC.1980.4308391

45. Herrera F, Martínez L (2000) A 2-tuple fuzzy linguistic representation model for computing with words. IEEE Trans Fuzzy Syst 8:746–752. https://doi.org/10.1109/91.890332
46. Rickard JT, Aisbett J, Yager RR (2015) Computing with words in fuzzy cognitive maps. In: 2015 Annual conference of the north american fuzzy information processing society (NAFIPS) held jointly with 2015 5th world conference on soft computing (WConSC), pp 1–6
47. Rickard JT, Aisbett J, Yager RR (2015) A new fuzzy cognitive map structure based on the weighted power mean. IEEE Trans Fuzzy Syst 23:2188–2202. https://doi.org/10.1109/TFUZZ.2015.2407906
48. Najafi A, Amirkhani A, Papageorgiou EI, Mosavi MR (2017) Medical decision making based on fuzzy cognitive map and a generalization linguistic weighted power mean for computing with words. In: 2017 IEEE international conference on fuzzy systems (FUZZ-IEEE). IEEE, Naples, Italy, pp 1–6
49. Gupta M, Garg K, Jain R (2021) Multi-criteria and fuzzy-based decision making: applications in environment pollution control for sustainable development. In: Big data analysis for green computing. CRC Press, pp 1–15
50. Frias M, Filiberto Y, Nápoles G, Falcon R, Bello R, Vanhoof K (2019) Comparative analysis of symbolic reasoning models for fuzzy cognitive maps. In: Bello R, Falcon R, Verdegay JL (eds) Uncertainty management with fuzzy and rough sets: recent advances and applications. Springer International Publishing, Cham, pp 127–139
51. Akram M, Niaz Z, Feng F (2023) Extended CODAS method for multi-attribute group decision-making based on 2-tuple linguistic Fermatean fuzzy Hamacher aggregation operators. Granul Comput 8:441–466. https://doi.org/10.1007/s41066-022-00332-3
52. Dikopoulou Z, Papageorgiou EI, Vanhoof K (2020) Retrieving sparser fuzzy cognitive maps directly from categorical ordinal dataset using the graphical lasso models and the MAX-threshold algorithm. In: 2020 IEEE international conference on fuzzy systems (FUZZ-IEEE), pp 1–8
53. Martínez L, Rodriguez RM, Herrera F (2015) The 2-tuple linguistic model: computing with words in decision making. Springer International Publishing, Switzerland
54. Pérez-Teruel K, Leyva-Vázquez M, Espinilla M, Estrada-Sentí V (2014) Computación con palabras en la toma de decisiones mediante mapas cognitivos difusos. Revista Cubana de Ciencias Informáticas 8:19–34
55. Pérez-Teruel K, Leyva-Vázquez M, Estrada-Sentí V (2015) Mental models consensus process using fuzzy cognitive maps and computing with words. Ingeniería y Universidad 19:173–188. https://doi.org/10.11144/Javeriana.iyu19-1.mmcp
56. Gonçalves G, Wanke P, Tan Y (2022) A higher order portfolio optimization model incorporating information entropy. Intell Syst Appl 15:200101. https://doi.org/10.1016/j.iswa.2022.200101
57. Nair A, Reckien D, van Maarseveen MFAM (2019) A generalised fuzzy cognitive mapping approach for modelling complex systems. Appl Soft Comput 84:105754. https://doi.org/10.1016/j.asoc.2019.105754
58. Anninou AP, Groumpos PP, Panagiotis P (2013) Modeling health diseases using competitive fuzzy cognitive maps. In: Papadopoulos H, Andreou AS, Iliadis L, Maglogiannis I (eds) Artificial intelligence applications and innovations. Springer, Berlin, Heidelberg, pp 88–95
59. Chen J, Gao X, Rong J, Gao X (2021) The dynamic extensions of fuzzy grey cognitive maps. IEEE Access 9:98665–98678. https://doi.org/10.1109/ACCESS.2021.3096058
60. Georgopoulos VC, Stylios CD (2008) Complementary case-based reasoning and competitive fuzzy cognitive maps for advanced medical decisions. Soft Comput 12:191–199. https://doi.org/10.1007/s00500-007-0194-7
61. Douali N, Csaba H, Roo JD, Papageorgiou EI, Jaulent M-C (2014) Diagnosis support system based on clinical guidelines: comparison between case-based fuzzy cognitive maps and bayesian networks. Comput Methods Program Biomed 113:133–143. https://doi.org/10.1016/j.cmpb.2013.09.012
62. Aamodt A, Plaza E (1994) Case-based reasoning: foundational issues, methodological variations, and system approaches. AI Commun 7:39–59. https://doi.org/10.3233/AIC-1994-7104

63. Salmeron JL (2010) Modelling grey uncertainty with fuzzy grey cognitive maps. Expert Syst Appl 37:7581–7588. https://doi.org/10.1016/j.eswa.2010.04.085
64. Concepción L, Nápoles G, Bello R, Vanhoof K (2020) On the behavior of fuzzy grey cognitive maps. In: Bello R, Miao D, Falcon R, Nakata M, Rosete A, Ciucci D (eds) Rough sets. Springer International Publishing, Cham, pp 462–476
65. Nápoles G, Salmeron JL, Vanhoof K (2019) Construction and supervised learning of long-term grey cognitive networks. IEEE Trans Cybern. https://doi.org/10.1109/tcyb.2019.2913960
66. Lavin EA, Giabbanelli PJ (2017) Analyzing and simplifying model uncertainty in fuzzy cognitive maps. In: 2017 winter simulation conference (WSC). IEEE, Las Vegas, NV, USA, pp 1868–1879
67. Zhang Y, Qin J, Zheng WX, Kang Y (2018) Extended evidential cognitive maps and its applications. J Franklin Inst 355:381–405. https://doi.org/10.1016/j.jfranklin.2017.10.032
68. Kang B, Deng Y, Sadiq R, Mahadevan S (2012) Evidential cognitive maps. Knowl-Based Syst 35:77–86. https://doi.org/10.1016/j.knosys.2012.04.007
69. Sumathi IR, Antony Crispin Sweety C (2019) New approach on differential equation via trapezoidal neutrosophic number. Complex Intell Syst 5:417–424. https://doi.org/10.1007/s40747-019-00117-3
70. Mkrtchyan L, Ruan D (2010) Belief degree-distributed fuzzy cognitive maps. In: 2010 IEEE international conference on intelligent systems and knowledge engineering. IEEE, China, pp 159–165
71. Ruan D, Hardeman F, Mkrtchyan L (2011) Using belief degree-distributed fuzzy cognitive maps in nuclear safety culture assessment. In: 2011 Annual meeting of the north american fuzzy information processing society. IEEE, El Paso, TX, USA, pp 1–6
72. Mkrtchyan L, Ruan D (2012) Using belief degree distributed fuzzy cognitive maps for energy policy evaluation. In: Lu J, Jain LC, Zhang G (eds) Handbook on decision making: Risk Management in Decision Making, vol 2. Springer, Berlin, Heidelberg, pp 39–67
73. Fahmi A, Aslam M, Riaz M (2019) New approach of triangular neutrosophic cubic linguistic hesitant fuzzy aggregation operators. Granul Comput. https://doi.org/10.1007/s41066-019-00177-3
74. Nápoles G, Grau I, Papageorgiou E, Bello R, Vanhoof K (2016) Rough cognitive networks. Knowl-Based Syst 91:46–61. https://doi.org/10.1016/j.knosys.2015.10.015
75. Nápoles G, Mosquera C, Falcon R, Grau I, Bello R, Vanhoof K (2018) Fuzzy-rough cognitive networks. Neural Netw 97:19–27. https://doi.org/10.1016/j.neunet.2017.08.007
76. Papageorgiou EI, Markinos A, Gemptos T (2009) Application of fuzzy cognitive maps for cotton yield management in precision farming. Expert Syst Appl 36:12399–12413. https://doi.org/10.1016/j.eswa.2009.04.046
77. Mourhir A, Rachidi T, Papageorgiou EI, Karim M, Alaoui F (2016) A cognitive map framework to support integrated environmental assessment. Environ Model Softw 77:81–94. https://doi.org/10.1016/j.envsoft.2015.11.018
78. Carvalho JP, Gregorio P (2019) A case study on wildfire propagation modelling using evolving rule based fuzzy cognitive maps. In: 11th conference of the european society for fuzzy logic and technology (EUSFLAT 2019). Atlantis Press, pp 573–579
79. Barriba I, Rodríguez-Díaz A, Castro JR, Sanchez MA (2018) An approach to fuzzy inference system based fuzzy cognitive maps. In: Aguilar L, Castañón-Puga M, Rodríguez-Díaz A (eds) Computer science and engineering—Theory and applications. Springer International Publishing, Cham, pp 151–166
80. Mamdani EH, Assilian S (1975) An experiment in linguistic synthesis with a fuzzy logic controller. Int J Man Mach Stud 7:1–13. https://doi.org/10.1016/S0020-7373(75)80002-2
81. Zdanowicz P (2017) Development and application of new rule based fuzzy cognitive maps. Ph.D., Coventry University
82. Zdanowicz P, Petrovic D (2017) New mechanisms for reasoning and impacts accumulation for rule based fuzzy cognitive maps. IEEE Trans Fuzzy Syst 26:543–555. https://doi.org/10.1109/TFUZZ.2017.2686363

83. Carvalho JP (2013) On the semantics and the use of fuzzy cognitive maps and dynamic cognitive maps in social sciences. Fuzzy Sets Syst 214:6–19. https://doi.org/10.1016/j.fss.2011.12.009
84. Jia Z, Zhang Y, Dong X (2020) An extended intuitionistic fuzzy cognitive map via dempster-shafer theory. IEEE Access 8:23186–23196. https://doi.org/10.1109/ACCESS.2020.2970159
85. Atanassov KT (1986) Intuitionistic fuzzy sets. Fuzzy Sets Syst 20:87–96. https://doi.org/10.1016/S0165-0114(86)80034-3
86. Luo C, Zhang N, Wang X (2019) Time series prediction based on intuitionistic fuzzy cognitive map. Soft Comput 24:6835–6850. https://doi.org/10.1007/s00500-019-04321-8
87. Papageorgiou EI, Iakovidis DK (2009) Towards the construction of intuitionistic fuzzy cognitive maps for medical decision making. In: 2009 9th international conference on information technology and applications in biomedicine. IEEE, Larnaca, Cyprus, pp 1–4
88. Dogu E, Albayrak YE, Tuncay E (2020) Multidrug-resistant tuberculosis risk factors assessment with intuitionistic fuzzy cognitive maps. J Intell Fuzzy Syst 38:1083–1095. https://doi.org/10.3233/JIFS-179470
89. Zadeh LA (1965) Fuzzy sets. Inf Control 8:338–353. https://doi.org/10.1016/S0019-9958(65)90241-X
90. Smarandache F (2019) NIDUS IDEARUM. Scilogs, VI: annotations on neutrosophy, Editions Pons. Pons Publishing, Brussels, Belgium
91. Smarandache F (2017) Nidus Idearum. Scilogs, III: Viva la Neutrosophia! Infinite Study, Brussels, Belgium
92. Rodríguez IXL, Diaz JAE, Salazar LDP (2020) Analysis of crimonogenic factors in femicide crimes. NSS 37:90–98
93. Al-subhi SH, Papageorgiou EI, Pérez PP, Mahdi GSS, Acuña LA (2021) Triangular neutrosophic cognitive map for multistage sequential decision-making problems. Int J Fuzzy Syst 1–23. https://doi.org/10.1007/s40815-020-01014-5
94. Kandasamy WBV, Smarandache F (2003) Fuzzy cognitive maps and neutrosophic cognitive maps. Xiquan, New Mexico, USA
95. Cabrera MAC, Chasiguano BJI, Pillajo ALS, Vázquez RD (2020) Prioritization of non-functional requirements in a mobile application for panic button system using neutrosophic decision maps. NSS 34:153–158
96. Anitha R, Gunavathi K, Shadrach FD (2020) Investigation on the musical features of carnatic ragas using neutrosophic logic. J Phys: Conf Ser 1706:012051. https://doi.org/10.1088/1742-6596/1706/1/012051
97. Rodríguez Lara A, Tello CR, Blacio JA, Guevara RH (2020) Neutrosophic interrelationship of key performance indicators in an accounting process. Neutrosophic Sets Syst 34
98. Ricardo J, Flores D, Díaz J, Teruel K (2020) An exploration of wisdom of crowds using neutrosophic cognitive maps. Neutrosophic Sets Syst 37. https://doi.org/10.5281/zenodo.4121955
99. Aguilar J (2016) Multilayer cognitive maps in the resolution of problems using the FCM designer tool. Appl Artif Intell 30:720–743. https://doi.org/10.1080/08839514.2016.1214422
100. Puerto E, Aguilar J, López C, Chávez D (2019) Using Multilayer fuzzy cognitive maps to diagnose autism spectrum disorder. Appl Soft Comput 75:58–71. https://doi.org/10.1016/j.asoc.2018.10.034
101. Christoforou A, Andreou AS (2017) A framework for static and dynamic analysis of multi-layer fuzzy cognitive maps. Neurocomputing 232:133–145. https://doi.org/10.1016/j.neucom.2016.09.115
102. Stach W, Kurgan L (2004) Parallel fuzzy cognitive maps as a tool for modeling software development projects. In: IEEE annual meeting of the fuzzy information, 2004. Processing NAFIPS '04, vol 1. IEEE, Canada, pp 28–33

# Project Scheduling a Critical Review of Both Traditional and Metaheuristic Techniques

**Pedro Yobanis Piñero Pérez, Iliana Pérez Pupo, Gaafar Sadeq Saeed Mahdi, Julio Madera Quintana, and Luis Alvarado Acuña**

**Abstract** Project planning is a problem usually discussed in the different project management standards as an essential problem to be addressed from the project initiation stage. It is a problem that has traditionally been treated by formal methodologies. But current trends in project development have a greater focus on agile methodologies. This situation causes greater variability in project plans. In the particular case of BIM methodologies, the approach is aimed at achieving the simulation of the production process through virtual construction. In this context, in this work, a critical analysis of different approaches that deal with the construction of project schedules is carried out. In particular, the problem is analyzed from a hybrid perspective. The approach proposed by project management standards and the approach to scheduling problems raised by computer science are analyzed. As a result of the analysis, a group of lines open to research are proposed that combine traditional tendencies with metaheuristics.

**Keywords** Agile project scheduling · Metaheuristics · Optimization · Project management

P. Y. Piñero Pérez (✉) · I. P. Pupo · G. S. S. Mahdi
Artificial Intelligence for a Sustainable Development Group, IADES, Havana, Cuba
e-mail: pppyob@gmail.com; iades@businessredmine.com

I. P. Pupo
e-mail: iliperezpupo@gmail.com; iades@businessredmine.com

G. S. S. Mahdi
e-mail: iades@businessredmine.com

J. M. Quintana
Vicerrectoría de Informatización, Universidad de Camagüey, Camagüey, Cuba
e-mail: julio.madera@reduc.edu.cu

L. A. Acuña
Departamento de Ingeniería de La Construcción, Universidad Católica del Norte, Antofagasta, Chile
e-mail: lualvar@ucn.cl



P. Y. Piñero Pérez et al. (eds.), *Computational Intelligence in Engineering and Project Management*, Studies in Computational Intelligence 1134,
https://doi.org/10.1007/978-3-031-50495-2_3

## 1 Introduction

In recent years, and with a growing trend, the development of new products and services based on project-oriented production models has manifested itself in society. Within this context, there are a group of elements that must be carefully planned to achieve the objectives of the projects with an adequate balance of cost, time and quality, among which are: technological changes, the complexities of working with multidisciplinary teams and restrictions on material and financial resources [1]. Two spheres of knowledge that exemplify this situation are the growing development of Building Information Modeling (BIM) technologies [2, 3] for construction projects and the development of software projects.

Particularly in the construction sector, BIM technologies aim to simulate the complete execution of projects in virtual environments. The use of BIM methodologies and tools is still poorly standardized and not widespread at a global level [4] and constitutes an open field in the development of new research [5]. At the current stage of BIM development, fundamental progress has been made in the integration and parameterization of the designs or plans of work with the planning activities. It is proposed that around 34% of the resources invested in a project are wasted and that approximately 95% of the information that is generated later is underused and not used as lessons learned in new projects [6]. In this same source, it is proposed that the main factor that affects the inappropriate use of resources is inefficient planning. In this context, the development of new algorithms and techniques that allow the optimal or quasi-optimal construction of project schedules and take into account resource limitations and variability in the ways of executing tasks remains a necessity.

A similar situation occurs in the development of software projects. In this area of knowledge, it is identified by several sources that, in medium-sized projects, approximately 26% need to be renegotiated and 31% are canceled [7, 8]. This scenario is characterized by a high dependence on human resources and their competencies, accompanied by tasks that can be executed in various forms or modes. Among the fundamental causes of project failure [9] are: insufficient training of human resources and frequent errors in the planning processes, whether in scope, time or logistics [10] and poor management and insufficiencies in the control and monitoring processes [11]. This research focuses on contributing to the solution of the planning problem, understanding this as one of the fundamental problems of project management.

In both scenarios, competitiveness in project development imposes the need for algorithms and tools for the rapid construction of schedules that optimize, as much as possible, the time and costs of project execution, considering that the tasks can be executed in a timely manner in multiple modes and the existence of restrictions on the use of renewable and non-renewable resources.

In this sense, standards and guides have been developed by different project management schools, among which are the A Guide to the Project Management Body of Knowledge (PMBOK) [12], *International Project Management Association* (IPMA), the ISO 21500 International Organization for Standardization standard [13] and the (SEI, *Software Engineering Institute*), with its proposal for Capability

Maturity Model Integration [14, 15]. In recent versions, these standards refer to the need for simulation techniques, data analysis and resource optimization. However, they do not propose specific techniques based on optimization models for schedule construction processes [12]. On the other hand, the standards explain the need to consider resource availability restrictions but do not take into account other restrictions related to the competencies of human resources or the treatment of variability with respect to the modes of execution of tasks.

In general, although different schools analyze planning from different perspectives, they all agree on the need to identify and plan tasks logically to achieve project objectives. This planning problem is known as "Task Scheduling Problem or Scheduling" called in the scientific community "PSP, Project Scheduling Problem". The first approaches to conceptualizing and solving these problems were developed in the 1950s [16]. PSPs are presented as a family of optimization problems that aim to organize a set of tasks respecting a group of precedence restrictions and limited use of renewable and non-renewable resources with the least possible time and cost [17].

All this, in the context of the development of agile methodologies, incorporates greater dynamism in the construction of schedules. That is why, in this research, the authors propose to carry out an exploratory study that allows for the characterization of the approach in the construction of chronograms of project management standards and the computational intelligence techniques used in the solution of these problems. The second section of the work presents the protocol used in the review of the information. Then, in Sect. 3, the planning problem is addressed from the perspective of the different management standards. Section 4 characterizes planning problems and their impact on project management. In Sect. 5, different metaheuristics reported in the literature for solving scheduling problems are analyzed. Finally, conclusions and future work are presented.

## 2 Systematic Review Protocol

The systematic review protocol used in the exploratory study is presented below. The study constitutes a critical analysis of the following two approaches: the process approach of project management schools and the approach that treats planning problems as optimization problems.

Review protocol for critical analysis:

1. Definition of a Bibliographic Reference Manager
2. Definition of sources or databases of scientific information such as: *Google Scholar, Semantic Scholar, Scopus,* and the *Web of Science* and work in research networks for access to scientific information associated with the object of the research.
3. Definition of the following key phrases for the search process:
   - "Project Management", "Project Scheduling", "Project Scheduling Problem"

- "Estimation of Distribution Algorithm", "Estimation of Distribution Algorithm & Scheduling", "Estimation of Distribution Algorithm & Project Scheduling Problem", "Estimation of Distribution Algorithm & Resource-Constrained Project Scheduling Problem", "Estimation of Distribution Algorithm & Multi-Mode Resource-Constrained Project Scheduling Problem", "Estimation of Distribution Algorithm & Resource-Constrained Multi-Project Scheduling Problem", "Estimation of Distribution Algorithm & Multi-Mode Resource-Constrained Multi-Project Scheduling Problem".
- "Metaheuristics & Multi-Mode Resource-Constrained Multi-Project Scheduling Problem", "Multi-Mode Resource-Constrained Multi-Project Scheduling Problem & Evolutionary".
- "Constraints Handling", "Estimation of Distribution Algorithm & Constraints Handling", "Constraints Handling & Evolutionary", "Constraints Handling & Metaheuristics".

4. Establishment of the following criteria for the analysis of the search:
   - *Publication behavior per year.*
   - *Identification of the main sources of information on the topic.*
   - *Main authors and their affiliations.*
   - *Analysis differentiated by types of publications: articles, conference proceedings, books, doctoral theses.*
5. Classification and filtering of publications into classics, extensions to theories, applications, tertiary reviews and others that facilitate the analysis of trends.
6. Detailed review of documentation and synthesis of trends.

## 3 Treatment of Planning Problems by Project Management Schools

The analysis of the main project management standards allowed the authors to characterize these standards from a planning perspective. In addition, the following concepts are considered important for the exploratory study:

- Project cut: Refers to the moment in time at which cuts are planned in the evaluation of the execution of a project. Generally, in production projects, weekly project cuts are planned, while in research projects, a greater amount of time is planned between cuts [18].
- Quasi-optimal: term used in optimization works that, due to their complexity, are solved using metaheuristics. A quasi-optimal solution represents a solution to an optimization problem that does not reach the value of the global optimum but is very close to it, can be found in a reasonable execution time, and satisfies the objectives of the problem.

- Effectiveness: in the research context, effectiveness reflects the ability of algorithms to achieve a balance between effectiveness and efficiency in finding solutions to optimization problems. Understanding effectiveness as the ability to find feasible solutions to the problem and efficiency as the computational cost to find them.
- Bottom-up estimation: The costs of the resources used to execute each task are estimated. These estimated costs are added together, resulting in the total cost of the project [12].
- Non-renewable resources: are resources that are limited to a certain amount for the entire life cycle of the project, and once used in the processing of one task, they cannot be assigned to another. They can be made available, as long as they are available, at any time during the execution of the project [19]. Example: budget, some types of raw materials, fuel, etc.
- Renewable resources: the availability of these resources is limited in each unit of time of the project. As a restriction, each renewable resource is available in a certain constant quantity in each unit of time, and its utilization cannot exceed that quantity in any of those units. Example: labor, tools, space, etc.
- Waiting slack: [20] is the period of time between the completion of a task and the start of its successor. This slack is introduced because the start of a task depends on the completion of all of its predecessors, and therefore a wait time is created between the task and its predecessors that have already finished.
- Free slack: [21] is the period of time between the completion of a task and the scheduled date to make a project cut. This slack is planned as part of the reserve time for risk management processes. During the execution of the project, this slack allows necessary adjustments to be made to the schedules in the event of unforeseen delays.

## *3.1 Project Planning as Seen from the PMBOK Guide*

PMBOK developed by the *Project Management Institute* [12], constitutes one of the most internationally recognized project management standards. These standard states that planning consists of organizing the set of tasks to achieve the project objectives; For each of these tasks, the start and end dates are estimated, taking into account time and resource restrictions, but noting that this process continues during the life of the project. It is iterative because, in each project check, the estimates are corrected based on progress. The iterative approach proposed by PMBOK denotes the need for the use of algorithms and tools that help to frequently replan the project, seeking optimal or quasi-optimal solutions.

In addition, these standard concentrates planning tasks in the knowledge area of project planning management, organizing them into the following processes: planning, definition of tasks, sequence of tasks, estimation of the duration of tasks, development of the schedule, and control of its changes.

In the PMBOK, the main result of time management is the construction of the schedule using several techniques, among which the following stand out: the critical path method, the critical chain, project management tools, schedule compression, and leveling of renewable and non-renewable resources. These techniques help build schedules that avoid collisions between tasks due to the use of resources and help detect those critical paths that represent sequences of tasks that, if delayed, can affect the fulfillment of project commitments. But it does not present an approach to optimizing resources and schedules, an element that constitutes an opportunity to improve this good practice guide. Furthermore, the process of building schedules in the PMBOK focuses on the planning of a single project and encourages the idea of bottom-up estimation of project times and required resources without a strong constraint approach but rather based on the usage of all renewable and non-renewable resources that are necessary.

### *3.2 Approach from the International Organization for Standardization (ISO)*

The ISO 21500 standard standardizes project management with an international reference standard that allows you to compete in any geographic area by using a "universal language" [13].

This standard identifies direction and management processes organized into process groups, with the time management knowledge area being the most related to planning problems. Like the PMBOK, ISO 21500 organizes planning tasks into the following processes: task sequencing, resource estimation, task duration estimation, and schedule development. However, it does not propose any techniques or tools to build it. In general, the ISO 21500 proposal, like the PMBOK, focuses on the planning of a single project at a time with the idea of using all the resources that are necessary without a strong focus on renewable resource restrictions and non-renewable resources during planning.

In particular, the Cuban Standards for Project Management (NC) [22] are completely aligned with the ISO 21500 standards and have the same difficulties that these present [23].

### *3.3 CMMI Approach*

The CMMI developed by the *Software Engineering Institute* (SEI) [14, 15], is an integrated model for improving the capacity of software project development organizations. It proposes five levels of maturity: initial, managed, defined, quantitatively managed and optimized. At level two, CMMI proposes generic and specific practices focused on solving the project planning problem by considering them independently, and only from the third level onwards does it propose practices that focus on the planning problem of multiple projects. In this standard, the areas that most strongly focus planning problems are:

- At level two, the project planning area in the specific practice SP 2.1 suggests the use of traditional tools such as the critical path (CPM, Critical Path Method) [24], the Program Evaluation and Review Technique (PERT) [25] and planning with limited resources.
- At level three, in the area of integrated project management, the practices associated with the integration of schedules are included.
- At level five, in the area of organizational performance management, there are generic and specific practices that promote the reduction of execution times and the optimal use of renewable and non-renewable resources. But they do not suggest the use of tools, nor do they propose ideas based on the use of schedule optimization techniques that represent projects with various execution modes.

However, such techniques do not fully address the limited resources and the stochastic and dynamic nature of multiproject systems. Nor do they consider the use of optimization techniques in the construction of schedules, nor are tools introduced or suggested that allow the optimization of renewable and non-renewable resources in the process of integrating the plans.

### *3.4 Analysis Regarding the Tools that Support the Standards*

The first tools that have to do with planning date back to 1917 with Gantt charts, which were introduced mainly for the US arms industry [26]. Then other instruments emerged that have set guidelines in project planning, such as the CPM in 1957 [24]. Currently, a wide variety of different computer tools are reported to support the standards set out above, among which are Microsoft Project [27], Redmine [28], Primavera [27], or the BusinessRedmine Integrated Project Management Package [29]. These tools allow visual modification of schedules, but essentially execute planning processes manually. As for the restrictions, they are limited to presenting alerts that allow users to modify the schedules manually until they are satisfied [10].

The main standards, guides for project management, recognize the importance of constructing optimal or quasi-optimal schedules, but do not propose metaheuristics or other optimization techniques that facilitate the planning process. In general, they apply time-consuming manual techniques in the generation of project plans [9]. The following section presents different metaheuristics used in solving project planning problems.

## 4 Characterization of Planning and Modeling Problems as an Optimization Problem

From the point of view of mathematical sciences and computer science, project planning problems are presented as optimization problems that aim to organize a set of tasks respecting a group of precedence restrictions and the limited use of renewable and non-renewable resources renewable, with the least possible time and cost [17, 30]. Conceptually, these problems raise the question of how to find a solution $x = (x_1, x_2, \ldots, x_n)^T$ in the solution space $\mathbb{R}^n$ that minimizes the function $f(x)$ subject to the restrictions posed in the Eqs. (1) and (2).

$$g_i(x) \leq 0, i = 1, \ldots, p \tag{1}$$

$$h_j(x) = 0, j = 1, \ldots, q \tag{2}$$

where $x \in \mathbb{R}^n$ represents a feasible solution that satisfies all the inequality and equality constraints of $p$ and $q$ respectively. The set of all solutions that satisfy the constraints of the problem is called the feasible region.

Under this approach, a fundamental element is the implementation of different strategies for the treatment of restrictions [31]. Coello and Mezura Montes in [32, 33] group these strategies into four fundamental categories:

1. *Penalty*: the idea is to add a penalty coefficient proportional to the sum of the violations of the constraints on all objective functions. It consists of reducing the fitness value of the solution obtained so that another solution with similar fitness that does meet the restriction has a better chance of surviving. The fundamental thing in this technique is to determine the penalty coefficient [34]. In this method, both the objective functions and the constraints must be normalized. There is a specific category within this so-called deadly penalty strategy, in this case, individuals who do not comply with the restrictions are eliminated.
2. *Feasible region methods*: consists of generating an initial feasible solution (or a set of feasible solutions) and moving from it without ever leaving the feasible region, trying to improve the value of the objective function. Internal penalty methods follow this principle [35].
3. *Repair methods*: in this case, it is about repairing those solutions that are not feasible using different heuristics, adapting them to each particular problem, and always replacing the solution that gave rise to it in the population. However, this is not always possible, and in some cases, repair operators can introduce noise into the search space [32], thus affecting the quality of the solutions obtained.
4. *Separation of constraints and objectives*: unlike the idea of converting constraints into objectives of the problem as in the penalty function, here they are kept separate during the selection process. The idea is to divide the search process into two phases: the first aims to find feasible solutions without taking into account the objective function, and after finding a number of feasible solutions, the second

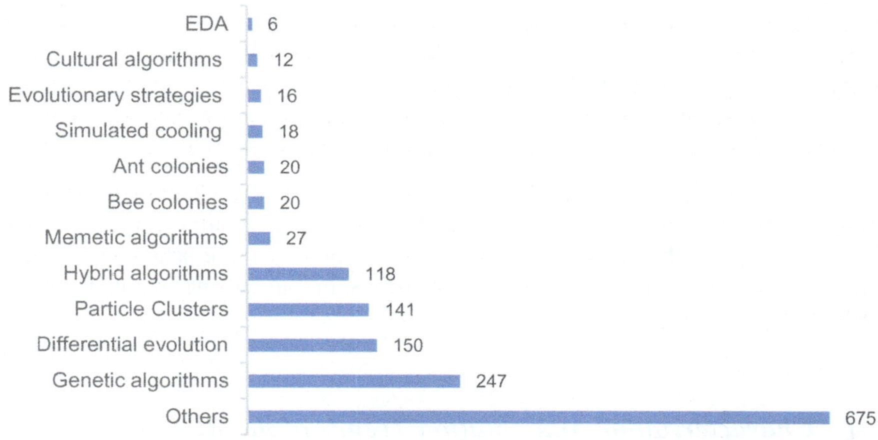


**Fig. 1** List of bibliographic references associated with the treatment of restrictions within evolutionary algorithms [31]

phase begins, whose purpose is to optimize the objective function of the problem based on the set of feasible solutions found in the first phase. According to [36], this approach may cause a lot of time to be spent in the search process in infeasible regions.

In this context, project planning problems are presented as complex problems of the NP-Complete type [24, 37]. Considering the complexity of these problems, several authors make use of metaheuristics such as: Particle Swarm Optimization (PSO) [38], Genetic Algorithms (GA, del inglés, *Genetic Algorithm*) [39] and Algorithms with Estimation of Distributions (EDA, del inglés, *Estimation of Distribution Algorithms*) [40]. A tertiary review carried out in [32] includes precisely the treatment of restrictions from different metaheuristics, see Fig. 1.

This study identifies that genetic algorithms represent the most used technique in isolation, while Algorithms with Distribution Estimation are the least worked on [31].

For a better characterization of the different strategies for the treatment of restrictions, the planning problems have been subdivided as follows:

- Limited resource project planning problem (RCPSP) [41].
- Resource Limited Multiple Project Planning Problem (RCMPSP) [42].
- Project planning problem with limited resources and tasks that are executed in multiple modes (MMRCPSP) [43].
- Multi-mode, multi-resource constrained multi-project planning problem (MMRCMPSP) [43].

These problems share the existence of restrictions that generally involve several variables. Examples of these restrictions are presented in:

- Precedence restrictions between tasks are given by logical dependencies in their execution or by limitations of renewable or non-renewable resources.
- Dependency restrictions between execution modes in which problems exist, where the selection in the early stages of certain execution modes conditions the execution modes of the tasks in the advanced stages of planning.

However, in solving these problems, the potential of EDA, which by its nature is designed to solve problems with correlated variables, is not sufficiently exploited. In the following subsections, the aforementioned problems are characterized, and the metaheuristics that have been used in their solution and the treatment of the restrictions in each case are presented.

## 4.1 Characterization and Solution Trends of the RCPSP Problem

The RCPSP is one of the basic and key problems in project planning. Its objective is, given a project, to establish the sequence of a set $J = \{1, \ldots, n\}$ of tasks of the project, minimizing the total time of its duration, subject to two types of restrictions: precedence relationships and the amount of renewable and non-renewable resources available to execute the tasks at each instant of time [44].

The RCPSP problem can be modeled as follows [41].

Let a project be represented by a directed graph without cycles $G = (J, E)$ where $J$ is the set of nodes (project tasks), $E = \{(h, j) : h \in A_j, j \in J\}$ is the set of arcs (precedence relations), and $A_j \subset J$ is the set of $j$´s predecessor tasks. Each task $j \in J$ has a duration $d_j$ with $d_1 = d_n = 0$ since $j = 1$ and $j = n$ are fictitious nodes, which are taken to represent where the project begins and ends. Each task $j$, has a start time $s_j$ (problem variable) and an end time $\overline{f}_j$ calculated from the sum of the start time plus the duration of the task. Furthermore, in the RCPSP, $R_k$ are considered units of renewable resources of the type $k \in K$. Each task $j$ to be processed consumes an amount $r_{jk}$ of renewable resources of type $k$. A feasible solution is given by the pair $(S, \overline{F})$, where $S = (s_1, s_2, \ldots, s_n)$ is the vector of the start times and $\overline{F} = (\overline{f_1}, \overline{f_2}, \ldots, \overline{f}_n)$ is the vector of the completion times of the project tasks, calculated without violating the precedence relationships and the amount of available resources.

Modeling this problem with binary variables, the decision variable is defined as follows:

$$x_{jt} = \begin{cases} 1 \text{ if task } j \text{ starts executing at instant } t \in [ES_j, LS_j] \\ 0 \qquad\qquad\qquad\qquad \text{in another case} \end{cases}$$

where $ES_j$ is the earliest possible start time and $LS_j$ is the latest possible start time of task $j$ calculated without considering resource limitation by sequencing methods. For more details on the calculation of these values, see [31]. The optimization problem is formulated as follows:

$$\text{minimize } Z = \sum_{t=ES_n}^{LS_n} tx_{nt} \tag{3}$$

$$\sum_{t=ES_j}^{LS_j} x_{jt} = 1 \forall j \in J \tag{4}$$

$$\sum_{t=ES_j}^{LS_j} tx_{jt} - \sum_{t=ES_i}^{LS_i} tx_{it} \geq d_i \forall (i, j) \in E \tag{5}$$

$$\sum_{j=1}^{n} \sum_{q=\max\{t, ES_j\}}^{\min\{t+d_j-1, LS_j\}} r_{jk} x_{jq} \leq R_k \forall k \in K, t = 1, \ldots, T \tag{6}$$

$$x_{jt} \in \{0, 1\} \forall j \in J, t \in \left[ES_j, LS_j\right] \tag{7}$$

Expression (3) is the objective function of the problem; it minimizes the start time of the fictitious task $j = n$ which is equal to the duration of the project. The constraint group (4) guarantees that all tasks are executed. Constraints (5) control the precedence relationships between tasks. The group of restrictions (6) controls that the number of renewable resources available at the instant $t$ at which any task can begin to be executed is not exceeded, where $T$ is an upper bound on the duration of the project. Constraints (7) force the decision variable to take binary values.

In solving this problem, authors such as [45, 46], apply tabu search techniques by combining information associated with the critical path with the neighbor generation process. The algorithms proposed by these authors only focus on the analysis of the time variable and do not consider costs in the optimization process.

Other authors have treated the RCPSP problem, considering that resources can be distributed across several sites, some located at fixed sites and others at mobile sites [38]. These works focus on minimizing the duration of tasks and resolving where each task should be executed, minimizing the movement time of resources. In particular, in [38] they use PSO-based techniques, while in [47] local search, simulated annealing and iterative local search methods are used. As an interesting element of these works, it can be considered that the execution of a task at multiple sites can have an analogy with the execution of the task in different modes, although the authors never consider the multimode problem. Another element that constitutes an opportunity for improvement in these works is that they only focus on optimizing the duration and do not include the cost.

GAs have been used to solve the RCPSP problem. One of the first works is [41], which impacted later works aimed at solving the RCPSP and MMRCPSP problems. Positive elements of this work that can be used in this research are: the design of the solutions, where each individual is represented by a list of tasks; and the evaluation criteria of each solution focused on the calculation of the total duration of the project. Another important element that can also be used is the treatment of restrictions, where Kolisch and Hartmann introduced crossover and mutation operators that did not violate the precedence relationships between tasks.

Debels and Vanhoucke [48] present the BPGA genetic algorithm, where classic genetic algorithm operators are used, and two populations are constructed considering the calculation of the critical path as explained below. In the first population, individuals are represented under the "as early as possible" approach, while in the second population, the solutions follow the "as late as possible" approach. It is about building roads with minimum clearance where the earliest calendar and the latest calendar coincide. In this algorithm, pairs of parents are selected from both populations and crossed using two-point crossing. An interesting element of this proposal is that it assigns priority to tasks, and it is understood that the tasks with the highest priority are those that are executed first. As a limitation of this proposal, it is noted that it does not address the impact of cost on the solutions or the multimode approach. However, an extension of this work for the multimode approach is presented by Vanhoucke and collaborators [49].

Other authors, such as [41] combine several meta-heuristics to solve the RCPSP. In this work, the authors propose the use of the *Multi-objective Variable Neighborhood Search* (MOVNS) algorithm, which starts with a solution S and generates a neighborhood for this solution based on small changes in it. To construct the neighborhood, the authors use two methods: the first consists of exchanging tasks in the sequence of the solution S, while the other is to cause a mutation of the solution by inserting a task in the sequence. This strategy can be considered important in the framework of this research. Although the approach used in the proposal is multi-objective, it does not take into account the cost of project execution and focuses only on the optimization of the project duration and the total start time of the tasks.

In the literature consulted, it was found that EDAs were introduced for the solution of the RCPSP by [50]. The authors propose the *Hybrid Estimation of Distribution Algorithm* (HEDA), which combines an EDA with local search. In their proposal, a procedure called *Probability Generation Mechanism* (PGM) is used to generate the population according to the learned probabilistic model, guaranteeing that each task appears only once in each solution. Then local search procedures are incorporated that consist of exchanging tasks as long as they do not violate the precedence restrictions. According to [51], the results shown in this work are not competitive and are only optimized with respect to the duration of the project.

In the work of [51] the *Random Key-Based Estimation of Distribution Algorithm* (RK-EDA) is proposed, which is an EDA designed to solve problems based on permutations, and they apply it to the solution of the resource allocation problem in the tasks of planning. This algorithm is based, like the UMDA, on the calculation of the probability of occurrence of the values in the prominent solutions, but in this case the authors set a variance value with respect to the mean and use this value to control the level of dispersion or search concentration. They then combine the calculated mean with the established variance and generate the new solutions using a Gaussian distribution function. According to the experimental results, this algorithm is not competitive in terms of optimizing the total project duration, especially for instances that contain a greater number of tasks. Furthermore, it does not take into account the cost of executing the project.

## 4.2 *Characterization and Solution Trends of the RCMPSP Problem*

The RCMPSP problem involves the simultaneous development of multiple projects with limited resources [52]. In this scenario, in addition to the variables considered in the previous case, a new variable appears: the priority of the projects [53, 54]. In this problem, we consider a set $I = \{1, \ldots, n\}$ of projects, where each one is formed by a set $J_i = \{1, \ldots, N_i\}$ of tasks, where $j = 1$ and $j = N_i$ are fictitious tasks that have duration $d_{i_1} = d_{i_n} = 0$, as they indicate when project $i$ begins and when it ends. Precedence relationships between tasks are established only between tasks in the same project.

This problem can be represented by several networks, one for each project $G_i = (J_i, E_i), i \in I$ or by a single network in which there are only two fictitious tasks that act as the start and end for all the projects. This problem considers the availability $R_k$ of renewable resources of type $k \in K$, which are available for all projects $i \in I$. The formulation of this problem is shown below [30, 55]:

$$\text{minimize } Z = \sum_{i=1}^{n} \sum_{t=ES_n}^{LS_n} t x_{int} \tag{8}$$

$$\sum_{t=1}^{T} x_{ijt} = 1 j \in J_i, i \in I \tag{9}$$

$$\sum_{t=1}^{T} t(x_{iht} - x_{ijt}) \geq d_{ij} \forall (j, h) \in E_i, i \in I \tag{10}$$

$$\sum_{i=1}^{n} \sum_{j=1}^{N_i} r_{ijk} \sum_{q=t-d_{ij}+1}^{t} x_{ijq} \leq |R_k| \forall k \in K, t = 1, \ldots, T \tag{11}$$

$$x_{ijt} \in \{0, 1\} \forall j \in J_i, i \in I, t = 1, \ldots, T \tag{12}$$

In the previous model, we saw that (8) is a time-dependent objective function. Constraints (9) control that all tasks in all projects are executed. Constraints (10) ensure that precedence relationships are not violated. Constraints (11) prevent the available renewable resources from being exceeded when task $j$ of project $i$ is executed. Finally, constraints (12) force the decision variables to take binary values.

In solving this problem, authors such as Dong and collaborators in [56] propose a GA-based method for planning that focuses the search on the analysis of project cost and time. Its greatest contribution is found in the treatment of the problem of multiple projects, for which the authors of this algorithm introduce the priority of projects as a base element for the distribution of shared resources.

Medrano [55] proposes the PSO-J and PSO-R algorithms for solving problems of the RCMPSP type where the particles move simultaneously in two sets that form the problem solution space: the set of real vectors that store the values of the task priorities of each solution and the set of integer vectors containing the resource allocation to each solution. This proposal also does not take into account the different modes and their influence on the balance between cost and time.

For their part, Schmidt and collaborators [57] use a customized GA to solve the RCMPSP and introduce a dynamic calculation of the priority of the tasks using 10 indicators and considering the priority of the projects. As a negative element, the authors themselves recognize in [58] the high computational cost of their proposal.

## 4.3 Characterization and Solution Trends of the MMRCPSP Problem

The MMRCPSP problem is also commonly represented as a $G = (J, E)$ network, similar to the RCPSP problem. Its particularity lies in the fact that tasks $j \in J$ are considered to have a set $M_j = \{1, \ldots, \mu_j\}$ of processing modes. This means that each task $j$ can consume $r_{jmk}$ resource units of type $k$, depending on the mode $m \in M_j$ in which it is executed. Also, in this problem $k \in \overline{K}$ types of resources in the non-renewable category are taken into account, whose availability in the horizon $T$ of the project is $\overline{R}_k$. A discrete linear modeling of this problem is proposed in [59] using binary variables as shown below: $x_{jmt} = 1$ if task $j$ is executed in mode $m$ and starts at time $t \in [\mathrm{ES}_j, \mathrm{LS}_j]$ and $x_{jmt} = 0$ otherwise. Where $\mathrm{ES}_j$ is given by the earliest start time of task $j$, which is calculated by solving the relaxed MMRCPSP (taking the duration of the tasks equal to the duration of the processing mode that contributes the shortest duration). $LS_j$ is the latest start time that a task $j$ can start, and is calculated by solving the relaxed MMRCPSP, but now taking the duration of each task $j$ equal to the duration of the processing mode that contributes the longest duration. The problem formulation is shown below:

$$\text{minimize} \sum_{m=1}^{\mu_n} \sum_{t=ES_n}^{LS_n} t x_{nmt} \tag{13}$$

$$\sum_{m=1}^{\mu_j} \sum_{t=ES_j}^{LS_j} x_{jmt} = 1 \forall j \in J \tag{14}$$

$$\sum_{m=1}^{\mu_h} \sum_{t=ES_h}^{LS_h} (t + d_{hm}) x_{hmt} \leq \sum_{m=1}^{\mu_j} \sum_{t=ES_j}^{LS_j} t x_{jmt} \forall (h, j) \in E \tag{15}$$

$$\sum_{j=1}^{n} \sum_{m=1}^{\mu_j} r_{jmk} \sum_{q=t}^{t+d_{jm}-1} x_{jmq} \leq |R_k| \forall k \in K, t = 1, \ldots, T \tag{16}$$

$$\sum_{j=1}^{n} \sum_{m=1}^{\mu_j} \hat{r}_{jmk} \sum_{t=ES_j}^{LS_j} x_{jmt} \leq \left|\hat{R}_k\right| \forall k \in \hat{K} \tag{17}$$

$$x_{jmt} \in \{0, 1\} j \in J, m \in M_j, t = ES_j, \ldots, LS_j \tag{18}$$

Expression (13) represents the minimization of the project duration. The group of constraints that represents (14) controls that all tasks are executed in a processing mode, and at some point, the group (15) of constraints guarantees that precedence relationships are not violated, and (16) are the constraints that control that the quantity of each renewable resource of type $k \in K$ available in each period $t$ is not exceeded. The value of $T$ is an upper bound on the duration of the project that is equal to the length of the critical path of the project that is calculated considering that the duration of each task is equal to the longest possible duration according to its processing modes. The group of restrictions (17) is the one that controls that the number of non-renewable resources available is not exceeded. The set of constraints (18) guarantees that each decision variable takes binary values $\{0, 1\}$. Finding a feasible solution for an MMRCPSP is an NP-Complete problem [37].

## 4.4 *Characterization and Solution Trends of the MMRCMPSP Problem*

The MMRCMPSP planning problem combines the multi-project and multimode approaches to address much more complex problems. It focuses on building schedules for projects that are executed simultaneously and whose tasks can be performed in multiple ways [60]. An example of this situation is presented in [61] where the author works on this problem by representing each project as a directed graph with a final node that represents the completion of all projects but does not have a single initial node.

In the proposal of [61] it is assumed that with the use of a maximum number of resources, a minimum duration of the tasks is achieved, and vice versa, with a minimum number of resources, the duration is maximum, but this situation is not the case. It behaves exactly like this for different types of projects. For example, in the case of software projects where there is a high degree of dependency between certain technical tasks and a need for high levels of human resource competencies, putting in a greater number of resources can generate delays in execution due to the effort required for the project's communication management and conflict management.

Other works, such as [62] handle the issue of multimode and multi-project as a single directed graph, in which each node is represented by the mode of execution of the task, the resources associated with the mode and the duration time.

In general, there are fewer works for the solution of the MMRCMPSP problem compared to the other explained problems [43, 63]. In the proposal by [64] it is proposed to decompose the MMRCMPSP problem into two phases. The first is aimed at identifying the interrelationships between the projects and analyzing the tasks that mark the dependency between the projects, as well as the influence on them of the different execution modes. In a second stage, they work on the specificities of each project, causing minor changes in planning. This work focuses on resolving the precedence restrictions of the tasks of the different projects, but its limitation is that it does not address the restrictions associated with the use of shared resources between projects. Furthermore, this work can be criticized for the fact that in the validation, they only carry out a theoretical analysis and do not include experimentation with any database, which makes its comparison with other approaches difficult.

Wang et al. [65] present a GA for multi-objective optimization in a multi-project environment. They consider the total duration and financing costs determined by the critical chain to be objectives. The positive element of this work is the simultaneous treatment of costs and time under a multi-objective approach, an element that will be considered in this research. But it considers only the tasks on the critical path in the costs and does not handle restrictions associated with the number of resources.

Beşikci et al. [66] propose the use of genetic algorithms for the construction of schedules. These authors, unlike others, base their analysis on the minimization of renewable and non-renewable resources required for the development of the project with a limited budget. As a negative element of this work, it is noted that they do not treat the time and cost variables simultaneously, ignoring the interrelationships that exist between these two variables during the execution of a project.

Zhou and collaborators [67] propose a two-phase GA to solve MMRCMPSP type problems applied to complex assembly line problems. The objectives are to minimize the assembly duration, waiting times and cost associated with the use of resources. As a negative element of this approach, it is noted that each objective is resolved independently, affecting the integrated analysis of the cost and time variables.

## 5 Algorithms Reported in the Bibliography in the Solution of the MMRCPSP Problem

The present research delves into the solution of the Project Planning Problem with Limited Resources and tasks that are executed in multiple modes (MMRCPSP) because it is considered that the solution to this problem can help face solutions for any of the other three problems presented in the introduction, as shown below:

- The Limited Resource Project Planning Problem (RCPSP) is the simplest problem when addressing tasks of a single project that can be executed in a single way, and, therefore, it is a particular case of the MMRCPSP.
- The Limited Resource Multiple Project Planning Problem (RCMPSP) arises in the management of project programs with shared resources. In the particular case of software projects, this problem is closely related to the implementation of production models such as "Software Product Lines" [68]. It is modeled as a single macro project, trying to harmonize the tasks of all the projects in a single schedule and also considering that each task can be executed in a single way; then it could also be modeled from the solution of the MMRCPSP problem.
- The Multiple Project Planning Problem with Multiple Modes and Limited Resources (MMRCMPSP) can also be transformed into the MMRCPSP problem, considering, as in the previous case, that it is a single macro-project where all the tasks of the subprojects must coexist in a single schedule and maintaining its multimode characteristic.

In the bibliography consulted, it is identified that, although the MMRCPSP constitutes a problem with the presence of correlated variables, there are few works that apply EDA (see Fig. 2) in the solution of these problems [40, 69].

This figure shows that 33% of the algorithms used to solve the MMRCPSP correspond to GA, while only 4% use EDA. Furthermore, it is identified that 29% use a hybridization of several meta-heuristic techniques.

Considering the objectives of this research, the systematic review deepened the analysis of the works reported in the bibliography (see Table 1). The analysis of the algorithms will be based on a group of criteria associated with the definition of effectiveness described in the introduction of this research, among which are: multi-objective approach, optimization with respect to time, optimization with respect to cost, treatment of restrictions and use during the search of the information associated with the correlation of the variables. Furthermore, in the particular analysis of each algorithm, the following criteria are delved into:

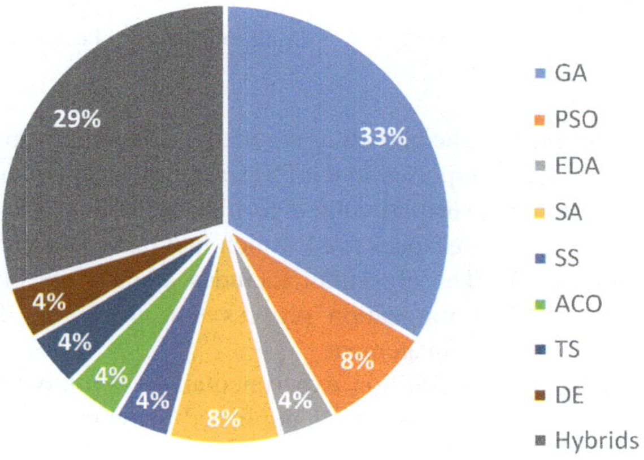


**Fig. 2** Main metaheuristics used to solve the MMRCPSP. *Source* Scopus and WOS

**Table 1** Abbreviations of the algorithms identified in the literature reported in the bibliography used in the validation

| Algorithm abbreviation | Reference | Optimization | Constraint handling | Variable correlation management |
|---|---|---|---|---|
| SS | [70] | Time | Penalty | No |
| DE | [71] | Time | Penalty | No |
| B&B | [72] | Time | Separación de las restricciones y los objetivos | No |
| MM-HGA | [73] | Time | Penalty | No |
| BPGA | [49] | Time | Penalty | No |
| $GA_{Coe\text{-}Van}$ | [74] | Time | Deadly penalty | No |
| $GA_{Van\text{-}Coe}$ | [75] | Time | Separation of constraints and objectives | No |
| $GA_{hart}$ | [76] | Time | Penalty | No |
| PBGA-EDA | [77] | Time | Penalty | Yes |
| $EDA_{Wan\text{-}Fan}$ | [78] | Time | Penalty | Yes |
| $EDA_{Sol\text{-}Elgendi}$ | [79] | Time | Penalty | Yes |
| BPEDA | [51] | Time | Penalty | No |
| CP-SAT | [80] | Time | Separación de las restricciones y los objetivos | No |
| MILP | [81] | Time | Separación de las restricciones y los objetivos | No |
| hGMEDA | [82] | Multi-objective | Penalty | Yes |
| Robust hGMEDA | [40] | Multi-objective | Penalty | Yes |
| SPEA2 | [83] | Multi-objective | Penalty | Yes |
| UMDA | [84] | Multi-objective | Penalty | No |
| FDA | [85] | Multi-objective | Penalty | Yes |

Criterion 1. The particular algorithm is included among those with the best results reported in the PSPLib library, regardless of the publication date.

Criterion 2. The particular algorithm includes PSPLib data for each of the database instances used, which facilitates comparison with other algorithms.

Criterion 3. The algorithm includes a multi-objective approach or proposes an optimization process that integrates different processes in project management.

Criterion 4. In addition, experimentation is promoted through the selection of works that have been published in the last five years.

These criteria will be decisive in selecting which of these algorithms can be used in future research.

Regarding the treatment of the correlation of variables during the search process, it is identified that 63.16% of the analyzed algorithms do not take into account this important element that can affect the effectiveness of the methods in the search for feasible solutions. Furthermore, it is evident that 78.95% of the analyzed algorithms use approaches based on penalty and that 100% of the works based on EDA apply this same approach.

Regarding the optimization approach used, it is identified that 73.68% of the consulted algorithms focus only on the optimization of the time objective.

From the above, it is concluded that the analyzed algorithms do not exploit the correlation of the variables and that, in the case of EDA, none of the consulted works handles the restrictions of the probabilistic model, which constitutes one of the fundamental motivations of this investigation.

Below is a more in-depth analysis of the algorithms presented in Table 1.

Sprecher and Drexl [72] have developed an exact branch and pruning (B&B) method based on the concept of a precedence tree, in which a task is considered eligible if all of its predecessors are planned but not necessarily completed. It is a complex algorithm that starts from the graph of constraints of the problem to be solved and sets arcs successively. According to the authors, their algorithm surpasses the exact methods existing at the time and manages to find, for the first time, optimal solutions for instances with more than 20 tasks. However, different meta-heuristic techniques have proven to be better for solving the MMRCPSP problem. This method does not take advantage of the correlation of the variables during the search process, an element that influences its efficiency and, due to its nature, has a high computational cost. It should be noted that the results of [72] are available in the PSPLib library and are reported as one of the best in solving the MMRCPSP problem. For this reason, it was selected for comparison with the algorithms proposed in this research.

Hartmann in [76] presents the $GA_{hart}$ considered one of the first implementations of its type for the solution of the MMRCPSP and which has also been, for many years, the best adaptation of a GA in the solution of this type of problem [86]. The genetic coding of this algorithm is based on a list of tasks ordered by precedence and another list for the assignment of execution mode for each task. According to the results shown in [76], the author states that the $GA_{hart}$ obtained results much superior to those obtained by the B&B of [87] except for a few tasks where the latter obtained better results. A positive element of this work is that it is reported in the PSPLib library as one of the works that has effectively completed all the experimentation cycles and is taken in the present investigation as one of the essential works for comparison.

Lova and colaboradores in [73] propose the MM-HGA algorithm, where the solution is represented by two lists: the first with the tasks and the other with the execution modes. The next generations are obtained by applying the two-point crossover and mutation operators, which are applied to the list of assigned modes. An interesting element is the use of local search methods to reduce duration times as part of the feasible solutions obtained by the GA. This idea was considered in the design of the algorithms proposed in this research. However, it is not compared with this algorithm because the results it reports are older than the last five years and are not the best reported in PSPLib.

Damak and colaboradores [71] propose a Differential Evolution (DE) in which a solution is represented by a list of mode assignments and another list of tasks that do not have to be ordered by precedence. Neighbor solutions are generated using mutation and crossover operators. According to the experiments carried out by [71], the DE algorithm showed results superior to those achieved by the SA [88] and PSO [89] algorithms. As a shortcoming of this method, it is noted that it was not compared with instances with a greater number of tasks, only with 20 tasks at most. This algorithm is older than five years and does not report its results in PSPLib, an element that makes its use for this experimentation difficult.

Peteghem and Vanhoucke present the BPGA algorithm [49] that constructs two populations, the first under the "as late as possible" approach and the other under the "as early as possible" approach. It is based on using one population to construct the other using the so-called Serial Solution Generation Scheme (SGS) [90]. A solution is also represented by two lists: a list of tasks ordered by priority and another list of mode assignments. Considering the criteria expressed above, this algorithm is not among the algorithms used in the experiment.

Coelho and Vanhoucke propose the $GA_{\text{Coe-Van}}$ presented by [74] in which a new approach is proposed that divides the MMRCPSP into two interrelated sub-problems: mode allocation and the single resource-limited project planning problem mode. The mode assignment sub-problem is solved using a Satisfiability Problem Solver (SAT) while the RCPSP is solved using a GA. A limitation of this approach that is evident in the work itself is that the size of the memory necessary to execute this algorithm grows exponentially with the size of the problem (number of tasks and/or number of non-renewable resource constraints). The results obtained were classified by [74] as satisfactory, although with a high computational cost. This work meets criterion 2 mentioned above, so it will be selected for experimentation.

Van Peteghem and Vanhoucke [70] propose the use of scattered search (SS), where each solution is represented by two lists: a random key (RK), which determines the sequence in which the tasks are executed, and another list of modes, which determines the duration and resources required for the execution of each task. Solutions are constructed using SGS serial [90] to decode these lists into a schedule. Then several methods are used to improve the solutions obtained. In this work, it is shown that the performance of this algorithm is sensitive to variations with respect to the amounts of renewable and non-renewable resources. It is pointed out as a deficiency of their proposal that the experimentation was carried out on databases created especially for this purpose and not on the public instances available for experimentation, such as those proposed by PSPLib.

Wang and Fang [78] propose a hybrid EDA with local search to solve the MMRCPSP. This algorithm uses two probabilistic models to generate a solution to the scheduling problem, one for the tasks and one for the modes. For a problem with $n$ tasks and $m$ execution modes, the probabilistic model to generate the solutions will be of size $n * n$, while the other will have size $n * m$. It is ensured that each task appears only once in the list through the use of a procedure called Permutation-based Probability Generation Mechanism (PGM). As a positive element of this proposal that may be of interest for research, the use of the two probabilistic models is noted. The results shown in this work are competitive, however, it is identified that the implemented improvement procedures have a high influence on the results and do not exploit the potential of the EDA in the treatment of the correlation of the variables and in the treatment of the restrictions. This algorithm is older than five years and does not report its results in the PSPLib library, so it will not be used in the experimentation.

Then Soliman and collaborators [79] present a hybrid EDA with random local search techniques to solve the MMRCPSP type planning problem. The local search technique is based on the "delete" and "insert" operators to allow better exploitation of the search space. First, they use a list of tasks and modes, then they evaluate all individuals using a task generation scheme called the Multi-Mode Serial Schedule Generation Scheme (MSSGS), and then they apply local search techniques to evaluate the selected individuals. However, checking the execution mode of each task in the two previous proposals iteratively implies a significant increase in the computational cost to evaluate the solutions obtained [51]. This algorithm is older than five years and does not report its results in the PSPLib library.

On the other hand, Vanhoucke and Coelho present the $GA_{Van\text{-}Coe}$ in [75] in which they extend the planning problem with two new classes of logical constraints that model the relationship between each pair of tasks through the use of OR operators and BI (bidirectional). Furthermore, they introduce a SAT solver used years ago by one of the authors [74] and a GA to find feasible solutions regarding compliance with the constraints of resource availability. The authors make comparisons based on several parameters, among which the following stand out: the number of optima found and the percentage of deviation from the optima. According to the experiments reported in [75], the $GA_{Van\text{-}Coe}$ algorithm achieved competitive results in terms of the effectiveness of the solutions found, but it did not surpass the solutions to the MMRCPSP reported by $GA_{Coe\text{-}Van}$ [74], for this reason, it will not be used in the experimentation.

Another alternative was the combination of GA with the univariate EDA PBIL (Population Based Incremental Learning) [91], presented in [77] as the PBGA-EDA algorithm. In this algorithm, we work with two lists of solutions [49], $POP_L$ and $POP_R$. The best individuals are selected using the tournament method and are then used to learn the probabilistic model regarding the execution modes. With the learned probabilistic model associated with the modes, new solutions are generated that are incorporated into the two lists. Then, crossover and mutation operators are applied, generating two lists that replace the previous population.

Years later, the author Ayodele herself, in [51] proposed the BPEDA (Bi-Population Estimation of Distribution Algorithm) with the aim of improving the results of the PBGA-EDA. In this new proposal, we do not work with GA but with the RK-EDA [92] for task planning. Each task has a randomly assigned RK value (for the first generation), which is nothing more than the ranking of the task within a solution. The best solutions are selected from a list, and then the probabilistic model is built with respect to the tasks and modes. Through the probabilistic models built, the new solutions from the other list are generated. That is, each of these lists is updated based on the probabilistic model built from the best solutions from the other list. This algorithm applies penalty techniques to the treatment of non-feasible solutions and does not introduce any element that manages the restrictions of the probabilistic model. The results of this work do not include PSPLib data for each of the instances used in the experimentation, which makes its use difficult for the comparison of the algorithms proposed in this research.

Another exact approach to solving the MMRCPSP is proposed by Schnell and Hartl in [80]. The authors present three formulations based on constraint programming (CP) combined with constraint satisfaction techniques. Furthermore, they introduce a new global cumulative constraint specially adapted to check the feasibility of the solutions with respect to compliance with the availability of renewable resources and to reduce the search domain. According to the experimental results presented, their approach achieves competitive results regarding the duration of the projects based on the j30.mm data. Although its results are not reported in the PSPLib library, they are available on the web, an element that facilitates their comparison with the proposals of this research.

Also [39], developed a hybrid algorithm called (Adapted TS-NSGA-II), based on the tabu search with the NSGA-II algorithm (Non-dominated Sorting Genetic Algorithm Two), to solve the MMRCPSP. This work has as a positive element the optimization of execution time and cost under a multi-objective approach: both objectives are evaluated from a linear combination in the objective function, which was taken into consideration for the present investigation. However, the algorithm presented in this work is not sufficiently described for its use and reproduction. Furthermore, the metrics used to evaluate the results are specific, an element that makes comparison with other algorithms difficult. For these reasons, this algorithm was not considered in the experimentation of this research.

The Robust hGMEDA was developed by [40] as an extension of the hGMEDA Distribution Estimation Algorithm [82] and is based on the hybridization between EDA and GA for the solution of the MMRCPSP with uncertainty in the duration time between tasks and validate it with the "n0.mm" instances of PSPLib. In the case of optimization regarding the duration of the project, the proposed algorithm is compared with the SPEA2 algorithm [83] and the hGMEDA algorithm [82]. According to the results reported in the work of [40], the Robust hGMEDA obtains superior results than the aforementioned algorithms and will also be used in experimentation.

Araujo and collaborators [81] propose an exact approach based on Mixed-Integer Linear Programming (MILP). The authors developed a parallel shear plane algorithm, with the objective of reducing the feasible region, which includes five families of cuts, including the reinforced Chvátal-Gomory method [93] that are based on the introduction of new restrictions to the relaxed problem (without resource and precedence restrictions) until the optimal solution of the new problem is integer. For a given day, feasible subsets of tasks and modes are enumerated to create a linear problem with the goal of finding the best combination given a resource constraint. According to the experiments carried out by [81], the results of their algorithm are competitive in terms of the effectiveness of the solutions found. However, this approach involves a systematic and exhaustive search for solutions and is too costly. This work meets criteria 2 and 4 mentioned above, so it will be used in the validation of the results of this research.

As a summary so far, it has been identified that there are few works that use EDA to solve the MMRCPSP problem and that none of the algorithms proposed in the consulted bibliography address the restrictions of the probabilistic model. Therefore, it is a motivation for this research to analyze the influence of the treatment of constraints in the probabilistic model on the performance of the EDA.

## *5.1 Characterization of EDA Algorithms in Solving Planning Problems Considering the Correlation of Variables*

An essential element that distinguishes each of the EDAs is the learning of the probabilistic model from the distribution estimation of a selected set of individuals (see step 6 of Algorithm 1).

**Algorithm 1: Pseudocode of the EDA algorithm family**

1. $g = 1$
2. *Generate* $R >> 0$ *points randomly*
3. *WHILE the stop criterion is not met DO*
    4. *Evaluate the population in the function* $f(x)$
    5. *Construct a set, CS, of S points, where* $S \leq R$ *according to a selection method*
    6. *Estimate the distribution of the selected set or* $p^{CS} = p(x, g)$ *from* $CS$
    7. *Generate R new points from* $p(x, g + 1) p^{CS} = p(x, g)$
    8. $g = g + 1$
9. *END WHILE*

The UMDA and FDA algorithms frequently used in the literature are analyzed in this section [94]. The UMDA algorithm has been selected for analysis considering its simplicity and because it has been used in different scenarios, regardless of the nature of the data, with satisfactory results [95]. This algorithm was proposed in [84]; It is characterized because in each iteration, the marginal distribution of each variable in the selected solution is estimated from their relative frequencies, see Algorithm 2.

**Algorithm 2: UMDA pseudocode**

1. *Initialize population* $D_0$ *with R solutions randomly*
2. *WHILE the stop criterion is not met DO*
    3. *Evaluate the population in the objective function*
    4. $D^S_{g-1} \leftarrow$ *Select* $S \leq R$ *solutions according to a selection method*

5. *Estimate the joint distribution of the selected solutions*

$$p_g(X) = p\left(X \middle| D_{g-1}^S\right) = \prod_{i=1}^{n} p_g(x_i) = \prod_{i=1}^{n} \frac{\sum_{q=1}^{S} \delta_q\left(X_i = x_i \middle| D_{g-1}^S\right)}{S} \quad (19)$$

6. $D_g \leftarrow$ *Generate R new solutions by sampling from* $p_g(X)$
7. $g \leftarrow g + 1$

8. *END WHILE*

This algorithm does not manage the dependency restrictions existing between the variables in the optimization problem and focuses its learning on the probabilistic analysis of each variable independently. Therefore, in each iteration, the joint probability distribution, $p_g(X)$ of $n$ variables can be calculated as the product of the individual distributions of each variable as follows:

$$p_g(X) = p\left(X \middle| D_{g-1}^S\right) = \prod_{i=1}^{n} p_g(x_i) \quad (20)$$

where $X$ represents an individual (candidate solution), $n$ the number of variables $x_i$, $g$ nnumber of iterations (generations), and $D_{g-1}^S$ represents the solutions selected in generation $g-1$. The marginal distribution of each variable is estimated from the marginal frequencies. That is to say:

$$p_g(x_i) = \prod_{i=1}^{n} \frac{\sum_{q=1}^{S} \delta_q\left(X_i = x_i \middle| D_{g-1}^S\right)}{S} \quad (21)$$

where

$$\delta_q\left(X_i = x_i \middle| D_{g-1}^S\right) = \begin{cases} 1 \text{ if in the } q\text{th case of} D_{g-1}^S,\ X_i = x_i \\ 0 \qquad\qquad \text{in another case} \end{cases}$$

The EDA algorithm is a theoretical proposal because the calculation of all the parameters necessary to specify the distribution is intractable [96]. In other words, the most complicated thing in EDA is how to estimate the probability distribution in problems where hundreds of variables are involved. As the number of variables increases, the calculation of all the parameters necessary to specify the joint probability distribution becomes intractable [97]. For example, storing a distribution of $n$ binary variables mean storing $2^n - 1$ parameters (probabilities). Such is the case with problems with multiple dependencies.

With the aim of making EDA a computationally tractable algorithm, the Factorized Distribution Algorithm (FDA) was created, which was published for the first time in the work of [25]. Its main contribution is that the interaction between variables can be reduced to a subset of them through finding an additively decomposable function (ADF). This function is used to simulate problems that can be decomposed into smaller subproblems and consequently, the distribution can be calculated in polynomial time [25, 32]. The value of the function in the ADF is the sum of the evaluation of several subfunctions defined on variable subsets of $X$. This function is defined as follows:

**Definition 1** ([98]) Let $s_1, s_2, \ldots, s_m$ be sets of indices with $s_i \subseteq \{1, \ldots, n\}$. Let $f_{s_i}$ be functions that depend only on the variables $x_q$ with $q \in s_i$. So is an additive decomposition of the function $f$.

$$f(x) = \sum_{i=1}^{m} f_{s_i}(x) \tag{22}$$

**Definition 2** ( *Stream intersection property*) [85] Given $s_1, s_2, \ldots, s_m$ the sets $d_i, b_i$ and $c_i$ are defined for $i = 1, \ldots, m$ where

$d_i := \bigcup_{q=1}^{i} s_q$, $b_i := s_i \backslash d_{i-1}$ and $c_i := s_i \cap d_{i-1}$ with $d_0 = \emptyset$

These sets are known in the literature as histories, residues and separators.

The straight intersection property is satisfied if:

$$b_i \neq \emptyset \forall i = 1, \ldots, m d_m = \{1, \ldots, n\} \forall i \geq 2 \exists j < i \text{such that} \, c_i \subseteq s_j$$

What we have seen so far allows us to state the following theorem:

**Theorem 1** (Factorization theorem) [98]: *If $f(x)$ is an additive decomposition and satisfies the straight intersection property, then the joint probability distribution can be factored, according to a more complex probability model simplified as follows*:

$$p(x) = \prod_{i=1}^{m} p\left(x_{b_i} | x_{c_i}\right) \tag{23}$$

*Precisely, Theorem allows us to approximate problems with multiple correlated variables with the FDA algorithm* (Algorithm 3). *The distribution reflected in* Eq. (23) *is valid for all iterations of the algorithm.*

**Algorithm 3: FDA Pseudocode**

1. *Calculate $b_i$ and $c_i$ from the decomposition of the additive function $f(x)$*
2. $g \Leftarrow 0$
3. *Generate R solutions randomly*
4. *DO* {
   5. *Evaluate the solutions in the objective function $f(x)$*
   6. *Select a set S of solutions where $|S| \leq R$ according to a selection method*
   7. *Estimate the conditional probabilities (from S) $p^S\left(x_{b_i} | x_{c_i}\right)$*
   8. *Generate R new solutions from $p(x, g + 1) \approx \prod_{i=1}^{m} p^s\left(x_{b_i} | x_{c_i}, g\right)$*

9. $g \Leftarrow g + 1$

10. } *Until the stopping criterion is met*

The FDA is selected because its intrinsic nature allows it to model problems with correlated variables through factorization, which simplifies the complexity of the construction of the probabilistic model. The dependency relationships between the tasks (variables) are considered criteria for the factorization and application of this algorithm, which constitutes one of the inputs to the project planning problem.

The computational complexity of the UMDA is $O(g \cdot n \cdot S)$, where $g$ represents the number of iterations, $n$ the number of variables, and $S$ represents the number of individuals selected from the population. For its part, according to [85], the computational complexity of the FDA depends on the factorization and the size of the population $R$ and is equal to $O\left(g \cdot m \cdot n^{|\hat{c}|} \cdot S\right)$, where $m$ denotes the number of factorizations and $|\hat{c}|$ is the average size of the factorization.

In general, both UMDA and FDA are not designed to handle the constraints in the probabilistic model. Several works from the literature consulted when using these algorithms treat restrictions as part of the solution evaluation process using penalty strategies.

Given this situation, the development of extensions to the UMDA and FDA algorithms that incorporate the treatment of restrictions within the probabilistic model constitutes an open line of research. Furthermore, characterize these two algorithms regarding their performance when faced with changes in MMRCPSP problems regarding the number of tasks, the number of modes, or the amount of renewable and non-renewable resources.

## 6 Conclusions

In the context of project management, planning problems are handled from two approaches: the process approach, described in project management schools, and the approach based on the treatment of planning problems as optimization problems. The main standards and guides for project management recognize the importance of building optimal or quasi-optimal schedules, but do not propose optimization techniques. In this sense, it is identified that there are open lines of research associated with the hybridization of these two approaches in the solution of planning problems.

In this context, problems of the MMRCPSP type are particularly interesting because their solutions can be adapted to the solutions of other problems such as: RCPSP, RCMPSP and MMRCMPSP. Despite the progress made in solving the MMRCPSP, a set of difficulties associated with the efficient treatment of restrictions still persist. In general, the MMRCPSP has been less treated compared to the other simpler planning problems, this is evident in the results reported in PSPLib.

In the bibliography consulted, the use of both exact and approximate approaches is reported for the solution of different planning problems, the first characterized by its high computational cost and the other by the use of meta-heuristics. Among the most commonly used meta-heuristics are genetic algorithms, despite their natural difficulty in dealing with problems with interrelated variables.

Particularly for the MMRCPSP problem, there are few works that apply Distribution Estimation Algorithms for its solution. Based on the nature of EDA and its ability to work with correlated variables, the application of this family of algorithms to project management is identified as an open line of research.

In the literature consulted, there are few works that address the issue of the treatment of restrictions in EDAs. Furthermore, the algorithms reported in the literature generally apply penalty-based strategies for the treatment of constraints. In the systematic review carried out, no works were found that handle restrictions from the probabilistic model, an element that constitutes a motivation for this research.

As future work, it is identified:

- Develop variants of algorithms with distribution estimation, exploiting the potential for working with related variables that allow the effective solution of planning problems.
- Identify strategies for optimization modeling, considering the priority of tasks and projects.
- Carry out an exploratory study on parallel and distributed programming techniques to reduce the complexity of complex project planning problems.

## References


1. Vrchota J, Řehoř P, Maříková M, Pech M (2021) Critical success factors of the project management in relation to industry 4.0 for sustainability of projects. Sustainability 13:281. https://doi.org/10.3390/su13010281
2. ISO (2018) ISO 19650-1:2018. Organization and digitization of information about buildings and civil engineering works. In: Including building information modelling (BIM)—Information management using building information modelling. Part 1: Concepts and Principles. In: ISO. https://www.iso.org/standard/68078.html. Accessed 10 Sep 2020
3. Mesároš P, Mandičák T, Behúnová A (2020) Use of BIM technology and impact on productivity in construction project management. Wirel Netw. https://doi.org/10.1007/s11276-020-02302-6
4. Loyola M (2019) Encuesta Nacional BIM 2019: Informe de resultados. University of Chile, Chile
5. González CFL (2019) Metodología BIM (building information modeling) aplicada a la prevención de riesgos laborales (PRL). J BIM Constr Manag 1:20–30
6. Dave B, Buda A, Nurminen A, Främling K (2018) A framework for integrating BIM and IoT through open standards. Autom Constr 95:35–45. https://doi.org/10.1016/j.autcon.2018.07.022
7. Lynch J (2018) Project resolution benchmark for IBEX financial corp. The Standish Group International Inc., United States
8. Johnson J (2018) CHAOS report: decision latency theory: it is all about the interval. The Standish Group International Inc., Boston, MA
9. Villavicencio NE (2016) Modelo integrado para la mejora de la productividad en organizaciones orientadas a proyectos de tecnologías de la información. Tesis para optar al grado de: Máster en Diseño, Gestión y Dirección de de Proyectos, Fundación Universitaria Iberoamericana

10. Pupo IP, Vacacela RG, Pérez PP, Mahdi GSS, Peña M (2020) Experiencias en el uso de técnicas de softcomputing en laevaluación de proyectos de software. Investigación Oper 41:108–119
11. Bermúdez NV, Abreu MP, Valareso SB, Pupo IP (2016) Experiencias en la integración de procesos en las organizaciones orientadas a proyectos de software. Revista Cubana de Ciencias Informáticas 10:171–185
12. Project Management Institute (2021) A guide to the project management body of knowledge, 7th edn. Project Management Institute Inc., Newtown Square, Pennsylvania
13. ISO (2012) ISO 21500: guidance on project management. In: ISO. https://www.iso.org/cms/render/live/en/sites/isoorg/contents/data/standard/05/00/50003.html. Accessed 16 Jan 2021
14. Chaudhary M, Chopra A (2017) CMMI for development: implementation guide, 1st edn. Apress, India
15. CMMI Product Team (2010) CMMI® for Development, Version 1.3. Software Engineering Institute, Carnegie Mellon University
16. Malcolm DG, Roseboom JH, Clark CE, Fazar W (1959) Application of a technique for research and development program evaluation. Oper Res 7:646–669. https://doi.org/10.1287/opre.7.5.646
17. Blazewicz J, Ecker KH, Pesch E, Schmidt G, Sterna M, Weglarz J (2019) Handbook on scheduling: from theory to practice, 2nd ed. Springer International Publishing
18. Vanhoucke M (2023) The data-driven project manager. In: Vanhoucke M (ed) The illusion of control: project data, Computer algorithms and human intuition for project management and control. Springer Nature Switzerland, Cham, pp 29–48
19. Ulusoy G, Hazır Ö (2021) An introduction to project modeling and planning. Springer International Publishing, Cham
20. Alves CGM de F, Meza LA (2023) A review of network DEA models based on slacks-based measure: evolution of literature, applications, and further research direction. Int Trans Oper Res. 30:2729–2760. https://doi.org/10.1111/itor.13284
21. Liu H, Qu S, Li R, Razaa H (2021) Bi-objective robust project scheduling with resource constraints and flexible activity execution lists. Comput Ind Eng 156:107288. https://doi.org/10.1016/j.cie.2021.107288
22. CubaIndustria (2020) Normas Cubanas Online. http://www.nconline.cubaindustria.cu:81/. Accessed 16 Jan 2021
23. Carter M, Price CC, Rabadi G (2019) Operations research: a practical introduction, 2nd edn. CRC Press, Broken Sound Parkway NW
24. Sarjono H, Kurnia VD (2022) A systematic literature review: optimization of implementation of project development in the company with PERT and CPM method. © IEOM society international. Telangana, India, pp 2478–2486
25. Ballesteros-Pérez P, Larsen GD, González-Cruz MC (2018) Do projects really end late? On the shortcomings of the classical scheduling techniques. JOTSE: J Technol Sci Educ 8:17–33. https://doi.org/10.3926/jotse.303
26. Marnada P, Raharjo T, Hardian B, Prasetyo A (2022) Agile project management challenge in handling scope and change: a systematic literature review. Procedia Comput Sci 197:290–300. https://doi.org/10.1016/j.procs.2021.12.143
27. Redmine (2020) Overview—Redmine. In: Redmine—Project management and application management. https://www.redmine.org/. Accessed 16 Jan 2021
28. Piñero Ramírez PE, Pérez Pupo I, Piñero Pérez PY, Marquez Ruiz Y, Fustiel Alvarez Y (2022) A software ecosystem for project management in BIM environments assisted by artificial intelligent techniques. In: Piñero Pérez PY, Bello Pérez RE, Kacprzyk J (eds) Artificial intelligence in project management and making decisions. Springer International Publishing, Cham, pp 191–212
29. Pritsker AAB, Waiters LJ, Wolfe PM (1969) Multiproject scheduling with limited resources: a zero-one programming approach. Manage Sci 16:93–108. https://doi.org/10.1287/mnsc.16.1.93
30. Gómez Sánchez M, Lalla-Ruiz E, Fernández Gil A, Castro C, Voß S (2023) Resource-constrained multi-project scheduling problem: a survey. Eur J Oper Res 309:958–976. https://doi.org/10.1016/j.ejor.2022.09.033

31. Rahimi I, Gandomi AH, Chen F, Mezura-Montes E (2023) A review on constraint handling techniques for population-based algorithms: from single-objective to multi-objective optimization. Arch Comput Methods Eng 30:2181–2209. https://doi.org/10.1007/s11831-022-09859-9
32. Mezura-Montes E, Coello Coello CA (2011) Constraint-handling in nature-inspired numerical optimization: Past, present and future. Swarm Evol Comput 1:173–194. https://doi.org/10.1016/j.swevo.2011.10.001
33. Zhu G, Feng Y, Chen Z, Lu H (2022) Hydrodynamic design of a circulating water channel based on a fractional-step multi-objective optimization. Ocean Eng 251:110959. https://doi.org/10.1016/j.oceaneng.2022.110959
34. Mohamed AW, Oliva D, Suganthan PN (2022) Handbook of nature-inspired optimization algorithms: the state of the art: volume II: solving constrained single objective real-parameter optimization problems. Springer Nature
35. Runarsson TP, Yao X (2005) Search biases in constrained evolutionary optimization. IEEE Trans Syst Man Cybern Part C (Appl Rev) 35:233–243. https://doi.org/10.1109/TSMCC.2004.841906
36. Balouka N, Cohen I (2019) A robust optimization approach for the multi-mode resource-constrained project scheduling problem. Eur J Oper Res. https://doi.org/10.1016/j.ejor.2019.09.052
37. Stiti C, Driss OB (2019) A new approach for the multi-site resource-constrained project scheduling problem. Procedia Comput Sci 164:478–484. https://doi.org/10.1016/j.procs.2019.12.209
38. Tao S, Dong ZS (2018) Multi-mode resource-constrained project scheduling problem with alternative project structures. Comput Ind Eng 125:333–347. https://doi.org/10.1016/j.cie.2018.08.027
39. Tian J, Hao X, Gen M (2019) A hybrid multi-objective EDA for robust resource constraint project scheduling with uncertainty. Comput Ind Eng 130:317–326. https://doi.org/10.1016/j.cie.2019.02.039
40. Ding H, Zhuang C, Liu J (2023) Extensions of the resource-constrained project scheduling problem. Autom Constr 153:104958. https://doi.org/10.1016/j.autcon.2023.104958
41. Sayah D (2023) Continuous-time formulations for multi-mode project scheduling. Comput Oper Res 152:106147. https://doi.org/10.1016/j.cor.2023.106147
42. Chen JC, Lee H-Y, Hsieh W-H, Chen T-L (2022) Applying hybrid genetic algorithm to multi-mode resource constrained multi-project scheduling problems. J Chin Inst Eng 45:42–53. https://doi.org/10.1080/02533839.2021.1983461
43. Brucker P, Drexl A, Möhring R, Neumann K, Pesch E (1999) Resource-constrained project scheduling: notation, classification, models, and methods. Eur J Oper Res 112:3–41. https://doi.org/10.1016/S0377-2217(98)00204-5
44. Servranckx T, Vanhoucke M (2019) A tabu search procedure for the resource-constrained project scheduling problem with alternative subgraphs. Eur J Oper Res 273:841–860. https://doi.org/10.1016/j.ejor.2018.09.005
45. Kumar M, Mittal ML, Soni G, Joshi D (2019) A Tabu search algorithm for simultaneous selection and scheduling of projects. In: Yadav N, Yadav A, Bansal JC, Deep K, Kim JH (eds) Harmony search and nature inspired optimization algorithms. Springer, Singapore, pp 1111–1121
46. Laurent A, Deroussi L, Grangeon N, Norre S (2017) A new extension of the RCPSP in a multi-site context: mathematical model and metaheuristics. Comput Ind Eng 112:634–644. https://doi.org/10.1016/j.cie.2017.07.028
47. Debels D, Vanhoucke M (2005) A bi-population based genetic algorithm for the resource-constrained project scheduling problem O. In: Gervasi, Gavrilova ML, Kumar V, Laganá A, Lee HP, Mun Y, Taniar D, Tan CJK (eds) Computational science and its applications—ICCSA 2005. Springer, Berlin, Heidelberg, pp 378–387
48. Peteghem VV, Vanhoucke M (2010) A genetic algorithm for the preemptive and non-preemptive multi-mode resource-constrained project scheduling problem. Eur J Oper Res 201:409–418. https://doi.org/10.1016/j.ejor.2009.03.034

49. Wang L, Fang C (2012) A hybrid estimation of distribution algorithm for solving the resource-constrained project scheduling problem. Expert Syst Appl 39:2451–2460. https://doi.org/10.1016/j.eswa.2011.08.095
50. Ayodele M (2018) Effective and efficient estimation of distribution algorithms for permutation and scheduling problems. Ph.D. thesis. Robert Gordon University
51. Tian M, Liu RJ, Zhang GJ (2020) Solving the resource-constrained multi-project scheduling problem with an improved critical chain method. J Oper Res Soc 71:1243–1258. https://doi.org/10.1080/01605682.2019.1609883
52. Yassine AA, Mostafa O, Browning TR (2017) Scheduling multiple, resource-constrained, iterative, product development projects with genetic algorithms. Comput Ind Eng 107:39–56. https://doi.org/10.1016/j.cie.2017.03.001
53. Wang Y, He Z, Kerkhove L-P, Vanhoucke M (2017) On the performance of priority rules for the stochastic resource constrained multi-project scheduling problem. Comput Ind Eng 114:223–234. https://doi.org/10.1016/j.cie.2017.10.021
54. Medrano BE (2013) Planificación de múltiples proyectos de desarrollo de software utilizando métodos metaheurísticos. Master Thesis, Universidad de La Habana
55. Dong N, Ge D, Fischer M, Haddad Z (2012) A genetic algorithm-based method for look-ahead scheduling in the finishing phase of construction projects. Adv Eng Inform 26:737–748. https://doi.org/10.1016/j.aei.2012.03.004
56. Schmidt T, Kühn M, Genßler PR (2017) Design of project-oriented calculation models for job priorities by using a customized genetic algorithm S. In: Wenzel, Peter T (eds) 17th ASIM proceedings. Kassel University Press, Kassel, Germany, pp 99–108
57. Kühn M, Schmidt T, Völker M (2019) Simulation-based optimization approach for efficient generation of sequencing heuristics for solving the stochastic resource-constrained scheduling problem M. In: Putz, Schlegel A (eds) Simulation in produktion und logistik 2019. Verlag Wissenschaftliche Scripten, Auerbach, pp 403–412
58. Talbot FB (1982) Resource-constrained project scheduling with time-resource trade offs: the nonpreemptive case. Manag Sci 28:1197–1210. https://doi.org/10.1287/mnsc.28.10.1197
59. Asta S, Karapetyan D, Kheiri A, Özcan E, Parkes AJ (2016) Combining Monte-Carlo and hyper-heuristic methods for the multi-mode resource-constrained multi-project scheduling problem. Inf Sci 373:476–498. https://doi.org/10.1016/j.ins.2016.09.010
60. Kuehn M, Zahid T, Voelker M, Zhou Z, Rose O (2016) Investigation of genetic operators and priority heuristics for simulation based optimization of multi-mode resource constrained multi-project scheduling problems (MMRCMPSP). In: Thorsten C, Frank H, Michael M, Oliver R (eds) ECMS 2016 proceedings. ECMS, pp 481–487
61. Pinha D, Ahluwalia R, Senna P (2016) The combinatorial multi-mode resource constrained multi-project scheduling problem. Int J Supply Oper Manag 3:1391–1412. https://doi.org/10.22034/2016.3.03
62. He N, Zhang DZ, Yuce B (2022) Integrated multi-project planning and scheduling—A multiagent approach. Eur J Oper Res 302:688–699. https://doi.org/10.1016/j.ejor.2022.01.018
63. Speranza MG, Vercellis C (1993) Hierarchical models for multi-project planning and scheduling. Eur J Oper Res 64:312–325. https://doi.org/10.1016/0377-2217(93)90185-P
64. Wang W, Wang X, Ge X, Deng L (2014) Multi-objective optimization model for multi-project scheduling on critical chain. In: Advances in engineering software, pp 33–39
65. Beşikci U, Bilge Ü, Ulusoy G (2015) Multi-mode resource constrained multi-project scheduling and resource portfolio problem. Eur J Oper Res 240:22–31. https://doi.org/10.1016/j.ejor.2014.06.025
66. Zhou Z, Angelidis E, Bohn D, Rose O (2017) A two-phase genetic algorithm to solve a multiobjective problem for complex assembly lines. In: Wenzel S, Peter T (eds) 17th ASIM proceedings. Kassel University Press, Kassel, Germany, pp 89–98
67. Sadowski C, Zimmermann T (2019) Rethinking productivity in software engineering, 1st ed. Springer Nature, Berkeley, CA
68. Kolisch R (2021) About the solution of multi-mode resource-constrained multi-project scheduling problem

69. Van Peteghem V, Vanhoucke M (2011) Using resource scarceness characteristics to solve the multi-mode resource-constrained project scheduling problem. J Heuristics 17:705–728. https://doi.org/10.1007/s10732-010-9152-0
70. Damak N, Jarboui B, Siarry P, Loukil T (2009) Differential evolution for solving multi-mode resource-constrained project scheduling problems. Comput Oper Res 36:2653–2659. https://doi.org/10.1016/j.cor.2008.11.010
71. Sprecher A, Drexl A (1998) Multi-mode resource-constrained project scheduling by a simple, general and powerful sequencing algorithm. Eur J Oper Res 107:431–450. https://doi.org/10.1016/S0377-2217(97)00348-2
72. Lova A, Tormos P, Cervantes M, Barber F (2009) An efficient hybrid genetic algorithm for scheduling projects with resource constraints and multiple execution modes. Int J Prod Econ 117:302–316. https://doi.org/10.1016/j.ijpe.2008.11.002
73. Coelho J, Vanhoucke M (2011) Multi-mode resource-constrained project scheduling using RCPSP and SAT solvers. Eur J Oper Res 213:73–82. https://doi.org/10.1016/j.ejor.2011.03.019
74. Vanhoucke M, Coelho J (2016) An approach using SAT solvers for the RCPSP with logical constraints. Eur J Oper Res 249:577–591. https://doi.org/10.1016/j.ejor.2015.08.044
75. Hartmann S (2001) Project scheduling with multiple modes: a genetic algorithm. Ann Oper Res 102:111–135. https://doi.org/10.1023/A:1010902015091
76. Ayodele M, McCall J, Regnier-Coudert O (2016) BPGA-EDA for the multi-mode resource constrained project scheduling problem. In: 2016 IEEE Congress on evolutionary computation (CEC). Vancouver, Canada, pp 3417–3424
77. Wang L, Fang C (2012) An effective estimation of distribution algorithm for the multi-mode resource-constrained project scheduling problem. Comput Oper Res 39:449–460. https://doi.org/10.1016/j.cor.2011.05.008
78. Soliman OS, Elgendi EAR (2014) A hybrid estimation of distribution algorithm with random walk local search for multi-mode resource-constrained project scheduling problems. IJCTT 8:57–64. https://doi.org/10.14445/22312803/IJCTT-V8P111
79. Schnell A, Hartl RF (2017) On the generalization of constraint programming and boolean satisfiability solving techniques to schedule a resource-constrained project consisting of multi-mode jobs. Oper Res Perspect 4:1–11. https://doi.org/10.1016/j.orp.2017.01.002
80. Araujo JAS, Santos HG, Gendron B, Jena SD, Brito SS, Souza DS (2020) Strong bounds for resource constrained project scheduling: preprocessing and cutting planes. Comput Oper Res 113:104782. https://doi.org/10.1016/j.cor.2019.104782
81. Hao X, Sun L, Gen M (2018) Solving robust resource constrained scheduling problem by multi-objective optimization method based on hybridization of EDA and GA. Procedia Manuf 17:1187–1194. https://doi.org/10.1016/j.promfg.2018.10.001
82. Zitzler E, Laumanns M, Thiele L (2001) SPEA2: improving the strength Pareto evolutionary algorithm. Eidgenössische Technische Hochschule Zürich (ETH), Institut für Technische Informatik und Kommunikationsnetze (TIK) 103. https://doi.org/10.3929/ETHZ-A-004284029
83. Mühlenbein H (1997) The equation for response to selection and its use for prediction. Evol Comput 5:303–346. https://doi.org/10.1162/evco.1997.5.3.303
84. Mühlenbein H, Mahnig T (1999) FDA-A scalable evolutionary algorithm for the optimization of additively decomposed functions. Evol Comput 7:353–376. https://doi.org/10.1162/evco.1999.7.4.353
85. Mika M, Waligóra G, Węglarz J (2015) Overview and state of the art. In: Schwindt C, Zimmermann J (eds) Handbook on project management and scheduling, vol 1. Springer International Publishing, Cham, pp 445–490
86. Hartmann S, Drexl A (1998) Project scheduling with multiple modes: a comparison of exact algorithms. Networks 32:283–297. https://doi.org/10.1002/(SICI)1097-0037(199812)32:4%3c283::AID-NET5%3e3.0.CO;2-I
87. Bouleimen K, Lecocq H (2003) A new efficient simulated annealing algorithm for the resource-constrained project scheduling problem and its multiple mode version. Eur J Oper Res 149:268–281. https://doi.org/10.1016/S0377-2217(02)00761-0

88. Jarboui B, Eddaly M, Siarry P (2009) An estimation of distribution algorithm for minimizing the total flowtime in permutation flowshop scheduling problems. Comput Oper Res 36:2638–2646. https://doi.org/10.1016/j.cor.2008.11.004
89. Kolisch R, Hartmann S (1999) Heuristic algorithms for the resource-constrained project scheduling problem: classification and computational analysis. In: Węglarz J (ed) Project scheduling: recent models, algorithms and applications. Springer, US, Boston, MA, pp 147–178
90. Baluja S (1994) Population-based incremental learning. A method for integrating genetic search based function optimization and competitive learning. Carnegie-Mellon Univ Pittsburgh Pa Dept of Computer Science, Pittsburgh, Pennsylvania
91. Ayodele M, McCall J, Regnier-Coudert O (2016) RK-EDA: a novel random key based estimation of distribution algorithm J. In: Handl, Hart E, Lewis PR, López-Ibáñez M, Ochoa G, Paechter B (eds) Parallel problem solving from nature—PPSN XIV. Springer International Publishing, Cham, pp 849–858
92. Dash S, Günlük O, Lee D (2020) Generalized Chvátal-Gomory closures for integer programs with bounds on variables. Math Program. https://doi.org/10.1007/s10107-020-01539-5
93. Krejca MS, Witt C (2020) Theory of estimation-of-distribution algorithms B. In: Doerr, Neumann F (eds) Theory of evolutionary computation: recent developments in discrete optimization. Springer International Publishing, Cham, pp 405–442
94. Dang D-C, Lehre PK, Nguyen PTH (2019) Level-Based analysis of the univariate marginal distribution algorithm. Algorithmica 81:668–702. https://doi.org/10.1007/s00453-018-0507-5
95. Madera J (2009) Hacia una generación más eficiente de algoritmos evolutivos con estimación de distribuciones: pruebas de independencia + paralelismo. Ph.D. thesis, Instituto de Cibernética, Matemática y Física
96. Kumar M, Soomro AM, Uddin W, Kumar L (2022) Optimal multi-objective placement and sizing of distributed generation in distribution system: a comprehensive review. Energies 15:7850. https://doi.org/10.3390/en15217850
97. Soto M (2003) A single connected factorized distribution algorithm and its cost of evaluation. Ph.D. thesis. Universidad de La Habana
98. Muhlenbein H, Mahnig T (1999) The factorized distribution algorithm for additively decomposed functions. In: Proceedings of the 1999 congress on evolutionary computation-CEC99 (Cat. No. 99TH8406) vol 1, pp 752–759

# Systematic Review of Augmented Reality (AR) and Bim for the Management of Deadlines, Costs and Quality

**Luis Alvarado Acuña, Boris Heredia Rojas, Hugo Pavez Reyes, Juan Huidobro Arabia, Pedro Yobanis Piñero Pérez, and Iliana Pérez Pupo**

**Abstract** The BIM (Building Information Management) methodology is widely used today in construction, engineering and architecture. For its part, augmented reality is an emerging technology that has been promoted during the last decade, specifically in the construction industry. Its use has been mainly focused on visualization and virtual construction. This technology has facilitated risk management in projects. In addition, it facilitates collaborative development with clients and project stakeholders. The objective of this study is to carry out a survey of the tools available for the integration and applicability of AR and BIM today. The aim is to review the main uses of augmented reality and its integration with other emerging technologies. The technological advantages and limitations of the introduction of these technologies in projects in the industrial and mining sectors in Chile were also reviewed.

**Keywords** AEC · AR · AR mobile · Augmented reality · BIM · Construction · Engineering · Project management

## 1 Introduction

In interdisciplinary construction projects, conflicts frequently arise due to a lack of information or a poor reconciliation of information from the different disciplines. Furthermore, there is often little information about the environment surrounding the project. On the other hand, it is common to find development environments with a low level of computerization in the management of project files. All of the above can

L. A. Acuña (✉) · B. H. Rojas · H. P. Reyes · J. H. Arabia
Departamento de Ingeniería de La Construcción, Universidad Católica del Norte, Antofagasta, Chile
e-mail: lualvar@ucn.cl

P. Y. Piñero Pérez · I. P. Pupo
Artificial Intelligence for a Sustainable Development Group, IADES, Havana, Cuba



P. Y. Piñero Pérez et al. (eds.), *Computational Intelligence in Engineering and Project Management*, Studies in Computational Intelligence 1134,
https://doi.org/10.1007/978-3-031-50495-2_4

negatively impact the project from the point of view of compliance with deadlines, costs and quality.

Among the main challenges faced by construction projects are:

- The need for quick decision-making on the ground in low-budget situations.
- Works with a high level of operational and physical interference.
- Assemblies of critical equipment in limited time windows.

The use of methodologies for project control combined with cutting-edge technologies can offer important solutions to the above. The BIM methodology has been widely used in the execution of projects, showing its benefits in terms of control as well as rapid access to information during and after their completion. In the area of innovative technologies, augmented reality (AR) is presented, which allows the linking and interaction of virtual environments and the physical world through mobile devices.

The use of BIM and AR [1–3] together opens the door to a new way of visualizing and developing projects. The present study carries out an analysis of the state of the art through the methodology of bibliometric analysis. With the information collected, a deliverable is proposed, which is put for validation by a panel of experts. The results are analyzed using the Delphi method.

The work methodology used in this study is the following:

1. Definition of the object of study, field of action and objective of the research.
   - Research object.
   - General objective of Investigation.
2. Definition of a bibliographic manager.
3. Definition of sources of academic information for the development of the review: Semantic Scholar, Google Scholar, Scopus and other meta-search engines academics based on open science.
4. Definition of key phrases to carry out searches.
5. Definition of the goals of bibliometric analysis in the form of, research questions and inclusion–exclusion criteria:
   - What has been the trend in publications per year?
   - Who are the main authors?
   - What are the affiliations and countries of the main authors?
   - How are the publications distributed, considering the types of documents: articles, books, theses and conference proceedings?
   - Exclusion of works published in spaces with a low level of arbitration.
   - Exclusion of text mining works not associated with the use of techniques of linguistic data summarization.
6. Sort and filter posts into the following set of categories:
   - Classics: refers to pioneering publications in linguistic summary of data, where the fundamental principles that set guidelines are set out in theory.

  - Extensions to theories: refers to publications that extend the theory raised in publications understood as classic, they do not set guidelines that significantly change the methods proposed previously, although they do develop contributions to knowledge.
  - Application results: publications that focus on the use of existing theory in concrete practical scenarios.
  - Tertiary reviews refer to articles reviewing trends and evolution in the topic in question.

7. Synthesize the main trends.
8. Carry out a detailed analysis of each of the works and characterize them with respect to:

  - Structure of the summaries that are generated (protoforms).
  - Methods or techniques for generating linguistic summaries of data.
  - The main validation techniques and methods used in the research.
  - Areas of application of the proposal.

Some Reference management systems are:

- Zotero Free/Storage in free line up to 300 MB/Storage space additional available, licence AGPL and based the Web.
- Mendeley Free/Storage in free line up to 2 GB/Storage space additional available, private solution and Based the Web.
- EndNote: private solution not available in Web environment.
- Qiqqa Free solution with GNU GPL license and available in web environment
- Regarding the acquisition cost, the best options are the free ones, but not all of them.

## 2 Literature Review

The BIM (Building Information Modeling) methodology is one of the most promising developments in the architecture, engineering and construction industries. According to [4] BIM is an acronym used for two concepts. On the one hand, there is the concept of Building Information Modeling, which is a parametric digital representation of the construction product (understood as slabs, walls, pillars, equipment, etc.) that includes its geometry and information. And on the other hand, there is the concept of Building information modeling, which is defined as a methodology/process to develop and use BIM models to support design, construction and operation decisions throughout the life cycle of a project and involves the integration and management of information provided and used by different project actors [5]. The implementation of BIM in an organization can be related to its level of maturity, which can range from level 0 to 3 [6].

- Level 0: unmanaged CAD documents, possibly in 2D, using paper as a data exchange mechanism

- Level 1: Management of CAD files in 2D or 3D format using some standard (possibly BS1192:2007) [7] as a collaboration tool in a common data environment. Business data is managed through finance without integration with the team.
- Level 2: independent data from each discipline managed in a 3D environment through BIM tools. Business data management through ERP (Enterprise Resource Planner) software.
- Level 3: This is the complete integration of BIM. All production areas work with a common model. The work is based on ISO standards that have been developed specifically for the use of BIM technologies.

Broadly speaking, augmented reality (AR) can be defined as "A real-world context that is dynamically overlaid with consistent location or context-sensitive virtual information." AR has three main characteristics [1]:

- a combination of virtual objects and real objects in a real environment,
- people working interactively in real time, and
- an alignment between real and virtual objects [8].

The first steps of AR in the world date back to 1968, when Ivan Sutherland created the first augmented reality system, which is also considered the first virtual reality system. This hardware consisted of a transparent optical screen mounted on the hunt, which had two 6DOF trackers that allowed head movements to be recorded, see Fig. 1. By then, the system could only display basic wireframe drawings [9].

One of the first applications of AR dates back to 1990, when it was used for the training of aviation pilots. As the years went by, its use in the medical field became immediate. It is also present in the areas of science and education. Later, after the 2000s, AR started to be used in the fields of engineering and construction [10].

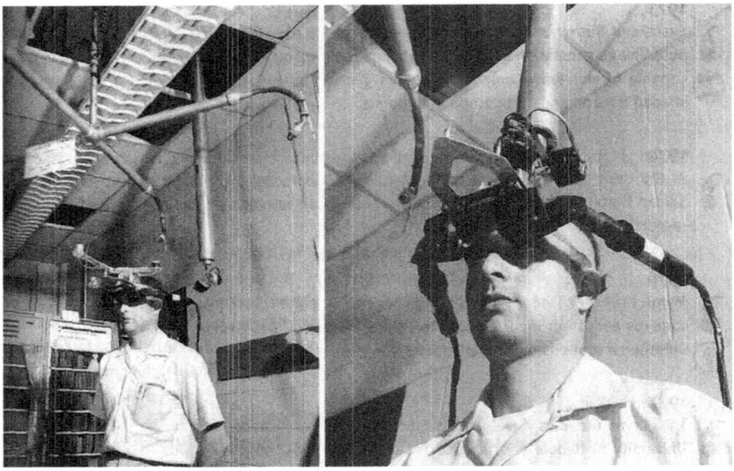

**Fig. 1** Sutherland system [10]

The adoption of BIM and AR is a field that is currently in the development phase globally. According by Panetta [11], AR and VR would be considered, within the top 10, as one of the most important strategic technologies worldwide. Under this same context, its main applications would be focused on the games and entertainment industries; however, other areas such as marketing, tourism, sports and education have been experiencing booms in recent times.

As far as the AEC industry is concerned, current projects are becoming more complex and challenging to control. Methodologies, such as BIM, were created to produce a 3D visualization of a project's information in order to present interested parties with a representation that is easier to understand, with the ultimate goal of generating collaborative work between them.

BIM has proven to be a powerful tool when making decisions within a project and carrying them out according to the proposed goals. However, for BIM to work as intended, it requires a large amount of highly accurate data; otherwise, it becomes a mere simulation and representation tool that, at the end of the project, does not reflect the final status of the project. This is where the integration of BIM with AR makes sense, since the latter can be used as a means of communication between the project site and the data contained in BIM [1, 5, 12].

In the last decade (2011–2021), there has been important development in the integration of BIM and AR. In 2013, Xiangyu Wang [13] proposed in a study a conceptual framework that would allow the integration of BIM and AR with the objective of achieving a contextual visualization of a 3D model (of a project) on site and in real time. The proposal considers a continuous interaction between BIM and AR, allowing field personnel to report project information to the personnel in charge of developing the BIM model and at the same time update the model and therefore its projection in the field. However, this study is from a conceptual point of view, and does not delve into its materialization [14].

Regarding the areas of interest and development of AR solutions in the world of AEC, Sara Rankohi [15] presents in her study that 26% of the literature reviewed referred to the use of AR for simulations and visualizations and 25% for communication and collaboration [16].

This study is consistent with what was carried out in 2020 by Juan Manuel Davila seven years later [17], which exposed the degree of adoption and development of BIM and AR in the world of the AEC. Using a scale of 1–5 (1 = not used, 2 = early testing, 3 = basic implementation, 4 = partial use, 5 = full implementation), it was found that in areas of "design review,", "design support," and "stakeholder participation," there is a greater development of AR and VR, with values that fluctuate between 2 and 3.

In the case of areas such as "construction support," the use and development of these tools fall between 2.08 and 1.8. In the same study, 32 companies were taken as references, and they were consulted, among other aspects, about the use of AR and VR tools as well as the level of investment in these technologies in the future.

## 2.1 *Software AR/BIM Review*

### 2.1.1 Unity Reflect

Unity Reflect is software from the Unity company, which is responsible for developing one of the most important video game engines worldwide. The Unity Reflect suite was launched at the end of 2019 and is based on Unity (video game software and engine), but utilities are added that allow connecting BIM with AR (and/or VR). In this way, it allows all members of an AEC project to be connected, providing coordination and communication solutions. These technologies provide an immersive and collaborative platform in real time. Unity Reflect is divided into two parts: Unity Reflect Review and Unity Reflect Develop.

The first part, called Unity Reflect Review, is the software in which the BIM model is imported to later be projected via AR on mobile devices. The transfer from BIM to AR occurs on Unity's own servers and can then be viewed directly on mobile devices with the software loaded. The software allows real-time communication between all parties involved in the project by incorporating notes, comments and modifications to the model presented in AR. Regarding BIM, Unity Reflect has compatibility with Revit, BIM 360, Navisworks, SketchUp, and Rhino.

For its part, Unity Reflect Develop allows the BIM model developer to make the adjustments they deem appropriate to the AR model before it is published to interested parties. Unity Reflect Develop provides an SDK package for the engineering developer to create custom applications and accelerate model development, tailoring it to the needs of the project. This includes an API to create custom plugins, import BIM data into Unity Editor, stream BIM data to a runtime application, and an open-source reference application [16].

### 2.1.2 Augín

Augín: software launched in 2019 by the company of the same name. It is a program that is installed on Android or iOS mobile devices and allows you to view BIM models in AR. Just like Unity Reflect, the BIM model is uploaded directly to Augín's servers, where it is processed to be later viewed on mobile devices. It is compatible with BIM formats with the IFC extension. Within the mobile application, the user can view the model, adjust it by layers, review information and make annotations, which are sent by the same system to the developer of the BIM model [18].

### 2.1.3 GammaAR

GammaAR: software released in 2020 by the company of the same name. It works similarly to Aguín: it is compatible with Android and iOS mobile devices, accepts

**Table 1** Theoretical propositions and their associated analysis factors

| Propositions | Analysis factors |
|---|---|
| 1. Project management: project management helps to achieve success in projects by reducing additional costs, delays and accidents | 1.1 Maturity level of the organization (in project management) |
| 2. BIM: The BIM methodology helps plan and manage engineering projects in terms of time and costs in graphic form | 2.1.BIM implementation level (3D@5D)<br>2.2. Scope, time, cost and Qa/Qc |
| 3. AR (with BIM): AR with BIM allows you to visualize and control construction projects on-site, streamlining the time used in Qa/Qc and decision-making | 3.1. Equipment and systems<br>3.2. Implementation |
| 4. Project evaluation and knowledge management: knowledge and knowledge management within an organization | 4.1. Feasibility analysis<br>4.2. SECI |

BIM files with the IFC extension, and allows the exchange of information between the parties involved in the field.

With all of the above, this study develops the following theoretical propositions and analysis factors, see Table 1.

## 3 State of the Art and Development of the Conceptual Model

For the collection of information and review of the literature, the following sources of information were considered:

- Web of Science
- Scopus
- Software development websites
- Hardware developer websites
- Google Books
- MeGIP UCN Thesis.

It is worth mentioning that the Web of Science (WoS) will be the main source of information related to papers and studies, while the software and hardware pages will be used to complement the technical information collected. As the first search, the combination of keywords "AR, BIM" is used, which yields a result of 309 documents. Of these, a filter is made by category, leaving only those of interest for the study in question. In total, there are 111 reference documents that are listed in the Table 2.

Using WoS tools, we have the following visualization of documents within the first 10 categories, see Fig. 2.

An interesting fact that WoS reveals from the search carried out is that the 101 publications found date from the years 2013 to 2023, with 74% of them concentrated

**Table 2** Number of documents per category

| Category (WoS) | No. related documents | % (of a total of 111 elements) |
|---|---|---|
| Engineering Civil | 51 | 45.95 |
| Construction building technology | 42 | 37.84 |
| Engineering multidisciplinary | 15 | 13.51 |
| Multidisciplinary sciences | 15 | 13.51 |
| Materials science multidisciplinary | 11 | 9.91 |
| Chemistry multidisciplinary | 10 | 9.01 |
| Physics applied | 10 | 9.01 |
| Computer science interdisciplinary applications | 8 | 7.21 |
| Engineering electrical electronic | 6 | 5.41 |
| Management | 6 | 5.41 |
| Computer science information systems | 5 | 4.51 |
| Engineering industrial | 4 | 3.60 |
| Chemistry analytical | 3 | 2.70 |
| Energy fuels | 3 | 2.70 |
| Geography physical | 3 | 2.70 |
| Instruments instrumentation | 3 | 2.70 |
| Remote sensing | 3 | 2.70 |
| Architecture | 2 | 1.80 |
| Computer science software engineering | 2 | 1.80 |
| Automation control systems | 1 | 0.90 |
| Chemistry physical | 1 | 0.90 |
| Computer science artificial intelligence | 1 | 0.90 |
| Computer science theory methods | 1 | 0.90 |
| Education scientific disciplines | 1 | 0.90 |
| Engineering chemical | 1 | 0.90 |
| Engineering mechanical | 1 | 0.90 |
| Geography | 1 | 0.90 |
| Operations research management science | 1 | 0.90 |
| Telecommunications | 1 | 0.90 |
| Thermodynamics | 1 | 0.90 |

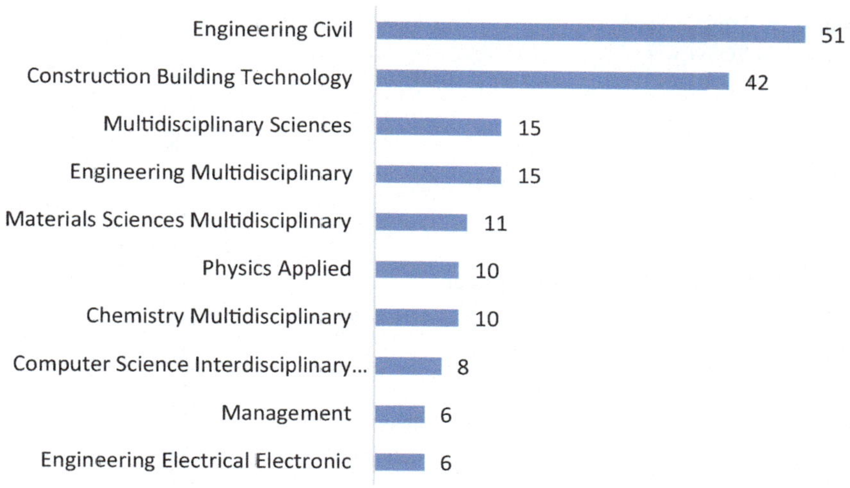


**Fig. 2** Number of posts in the top 10 categories

in the last 3 years (2020–2023). This speaks of a relatively new area of research that has gained strength in recent years.

An analysis was then performed with the VOSviewer software. To do this, the search was limited to occurrences of the keywords on at least two occasions. This showed that of a total of 670 keywords, only 100 met the indicated criteria.

The Table 3 shows the 20 keywords with the highest number of links based on the analysis carried out.

From the table, and taking into account the approach we want to give to the research, the following keywords are taken as reference: Augmented reality, Bim, Framework, Management, Augmented reality (AR), Building Information Modeling (BIM), AEC, Implementation and Safety.

### *3.1 Bibliometric Analysis—AR-BIM Software*

Similar to the one already described, a search was carried out with the keywords AR, BIM and software. This is in order to find the main software that is being used in the implementation of AR with BIM. Without performing any type of filter, the WoS search returns only 22 papers, which are detailed by category in Fig. 3.

Processing the information in VOSviewer does not obtain relevant information for the study.

Based on this analysis, the main papers that will be used for this study were selected, see Table 4.

**Table 3** Top 20 analysis keywords in VOSviewer

| Keyword | Total link | Strength |
|---|---|---|
| Augmented reality | 81 | 287 |
| BIM | 77 | 227 |
| System | 56 | 128 |
| Framework | 54 | 113 |
| Design | 48 | 92 |
| Visualization | 45 | 121 |
| Management | 44 | 112 |
| Building information modeling | 40 | 66 |
| Construction | 38 | 80 |
| Architecture | 36 | 69 |
| Building information | 35 | 77 |
| Information modeling bim | 33 | 53 |
| Virtual reality | 31 | 56 |
| Augmented reality (AR) | 31 | 53 |
| Mixed reality | 27 | 41 |
| Building information modeling (BIM) | 26 | 52 |
| AEC | 26 | 33 |
| Model | 25 | 39 |
| Implementation | 25 | 38 |
| Safety | 23 | 39 |

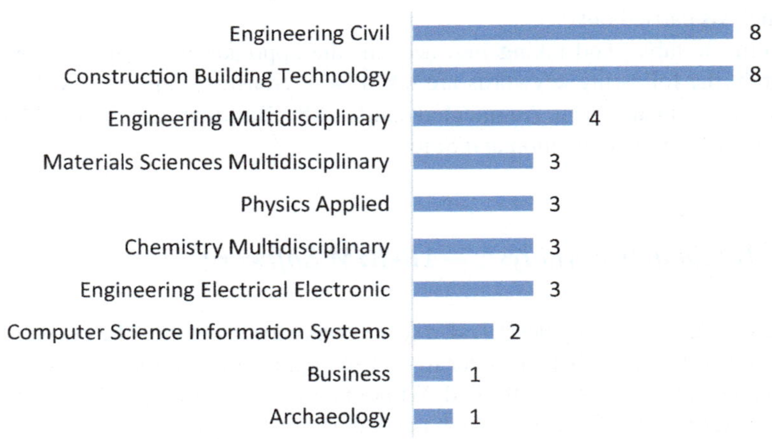


**Fig. 3** AR-BIM-software—categories of 22 resulting papers

**Table 4** Bibliographic summary of selected references

| | Q | Quantity | Areas of knowledge |
|---|---|---|---|
| WOS | Q1 | 11 | – Computer science, engineering<br>– Engineering architecture building and construction<br>– Science & technology<br>– Other topics; energy & fuels<br>– Engineering, Control and Systems Engineering<br>– Computer science artificial intelligence, information systems |
| | Q2 | 3 | – Engineering civil and structural<br>– Computer Science; Engineering<br>– Chemical Engineering,<br>– Engineering, Material Science |
| SCOPUS | Q1 | 2 | – Geography, planning and development<br>– Computer Science; Engineering |
| | Q2 | 1 | – Computer science, engineering, mathematics |
| | Q3 | 1 | – Business, management and accounting |
| Others | | 10 | – Social sciences, computer science<br>– Environmental Science (miscellaneous)<br>• Energy, engineering, social science<br>• Energy, Environmental Science |

## 4 Validation of the Theoretical Model

One of the objectives of this study is the validation of the proposed theoretical model. To do this, it is necessary to place this model under scrutiny through surveys and interviews with experts in the work area of the study in question. It is in this aspect that the Delphi methodology, which allows for consensus among experts through a qualitative perspective, allows the participation of experts in the topic investigated, who, through different iterations associated with filling out surveys and/or interviews, have the function of guiding, validating and, if required, modifying the presented model. Thus, achieving a more robust model validated by the panel of experts.

By definition, the Delphi method is an iterative process used to collect and distill different judgments from experts with extensive experience on a topic through the use of questionnaires and/or surveys designed to obtain feedback from them. These questionnaires are designed with a focus on the problem, opportunities, solutions, or forecasts. Each subsequent questionnaire is developed based on the results of the previous one. The process ends when the research question is answered through the consensus of the expert panel [19].

## 4.1 Questionnaire Preparation

From the research carried out through the bibliographic review using the methodology of bibliometric analysis, a theoretical model was obtained whose objective is to propose a model for incorporating AR from BIM, applying it to brownfield type projects (although it does not exclude the greenfield type) of engineering and construction for the management of deadlines, costs and quality.

This proposed model was shared with a panel of experts, to whom, in addition to presenting said model, a series of questions were asked with the ultimate objective of receiving the corresponding feedback and validating the model. It should be noted that this exercise is carried out individually by each of the experts, see Table 5.

Questions 1–4 aim to collect information about the respondent's profile.

Questions 12 through 29 use the five-point Likert scale (5—strongly agree, 4—somewhat agree, 3—neither agree nor disagree, 2—somewhat disagree, 1—strongly disagree). At the end of each question, there is the possibility for the expert to present his or her comments in relation to what was consulted, with the objective of validating or improving the model.

**Table 5** Analysis questions and factors

| No, questions | Proposition | Analysis factor |
|---|---|---|
| 5-6-7-22-23 | Project Management: Project management helps to be successful in projects. Reducing additional costs, delays and accidents | FA1.1 |
| 9-11-13 | BIM: BIM methodology helps plan and manage engineering projects in time and costs, graphically | FA2.1–2.2 |
| 8-9-10-15-16-17-18-19-20-22-23-24-29-30 | AR with BIM: AR with BIM allows you to visualize and control construction projects on-site, streamlining the time used in QA/QC and decision-making | FA3.1–3.2 |
| 12-14-16-21-23-24-25-26-27-28 | The evaluation and management of knowledge within an organization | FA4.1–4.2 |

## 4.2 Expert Panel Selection

For the present study, experts in the areas of BIM and/or AR with at least 5 years of experience were considered. Regarding the area of professional performance, engineering, construction and research and development were covered. In summary, the experts had to have one of the following characteristics:

- Experience in BIM
- AR and BIM implementation experience
- Knowledge of AR applied to BIM
- Experience in project development in AR and BIM
- Project leadership experience.

## 4.3 First Round of Consultations

In the present study, a total of two rounds of questions were asked. A total of five experts participated in the first round. Contact with each of the participants was made through a standard formal letter. Together with the form letter, the questionnaire is delivered in .docx format, as well as a link to the Google Forms platform. In this way, the instructions in the letter indicated that the questionnaire had to be answered directly through Google Forms, while the .docx file would serve to complement the answers with comments. The first round of queries is detailed in Table 6.

## 4.4 Results of the First Round of Consultations

From the point of view of a quantitative analysis, the consistency coefficient is considered acceptable if the value obtained exceeds 70%, and questions with values below this limit are eliminated due to their low consistency or small consensus among experts.

$$Cc = [1 - Vn/Vt] * 100 \tag{1}$$

where:

Cc corresponds to the consistency coefficient expressed as a percentage
Vn corresponds to the number of experts in disagreement
Vt belongs to the total number of experts.

According to the Table 7, the results of the first round of questions are presented. For the analysis, the answers of the 5 experts to questions Q12–Q29 are taken, and the coefficient Cc is calculated.

**Table 6** List of first round questions

| No, questions | Delphi questions round 1 |
|---|---|
| P1 | Name and surname |
| P2 | Title/studies |
| P3 | Indicate the sector in which you work or have specialized |
| P4 | Please indicate project management or leadership experience |
| P5 | Do you have experience in project development using BIM? |
| P6 | How many years of experience do you have in project development using BIM? |
| P7 | For multidisciplinary projects, indicate which BIM software you use (specify if more than one) |
| P8 | Do you have experience with projects implementing AR in BIM? If the answer is no, go to question 10 |
| P9 | What software have you used to implement AR in BIM? |
| P10 | In what year did you have your first experience implementing AR and BIM? |
| P11 | The use of AR and BIM—under what type of implementation was it? (You can select more than one.) |
| P12 | A correct integration of BIM and AR requires the continuous participation of the client in all phases of engineering design |
| P13 | The topographic survey (stage II) is essential for the development of the project design and its subsequent integration with AR (stage III) |
| P14 | Clients who are experiencing their first experience using AR will require basic training from the engineering company |
| P15 | The tools proposed to import BIM models into AR have a low economic, technological and human cost for the engineering office or company |
| P16 | During the bidding and acquisition phases of services or contracts, the use of AR is beneficial for the project, facilitating its understanding in terms of volume and location in space |
| P17 | The Unity Reflect software, unlike the other solutions proposed, allows editing of the 3D model, thus delivering a more finished result. For the industrial-type project approach, this does not represent a benefit for the client over the Augin and Gamma AR solutions |
| P18 | In all the AR solutions presented, the alignment of the model using the mobile device can be done manually on a surface, or more precisely, with reference points. This does not represent a problem from the point of view of the precision with which the model is projected in the real world |
| P19 | With the technology currently available, the level of precision of GPS on mobile devices (tablets and cell phones) is insufficient to position projects on site using AR. In this way, field alignment is presented as an option in terms of precision. In this regard, engineering must establish reference points on site for a correct visualization of the model in AR |

(continued)

**Table 6** (continued)

| No, questions | Delphi questions round 1 |
|---|---|
| P20 | In relation to the previous point, in situations where the project is inserted in a very congested area or some interference is required to be visualized with a high degree of precision (drilling in large diameter flanges, positioning of bolts, etc.), the AR solution, based on the software proposed, should not be seen as a recommended solution, especially for decision-making |
| P21 | During the bidding and acquisition phases of services and contracts, the use of AR is beneficial for the project, facilitating its understanding in terms of volume and location in space |
| P22 | The use of AR helps in the early detection of security risks to people and equipment, in the different stages of the project and subsequent operations |
| P23 | The use of AR helps in the early detection of security risks to people and equipment, in the different stages of the project and subsequent operations |
| P24 | The implementation of AR through BIM helps strengthen so-called "stakeholder engagement" |
| P25 | The physical representation of the project achieved through the implementation of AR helps detect interferences in stages prior to the purchase of equipment and construction |
| P26 | The implementation of BIM and AR can help a project even after its completion, during its productive life. Using AR, it will be possible to visualize elements underground or not directly visible (water lines, power lines, piping, structures, etc.) |
| P27 | The use of easily accessible and widespread technology, such as mobile devices (cell phones and tablets), simplifies the client's access to the representation of the BIM model in AR. The current state of the art of BIM and AR achieves the objectives of facilitating customer understanding, visualizing interferences and allowing security risks to be identified |
| P28 | The proposed model is highly applicable and recommended for minor and low-complexity projects (<USD10,000,000) |
| P29 | For intermediate or larger projects of medium or high complexity, the use of BIM and AR must be limited and evaluated depending on the particularity of the project |
| P30 | Given the state of the technology and the reference software proposed, under what scenarios would you recommend the use of AR and BIM? (You can check more than one option.) |

From the result, questions P14, P15, P16 and P28 have a Cc of 60%, while questions P18, P19 and P29 are much lower, with an acceptance level between 40 and 20%. For its part, an analysis is carried out using Cronbach's Alpha coefficient. Cronbach's Alpha is a statistical measure that is generally used as a measure of the internal consistency or reliability of a psychometric instrument (using Likert scales). This calculation expresses the internal consistency based on the covariance between the questionnaire items, so that the higher the covariance score, the higher the alpha score [20]. The following formula is used in this study to obtain the calculation of the coefficient [20]:

**Table 7** Cc coefficient, 1st round questions

| Expert/ask | Ex1 | Ex2 | Ex3 | Ex4 | Ex5 | Cc (%) |
|---|---|---|---|---|---|---|
| P12 | 5 | 4 | 5 | 5 | 3 | 86 |
| P13 | 5 | 5 | 4 | 2 | 4 | 86 |
| P14 | 3 | 5 | 2 | 5 | 4 | 71 |
| P15 | 5 | 5 | 3 | 2 | 4 | 71 |
| P16 | 5 | 3 | 1 | 5 | 4 | 71 |
| P17 | 5 | 4 | 4 | 5 | 4 | 100 |
| P18 | 3 | 2 | 2 | 5 | 2 | 43 |
| P19 | 4 | 5 | 5 | 4 | 4 | 100 |
| P20 | 2 | 2 | 4 | 2 | 3 | 43 |
| P21 | 4 | 3 | 5 | 5 | 4 | 86 |
| P22 | 5 | 4 | 5 | 5 | 5 | 100 |
| P23 | 5 | 4 | 5 | 5 | 4 | 100 |
| P24 | 5 | 5 | 5 | 5 | 4 | 100 |
| P25 | 4 | 4 | 5 | 5 | 4 | 100 |
| P26 | 5 | 4 | 5 | 5 | 5 | 100 |
| P27 | 5 | 3 | 4 | 5 | 5 | 86 |
| P28 | 4 | 3 | 5 | 4 | 3 | 71 |
| P29 | 2 | 3 | 5 | 4 | 3 | 57 |

$$\alpha = k\frac{\left(\sum s_i^2/s_t^2\right)}{k-1} \tag{2}$$

where:

- *k: number of items in the instrument*
- $s_i^2$: *variance of scores in item i*
- $s_t^2$: *variance of the total scores of the questionnaire or test.*

To simplify the calculation, SPSS statistical software is used. According to Nunnally [21], the minimum accepted score requires values equal to or greater than 0.80. Below are the results obtained directly with the aforementioned software were 0.802 Alfa de Cronbach ronda 1 considering 14 elements.

Table 8 shows a statistical summary of the questionnaire. For all of the above, and to adjust a result around 0.8, questions P13, P14, P15 and P19 are removed from the analysis.

With all the quantitative analysis carried out, the results were qualitatively analyzed to make sense of what was presented.

We begin by analyzing those questions that had a Cc below 40%: Q18, Q20 and Q29. In general terms, it is interpreted that the questions were poorly formulated and led to confusion among the experts. This emerges because question 18 was asked with a negative connotation, where all experts perceive and identify the proposed AR

**Table 8** Statistical results round 1

| Questions | Mean scale if element has been removed | Variance scale if element has been removed | Total correlation of elements | Alfa de Cronbach if element has been removed |
|---|---|---|---|---|
| P12 | 52.6 | 35800 | 0.645 | 0.772 |
| P16 | 53.4 | 41300 | −0.28 | 0.866 |
| P17 | 52.6 | 38800 | 0.645 | 0.782 |
| P18 | 54.2 | 32200 | 0.649 | 0.767 |
| P19 | 54.4 | 43300 | −0.051 | 0.826 |
| P20 | 52.8 | 34200 | 0.879 | 0.754 |
| P21 | 52.2 | 39200 | 0.732 | 0.782 |
| P22 | 52.2 | 37300 | 0.882 | 0.770 |
| P23 | 52.4 | 41700 | 0.277 | 0.800 |
| P24 | 52.6 | 38300 | 0.723 | 0.778 |
| P25 | 52.2 | 39200 | 0.732 | 0.782 |
| P26 | 52.6 | 36800 | 0.544 | 0.780 |
| P27 | 53.2 | 36200 | 0.656 | 0.772 |
| P28 | 53.6 | 38800 | 0.239 | 0.810 |
| P29 | 52.6 | 35800 | 0.645 | 0.772 |

solutions as having "low precision" and rather having a "representative or visual" approach. In this way, question 18 is reformulated in the second round of questions as a statement about the "imprecision" of the AR solutions presented.

Regarding questions Q20 and Q29, they are directly related to Q18. That is to say, in these questions, it is once again emphasized that the solutions proposed are imprecise (compared, for example, with topography or satellite positioning) and are useful if they are used as visualization elements. Despite the above, the experts in the first round indicate the opposite (they indicate AR is a precision technology), which generates inconsistency with the answer given in Q18. In this way, questions Q20 and Q29 are rephrased in a more direct way.

Regarding the questions with a Cc of 60%, in Q14, Q15, and Q16 there is an expert in disagreement (note 1 or 2), while in Q28 an expert is at an intermediate level of acceptance (note 3).

As in the analysis of those questions with a Cc under 40%, it is interpreted that the proposed model was not clear, and it is necessary to reformulate and clarify the information presented to the experts.

## 4.5 Second Round of Consultations

For the second round, there is the support of three of the five experts initially consulted. On this occasion, the same proposed model is sent, but with the following considerations:

- The objective and meaning of the proposed model are explained in more detail.
- The questions, explanation and proposed model are complemented with audiovisual material, which presents an example of a BIM model being visualized through AR on a mobile device using one of the applications proposed in the model.
- Questions P18, P20 and P29 are reformulated, eliminating the negative connotation expressed in the first round of questions.
- Each of the experts is given (separately) their answers from the first round (Table 9).

The reformulated questions are detailed below.

**Table 9** List of questions (modified) second round

| No, questions | Questions (modified) Delphi round 2 |
|---|---|
| P18 | In all the AR solutions presented, the alignment of the model using the mobile device can be done manually on a surface, or more precisely, with reference points. This does not represent a problem from the point of view of the precision with which the model is projected in the real world |
| P20 | In relation to the previous point, in situations where the project is inserted in a very congested area, or it is required to visualize with a high degree of precision some interference (drilling in large diameter flanges, positioning of bolts, etc.), the AR solution alone, based on the proposed software, is presented as an imprecise tool. It is recommended to complement it with precision technologies such as GPS and/or total stations |
| P29 | For intermediate or larger projects of medium or high complexity, the use of BIM and AR should be used as a visualization tool, complementary to technologies and tools such as GPS and/or total stations |

## 4.6 Results Second Round of Consultations

Analyzing the Cc for the second round of consultations, the results are in Table 10.

It appears that question P18 lowers its acceptance: two of the experts increase their approval, while one lowers it. In the case of P20 and P29, these have 100% acceptance. Again, a contradiction is seen since, as indicated during the first round, these questions are related.

**Table 10** Statistical results round 2

| Expert/ask | Ex1 | Ex4 | Ex5 | Cc (%) |
|---|---|---|---|---|
| P12 | 5 | 5 | 3 | 67 |
| P13 | 5 | 5 | 4 | 100 |
| P14 | 3 | 5 | 5 | 67 |
| P15 | 2 | 2 | 4 | 33 |
| P16 | 4 | 3 | 4 | 67 |
| P17 | 3 | 4 | 4 | 67 |
| P18 | 4 | 2 | 4 | 67 |
| P19 | 4 | 4 | 4 | 100 |
| P20 | 4 | 4 | 4 | 100 |
| P21 | 5 | 5 | 5 | 100 |
| P22 | 5 | 5 | 5 | 100 |
| P23 | 5 | 5 | 5 | 100 |
| P24 | 5 | 5 | 5 | 100 |
| P25 | 5 | 5 | 5 | 100 |
| P26 | 5 | 5 | 5 | 100 |
| P27 | 4 | 5 | 5 | 100 |
| P28 | 4 | 5 | 4 | 100 |
| P29 | 5 | 4 | 5 | 100 |

In the case of P15, which has the lowest Cc with 30%, it is a question focused on cost terms. A change in perception among the respondents is seen when experts identify the solution as high cost.

For P12, the experts maintain their perception; however, the Cc is less than 70% due to the decrease in the number of respondents. Based on the above, this question is considered accepted.

For Q14, the perception is positive, but the question suffers from the same problem as Q12: the sample is very small. Based on the above, this question is considered accepted.

With P16, the experts' perception changes. This question asserts that for industrial-type projects, Unity Reflect does not represent a benefit over Augin and Gamma AR. This alludes to the fact that Unity Reflect is more robust in terms of graphic representation (reflections, lighting, shadows, etc.), but at the same time more expensive. In this second round, the experts perceive this assertion negatively. Therefore, it is eliminated from the final deliverable.

For P17, it goes from having a Cc of 100 to 67%. From this result, it is interpreted that not all the software presented is easy to access and understand for the client. Here, it is assumed that Unity Reflect is the most elaborate and complex software to manage.

Due to the above, this aspect is modified in the final deliverable. With all of the above, the rounds of questions end, and therefore a final deliverable for this investigation is prepared.

## 5 Result Analysis

In the present study, an analysis of the state of the art of the technology that allows the integration of BIM and AR was carried out [2, 3]. This analysis is mainly focused on industrial-type projects in a brownfield environment. From the point of view of the client or stakeholder, the proposed software is easy to use, allowing the client to view the project in the field, filter by layers and provide feedback to the designer or engineering company, all of this through mobile devices (tablet, cell phone). On the part of the designer or engineering company, there is great compatibility with program files that work under the BIM methodology.

It is hypothesized that the use of BIM with AR will improve the understanding of the project by interested parties or stakeholders. This will bring with it fluid communication between all parties involved in the project. The main benefits sought are:

- Identification of interferences with the operation and adjacent structures or elements.
- Identification of potential security risks.
- Increase the understanding of the interested parties (stakeholders) and parties belonging to the project (vendors, tenders, services, insurance involved, etc.).
- Reduce decision-making time during project development.
- Establish and strengthen so-called "stakeholder engagement".

The three solutions presented to implement AR in BIM are: Aguín, Gamma AR and Unity Reflect.

In general terms, the three solutions work similarly, differing mainly in the compatibility they have with engineering software.

The workflow begins with the import of the design to the "Cloud Computing" service provided by each software. In this instance, the transformation of a BIM software file to AR is generated. It should be noted that all of them have compatibility up to 3D designs, leaving 4D out of the options [22].

Once the model is processed, it is ready to be viewed on Android and iOS mobile devices. Regarding mobile device compatibility, it is reduced to those Android devices that are compatible with ARCore and, in the case of iOS that are compatible with ARKit.

Within the mobile device application, the user must place the model on the site. For this, there are two options: on a flat platform without references or with references that connect the model with the place where the project will be located. Once placed, the user will be able to view the project by layers, sections, scale and make comments, which will be sent to the designer or engineering company.

A particular feature of Unity Reflect is that it offers the possibility of editing the 3D model through the Unity Review program. This feature is especially useful in projects that have a high architectural component.

This proposed implementation gives the user simple access to project visualization using AR technology, while for the designer or engineering company, it requires low implementation costs, both at the software level and at the staff training level. On the other hand, this scheme does not have great precision (it does not have geo-referencing), its main objective being to visualize and detect interference early.

From an economic point of view, the proposed solutions are detailed below, see Table 11.

The experts consulted reaffirm the idea that BIM and AR currently serve mainly as a visualization tool that improve communication between the parties involved in a project. This is reflected throughout the consultations made, and more specifically in the result of question 11 of the questionnaire, where they ask about their experience, and then in question 30, where they ask about how they see the use of this convergent technology.

The integration of AR and BIM can offer several benefits in terms of cost, time and quality in construction and project management. In summary, some of these benefits include:

**Table 11** Solutions economic point of view analysis

| Software | Compatibility | Cloud computing | Mobile compatibility | Price |
|---|---|---|---|---|
| –Gamma AR | Revit<br>Navisworks<br>IFC | Yes | IOS<br>Android | License per user: 700 EUR/year<br>License per client: free |
| –Augin | Revit<br>Tekla<br>Archicad<br>BricsCAD<br>Sketchup<br>IFC | Yes | IOS<br>Android | License for projects up to 100 MB: 199.90 USD/year<br>License per Client: free |
| | | | | License per projects up to 1 GB: 299.90 USD/year<br>License per Client: free |
| –Unity reflect develop<br>–Unity reflect review<br>–Unity reflect collaborate | Revit<br>Navisworks<br>Sketchup<br>Rhino | yes | iOS<br>Android | License per user: 690 USD/year<br>License per Client: 249 USD/yyear |

- Reduced errors and rework: AR and BIM can help identify and correct design errors before construction begins. This can avoid additional costs and project delays due to design changes and the need to make corrections.
- Improved collaboration and communication: AR visualization of BIM models allows work teams involved in a project to collaborate more efficiently throughout its stages. By working together on a shared model, problems can be identified and resolved more effectively, which in turn reduces project time and costs.
- Increased accuracy and quality: The use of AR and BIM can improve the accuracy and quality of construction, as problems can be detected and resolved before they become costly problems on the construction site. Additionally, AR visualization of BIM models can help workers better understand the project design and requirements, which can improve the quality of work. Saving time and money: AR visualization of BIM models can speed up the design and construction process by allowing quick and easy visualization of the models. This can reduce production costs while meeting deadlines, which translates into an eventual reduction in overall project costs.

Literature and experts expose some weaknesses in BIM and AR. Precision is a point to be discussed since, as indicated above, there are many actors and elements that participate and define the level of precision of the final model and its projection in the field. Another point is the cost and rapid understanding of the use of this technology, both for the developers of the AR model and for the users viewing the product. Based on the above, the areas for improvement in this technology can be summarized through the following points:

- System integration: One of the challenges in implementing AR and BIM is the integration of different software and hardware systems. To make the most of these technologies, companies need to be able to integrate and synchronize different systems to create a seamless, integrated workflow.
- Accessibility and ease of use: Although AR and BIM technology have advanced significantly, there is still a learning curve for their use, and some users may have difficulty accessing the technology. It is important that software and hardware developers focus on improving accessibility and ease of use to ensure that technology is available to all users.
- Accuracy and quality of data: The accuracy and quality of the data used to create BIM and AR models are critical to the effectiveness of these technologies. To improve data accuracy and quality, it is important that construction professionals have a greater awareness of the importance of accurate data and that standardized processes and procedures are implemented to ensure data accuracy and quality.
- Cost: The cost of AR and BIM technology remains a barrier to widespread adoption in the construction industry. As technology advances and more affordable solutions are developed, the adoption of AR and BIM technology is likely to accelerate.

## 6 Conclusions

Much of the information contained in the studies tends to confuse or conflate the worlds of AR, VR and MR. It must be remembered that this study focuses its efforts on compiling the state of the art of AR that is specific to industrial areas where particular PPE (personal protective equipment) is required for environments with the presence of gases and particulate matter (among others). VR and AR require bulkier hardware than AR. Particularly, VR causes the user to lose a complete visual notion of their surroundings while it is being used, which negatively impacts safety issues. For its part, VR generally requires hardware mounted on the user's head, making it incompatible with PPE such as airtight glasses and half-face or full-face respirators. In this way, AR allows the use of all types of PPE in addition to allowing the user to be completely aware of their surroundings.

For the preparation of the deliverable, the programs that integrate BIM and AR were used as a reference based on what was provided by the literature as well as what was consulted by engineering development companies in the northern part of Chile: Unity Reflect, Augín and Gamma AR.

During the validation with the experts, topics such as cost, the accessibility of AR/BIM for both the contractor and client company, the real usefulness of BIM/AR in the field of AEC and especially in brownfield type projects, and the perception of precision of the tools present today were discussed, among other aspects.

In general terms, experts validate AR/BIM integration as a useful tool when addressing projects throughout all their stages, in aspects such as communication with the client and early detection of interferences. Within the proposed deliverable, some deficiencies or aspects to improve in AR/BIM emerge, such as the cost of implementation, its level of pressure, and the need to train the client prior to its use. Regarding the precision of AR/BIM, locking a BIM model at a particular point within a brownfield project is where the most doubts arise. Based on the literature and the information provided by experts, there are areas for improvement, specifically in the hardware. Mobile devices have an acceptable level of precision for use in everyday applications such as positioning on maps or even video games, using global positioning systems (GPS, GLONASS) and/or triangulation of mobile phone antennas. However, for engineering and construction projects where high precision is required, it is insufficient or unreliable. The software that currently exists requires that the mobile device be equipped with a powerful power section at the hardware level. Developers use different abilities of mobile devices to interlock a BIM model: global positioning, antenna triangulation, cameras, depth sensors, gyroscopes, accelerometers, and, of course, mobile device processing. The use of some sensors, such as depth sensors, may be optional; however, the rest of the technical aspects are necessary. For a smooth and relatively accurate experience, a recent, high-end mobile device is desirable. The above somewhat limits the democratization of this type of technology.

It is important to consider that the precision of the model projected in AR is also conditioned by the quality of the data coming from the BIM model. It is relevant, then, that the taking of points on the ground via topographic survey and the processing and interpretation of these via BIM make sense and are consistent with the objective of the project.

All in all, BIM and AR are understood as "convergent technologies" [23], that is, different technologies that are independent but come together on a single platform to deliver a particular solution or product.

Technological convergence aims to improve the efficiency and accessibility of technological products and services. For example, a smartphone is an example of convergent technology, as it combines the features of a cell phone, a camera, a music player, a computer, and Internet access into a single device.

Technological convergence affects not only devices, but also the networks and services that support them. Converged technologies have the potential to create new business opportunities and improve the user experience by enabling access to a wide variety of services and applications on a single platform.

Focusing more specifically on the topic at hand, combining BIM and AR creates a convergent technology that allows AEC professionals to view and work with BIM models more efficiently and accurately. It could then be said that the convergence of these technologies has the potential to transform the way in which AEC projects are built and managed. Indeed, referring to the hypotheses of the present study, the combination of these technologies can improve efficiency and precision in construction and building management, which in turn can reduce costs and improve the quality of construction projects.

## References

1. Ibarra-Araya A, Heredia-Rojas B, Alvarado-Acuña L, Jamett-Aranda I (2022) Integration of augmented reality and BIM in early stages to improve design management in engineering and construction projects. In: Piñero Pérez PY, Bello Pérez RE, Kacprzyk J (eds) Artificial intelligence in project management and making decisions. Springer International Publishing, Cham, pp 363–380
2. Amaro NM, Piñero Pérez PY, Montesinos DFP, Pupo IP (2022) Tendencies in augmented reality. In: Piñero Pérez PY, Bello Pérez RE, Kacprzyk J (eds) Artificial intelligence in project management and making decisions. Springer International Publishing, Cham, pp 345–362
3. Amaro NM, Pupo IP, Pérez Montesinos DF, Piñero Pérez PY (2022) Trends in photogrammetry and its integration with artificial intelligence. In: Piñero Pérez PY, Bello Pérez RE, Kacprzyk J (eds) Artificial intelligence in project management and making decisions. Springer International Publishing, Cham, pp 417–431
4. Arellano Escobar K, Andrade Valles AI, Castillo Campoverde TO, Herrera RF, Arellano Escobar K, Andrade Valles AI, Castillo Campoverde TO, Herrera RF (2021) Evaluación del uso de BIM en las primeras fases de aplicación. Revista ingeniería de construcción 36:311–321. https://doi.org/10.7764/ric.00005.21
5. Vasconcelo Mir YO, Pérez Pupo I, Piñero Pérez PY, Alvarado Acuña L, Graffo Pozo A (2023) Learning of conversational systems based on linguistic data summarization applications in BIM environments. In: Rivera G, Cruz-Reyes L, Dorronsoro B, Rosete A (eds) Data analytics and computational intelligence: novel models, algorithms and applications. Springer Nature Switzerland, Cham, pp 241–267
6. Dastbaz M, Gorse C, Moncaster A (2017) Building information modelling, building performance, design and smart construction. Springer International Publishing, Cham

7. Lucena González C, Villena Manzanares F (2019) El marco normativo regulador en torno a la Metodología BIM enfocada a la prevención de riesgos laborales. Normative regulatory framework around the BIM Methodology focused on the Prevention of Labor Risks
8. Akçayır M, Akçayır G, Pektaş HM, Ocak MA (2016) Augmented reality in science laboratories: the effects of augmented reality on university students' laboratory skills and attitudes toward science laboratories. Comput Hum Behav 57:334–342. https://doi.org/10.1016/j.chb.2015.12.054
9. Cao J, Lam K-Y, Lee L-H, Liu X, Hui P, Su X (2023) Mobile augmented reality: user interfaces, frameworks, and intelligence. ACM Comput Surv 55:189:1–189:36. https://doi.org/10.1145/3557999
10. Renganayagalu S (2019) Virtual Reality as a future training medium for seafarers: potential and challenges, pp 79–86
11. Panetta K, Wan Q, Agaian S, Rajeev S, Kamath S, Rajendran R, Rao SP, Kaszowska A, Taylor HA, Samani A, Yuan X (2020) A comprehensive database for benchmarking imaging systems. IEEE Trans Pattern Anal Mach Intell 42:509–520. https://doi.org/10.1109/TPAMI.2018.2884458
12. Piñero Pérez PY, Bello Pérez RE, Kacprzyk J (2022) Artificial intelligence in project management and making decisions. Springer International Publishing, Cham
13. Wang X, Kim MJ, Love PED, Kang S-C (2013) Augmented Reality in built environment: classification and implications for future research. Autom Constr 32:1–13. https://doi.org/10.1016/j.autcon.2012.11.021
14. Mendoza-Ramírez CE, Tudon-Martinez JC, Félix-Herrán LC, de Lozoya-Santos J, Vargas-Martínez A (2023) Augmented reality: survey. Appl Sci 13:10491. https://doi.org/10.3390/app131810491
15. Rankohi S, Waugh L (2013) Review and analysis of augmented reality literature for construction industry. Visualiz Eng 1:9. https://doi.org/10.1186/2213-7459-1-9
16. El Kassis R, Ayer SK, El Asmar M (2023) Augmented reality applications for synchronized communication in construction: a review of challenges and opportunities. Appl Sci 13:7614. https://doi.org/10.3390/app13137614
17. Davila Delgado JM, Oyedele L, Beach T, Demian P (2020) Augmented and virtual reality in construction: drivers and limitations for industry adoption. J Constr Eng Manag 146:04020079. https://doi.org/10.1061/(ASCE)CO.1943-7862.0001844
18. Lopes PDS, Cheungb AB (2023) Estudo de caso: Aplicação da realidade aumentada utilizando o software Augin. Universidade Federal de Mato Grosso do Sul
19. Skulmoski GJ, Hartman FT, Krahn J (2007) The Delphi method for graduate research. J Inf Technol Educ: Res 6:1–21
20. Rodríguez-Rodríguez J, Reguant-Álvarez M (2020) Calcular la fiabilidad de un cuestionario o escala mediante el SPSS: el coeficiente alfa de Cronbach. REIRE Revista d'Innovació i Recerca en Educació 13:1–13. https://doi.org/10.1344/reire2020.13.230048
21. Nunnally JC (1978) An overview of psychological measurement. In: Wolman BB (ed) Clinical diagnosis of mental disorders: a handbook. Springer, US, Boston, MA, pp 97–146
22. Cardoch-Álvarez J, Alvarado-Acuña L, Heredia-Rojas B, Acevedo-Acevedo S (2022) Design of a proposal for the implementation of scope, schedule and cost management in fast-track projects using BIM methodology in 3D, 4D and 5D. In: Piñero Pérez PY, Bello Pérez RE, Kacprzyk J (eds) Artificial intelligence in project management and making decisions. Springer International Publishing, Cham, pp 381–399
23. Ardashkin IB, Chmykhalo A.yu., Makienko MA, Khaldeeva MA (2018) Smart-technologies in higher engineering education: modern application trends

# Assessing Adoption Archetypes of Advanced Technologies in Industrial Clusters

**Oscar D. Quiroga and Germán H. Rossetti**

**Abstract** The effective adoption of advanced Industry 4.0 (I4.0) technologies is a crucial research area. While some authors propose readiness or maturity models to analyze the level of adoption of I4.0 technologies, others examine their impact on company performance. However, there is a dearth of studies offering evidence on the implementation patterns of I4.0 technologies in general, and the adoption of advanced technologies related to artificial neural networks, deep learning, and computational intelligence, in particular, within various manufacturing clusters. This gap has prompted working hypotheses to investigate the current implementation and adoption patterns, as well as the degree of adoption or stage of maturity of these technologies in industrial clusters. This paper proposes a qualitative-quantitative study to determine the degree of adoption of advanced I4.0 technologies. It aims to develop an analysis that describes the archetypes characterizing the implementation patterns of I4.0 technologies and the adoption of advanced technologies related to machine learning applications in Argentinian clusters.

**Keywords** Adoption · Archetypes · Implementation patterns · Machine learning

## 1 Introduction

Industry 4.0 (I4.0) is a worldwide phenomenon that combines advanced technologies, revolutionizing economies globally through the smart factory concept. I4.0 offers benefits such as preventing epidemics, reducing waste, improving supply chain forecasting, and delivering personalized products [1, 2]. Additionally, smart factories represent a significant advancement in managing production procedures, and a

O. D. Quiroga · G. H. Rossetti (✉)
Universidad Nacional del Litoral. Fac. de Ing. Química (FIQ-UNL), 3000 Santa, Fe, Argentina
e-mail: german.rossetti1965@gmail.com; groseti@fiq.unl.edu.ar

O. D. Quiroga
e-mail: oquiroga@fiq.unl.edu.ar

 
P. Y. Piñero Pérez et al. (eds.), *Computational Intelligence in Engineering and Project Management*, Studies in Computational Intelligence 1134,
https://doi.org/10.1007/978-3-031-50495-2_5

key aspect is their utilization of comprehensive preventive maintenance programs to minimize production downtime [3, 4].

This stands as a highly promising development within machine learning applications, and the adoption of these advanced technologies related to artificial neural networks, deep learning and computational intelligence [5], given that system failures and defects decrease productivity and prolong production durations. By incorporating data analysis in decision-making processes, smart factories skillfully extract real-world data for use in a digital environment. I4.0 has emerged as an ideal environment for the use of cutting-edge artificial intelligence (AI) and machine learning services in process control and optimization. This is due to the emergence of massive process monitoring data, the adoption of hybrid IoT architectures, and the availability of advanced data repositories with data analytics capabilities [6]. The main reasons for adopting these advanced technologies are increasing efficiency and customization. Product innovation that is compatible with software is a priority for I4.0 technology developers [7].

The implementation of I4.0 technologies is an important area of research. Some authors have suggested models to analyze the level of adoption of I4.0 technologies [8], while others have examined their impact on company performance. Not all sectors are adopting these empowering technologies with equal ease and it is necessary to understand the reasons for these variations. On the one hand, companies are unable to correlate I4.0 with their business models and, on the other hand, they find it challenging to self-assess and gauge their current level of readiness. To surmount uncertainty and dissatisfaction, companies must adopt novel tools to guide and assist them. Therefore, examining I4.0 across different stages and utilizing maturity and readiness models can be advantageous [8, 9]. However, there is a shortage of studies that provide evidence on how advanced technologies are implemented in different manufacturing clusters in Argentina and Latin America [10]. This gap has led to working hypotheses to investigate the current implementation and adoption patterns, as well as the level of adoption or stage of maturity of these technologies in industrial clusters.

This work proposes a new research framework based on two dimensions: the core and the smart technologies of I4.0. Each dimension has three sub-dimensions: basic technologies and advanced technologies, which are related to seven design principles (DPs). These principles form the basis for the Industrial Cluster Maturity Assessment Tool (ICMAT), also called ICMAT—Questionnaire, and serve as the units of analysis (UAs) for a multiple—case study methodology presented in Sect. 2. The ICMAT questionnaire contains three types of questions related to I4.0 core technologies, I4.0 smart technologies, and characterization. The first two types include a 5-point adoption scale that indicates the level of adoption and maturity of I4.0 technologies. The ICMAT is described in Sect. 3 and serves as a guide to establish points of interest and develop specific questions, and its main characteristics are presented in Sect. 4. The results of the analysis reveal five archetypes of adoption of I4.0 technologies, which are discussed in Sect. 5, along with a brief comparative analysis. Finally, Sect. 6 presents the conclusions of this work.

## 2 Methodology of Multiple—Case Studies for Industrial Clusters

In a previous study, Quiroga [10] proposed a qualitative methodology using multiple—case studies and units of analysis (UAs) [11] to investigate the implementation of advanced technologies related to I4.0 in several companies within a region. To continue this research, this study updates the methodology by using industrial clusters as case studies instead of individual companies. The study includes seven cases, each comprising of eight UAs, with seven UAs directly connected to the Design Principles (DPs) described in Sect. 3.

The remaining UA, the Main Analysis Unit (MUA), characterizes each industrial cluster. Table 1 provides a description of each of the seven industrial clusters. The current phase of our research project centers on evaluating the level of maturity and adoption of advanced I4.0 technologies within industrial clusters situated in Argentina. To achieve this objective, we conducted an extensive analysis of eighty-three (83) companies, employing a structured research framework. Based on various parameters such as the size and nature of the industrial clusters, we classified these companies into seven distinct industrial clusters. We then administered a detailed questionnaire, as outlined in Sect. 4, to obtain a comprehensive understanding of their current state of adoption. Tables 1 and 2 provide an overview of the industrial clusters and the sizes of the companies, respectively.

**Table 1** Industrial clusters of the companies under study

| Industrial cluster | Percentage (%) |
|---|---|
| (1) Food industry (23/83) | 27.8 |
| (2) Energy (8/83) | 9.6 |
| (3) Metalworking (8/83) | 9.6 |
| (4) Chemical (8/83) | 9.6 |
| (5) Software and technology (9/83) | 10.8 |
| (6) Metallurgical (12/83) | 14.5 |
| (7) Services (15/83) | 18.1 |

**Table 2** Size of the companies under study

| Size | Quantity | Percentage (%) |
|---|---|---|
| SME (<100 employees) | 20 | 23.7 |
| Large (>100 employees) | 63 | 76.3 |

**Table 3** Meaning of I4.0 technologies

| Industry 4.0 technologies [10, 12] | |
|---|---|
| (1) Sensors—Actuators (SCADA, ERP) | (7) Cybersecurity |
| (2) Communication and networking (SCADA, ERP) | (8) Simulation |
| (3) RFID-RTLS (SCADA, ERP) | (9) Additive manufacturing |
| (4) Embedded systems (SCADA, ERP) | (10) Virtual–Augmented reality (VR-AR) |
| (5) Cloud technologies (cloud services) | (11) Mobile technologies |
| (6) Machine learning and data analysis [3–6] | (12) Adaptive—Industrial robotics |

# 3 Research Framework

This section proposes a research framework based on two dimensions: (A) core technologies of I4.0 and (B) smart technologies of I4.0. Each dimension comprises three sub-dimensions. These sub-dimensions are linked to seven Design Principles—DP(1–7).

## 3.1 *Design Principles*

Table 4 outlines seven Design Principles—DP(1–7) and their respective implications in terms of identifying the presence of technologies associated with the smart factory concept, which are depicted in Table 3. These DP(1–7) provide the foundation for the development of the new ICMAT—Questionnaire.

## 3.2 *The Framework*

The research framework shown in Fig. 1 illustrates the interconnections between the dimensions and subdimensions detailed in Table 5. Table 6 shows the association between the subdimensions and the Design Principles—DP(1–7). Figure 1 also shows the connection between Tables 3 and 4, and the linkages between the DP(1–7) and the I4.0 Technologies, which are connected to the Industrial Cluster Maturity Assessment Tool (ICMAT)—Questionnaire. In summary, the figure and tables provide a comprehensive overview of the research framework and the relationships among its components.

**Table 4** The design principles—DP(1–7) and their meanings

| Design principles—DP(1–7) and their meanings |
|---|
| DP(1) is related to the concept of process integration [10, 12], and DP(1) consists of verifying the level of adoption (VLA) of the following technologies (according to Table 3): (1) Sensors—Actuators. (2) Communication—networking. (3) RFID-RTLS. (5) Cloud. (7) Cybersecurity |
| DP(2) is related to the concept of Agility [10, 12]. DP(2) consists of VLA of the following technologies (according to Table 3): (2) Communication—networking. (3) RFID-RTLS. (5) Cloud. (6) Machine learning & Data analysis. (8) Simulation. (9) Additive Manufacturing. (10) VR-AR. (11) Mobile Technologies. (12) Adaptive robotics—Industrial robotics |
| DP(3) is related to the concept of Interoperability [10, 12]. DP(3) consists of VLA of the following technologies (according to Table 3): (2) Communication—networking. (7) Cybersecurity |
| DP(4) is related to the concept of real time data management [10, 12]. DP(4) consists of VLA of the following technologies (according to Table 3): (1) Sensors—Actuators. (3) RFID-RTLS. (6) Machine learning & data analysis |
| DP(5) is related to the concept of Automation [10]—Decentralization [12]. DP(5) consists of VLA of the following technologies (according to Table 3): (1) Sensors—Actuators. (2) Communication—networking. (3) RFID-RTLS. (4) Embedded systems. (6) Machine learning & data analysis |
| DP(6) is related to the concept of virtualization [10, 12]. DP(6) consists of VLA of the following technologies (according to Table 3): (8) Simulation. (10) VR-AR |
| DP(7) is related to the concept of Flexibility [10]—Service Orientation [12]. DP(7) consists of VLA of the following technologies (according to Table 3): (5) Cloud. (6) Machine learning & data analysis |

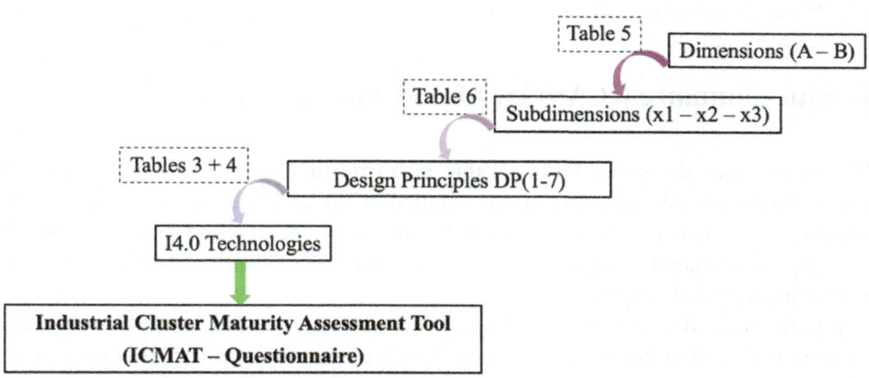


**Fig. 1** Research framework

**Table 5** The dimensions and sub dimensions of the framework

| Dimensions (A–B) | Subdimensions (x3-x2-x1) | |
|---|---|---|
| (A) Core technologies of I4.0 | A(x3) | Base technologies III (Level III) |
| | A(x2) | Base technologies II (Level II) |
| | A(x1) | Base technologies I (Level I) |
| (B) Smart technologies of I4.0 | B(x3) | Advanced technologies III (Level VI) |
| | B(x2) | Advanced technologies II (Level V) |
| | B(x1) | Advanced technologies I (Level IV) |

**Table 6** Relationship among the Design Principles—DP(1–7), the Dimensions (A–B) and the Subdimensions (x1-x2-x3) of the framework

| Design principles DP(1–7) | Dimension A core technologies of industry 4.0 | | | Dimension B smart technologies of industry 4.0 | | |
|---|---|---|---|---|---|---|
| | Subdimensions (x1-x2-x3) | | | | | |
| | A(x1) | A(x2) | A(x3) | B(×1) | B(×2) | B(×3) |
| Process integration DP(1) | * | | | + | | |
| Agility DP(2) | * | | | + | | |
| Interoperability DP(3) | | * | | | | |
| Real time data management DP(4) | | | * | | | |
| Automation-decentralization DP(5) | | | | | + | |
| Virtualization DP(6) | | | | | + | |
| Flexibilization-service orientation DP(7) | | | | | | + |

## 4 Questionnaire (ICMAT): Main Characteristics

Table 4 presents the seven DP(1–7) that underpin the ICMAT—Questionnaire, as well as their respective interpretations detailed in Table 3. The DP(1–7) establish the key points of focus and facilitate the formulation of specific and precise questions that help to evaluate the degree of adoption or state of maturity of I4.0 technologies across different industrial clusters.

Figure 2 provides an overview of this evaluation process. The DP(1–7) emphasize the importance of technologies associated with the smart factory or smart manufacturing concepts, which are recognized as the core of I4.0. Therefore, companies that have implemented such technologies can be deemed to possess a high level of adoption or maturity in relation to I4.0. This approach aims to assess the degree of adoption or level of maturity of intelligent I4.0 technologies through a questionnaire. The questionnaire classifies the adoption stage of each technology as basic, medium, or advanced and also covers the use of new paradigms. The majority of the questions focus on estimating the degree of adoption or maturity stage of these technologies and paradigms.

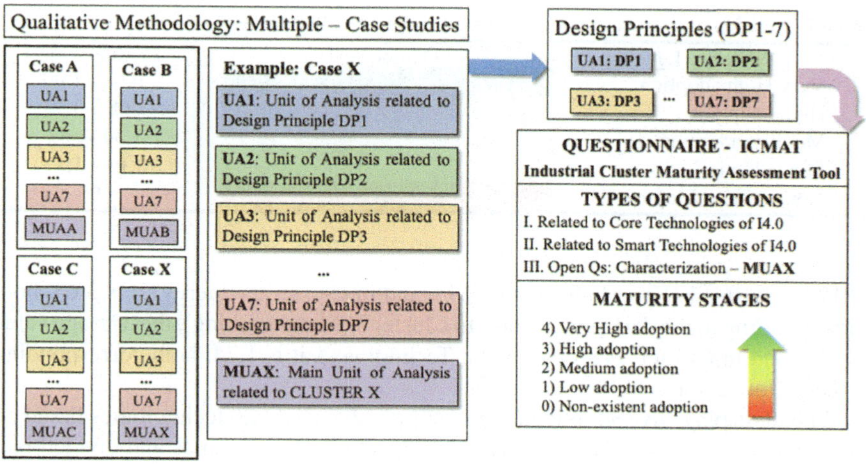


**Fig. 2** Relationship between the qualitative methodology and the ICMAT

The ICMAT—Questionnaire contains three types of questions: (I) questions related to the core technologies of I4.0, (II) questions related to smart technologies of I4.0, and (III) characterization questions that are open-ended. The first two types of questions are closed and include a 5-point adoption scale, ranging from non-existent adoption to very high adoption, to indicate the level of adoption and degree of implementation or maturity stage of core and smart technologies. The third type of question provides information about the company's focus, size, main product, and cluster or sector, which helps to identify the type of company and cluster. The following section examines the responses of companies to determine if the established industrial clusters are related to each other through common characteristics.

## 5 Analysis of Results

In order to facilitate the analysis of the results from the ICMAT—Questionnaire applied to 83 companies, a color scale (as shown in Table 7) was proposed as a qualitative tool to describe the five levels of maturity or degrees of adoption of I4.0 technologies. This allows for the differences among the clusters to be more easily distinguished. To establish the maturity level of each technology linked to each DP(1–7) in Table 4 and determine the corresponding color, the scale of values indicated in Table 7 was utilized.

Table 8 presents the adoption of I4.0 technologies using the color scale provided in Table 7. The 83 companies were grouped into seven clusters, and the results were averaged for each cluster, with each color assigned based on the value scale in Table 7. For instance, the average value for the technologies (1) Sensors-Actuators (SCADA,

**Table 7** Relation of the maturity levels with the color and value scales

| Maturity Levels | Color Scale | Value Scale |
|---|---|---|
| Very High adoption | | 3.34 – 4.00 |
| High adoption | | 2.68 – 3.33 |
| Medium adoption | | 1.34 -2.67 |
| Low adoption | | 0.10 – 1.33 |
| Non-existent adoption | | 0 |

ERP) in Cluster 1 (Food Industry) is 3.38, and in Cluster 2 (Energy) it is 3.64, both corresponding to the dark green color. In Cluster 3 (Metalworking), the average value is 1.56, and in Cluster 5 (Software and Technology), it is 1.74, both corresponding to the yellow color.

Table 8 displays five adoption archetypes of advanced I4.0 technologies based on the results of the ICMAT administered in 83 industrial companies and divided into seven different clusters.

The adoption archetypes are explained in detail below:

Archetype I. The Energy Cluster (2) has a majority with very high to high adoption, predominantly in advanced technologies of Machine Learning and Data Analysis. It shows a very high adoption in the cases of DPs (1, 3, 4, 5, 7). It indicates low or no adoption of simulation technologies and VR-AR, both linked to DP(6).

Archetype II. The Food Cluster (1) corresponds to the case of very high-high-medium adoption. Very High–high adoption in the cases of DPs (1, 5). It shows high-medium adoption in the cases of DPs (2, 3). Indicates low or no adoption in additive manufacturing, simulation, embedded systems and VR-AR technologies.

Archetype III. Clusters: (4) Chemical and (6) Metallurgical, correspond to similar patterns of high-medium adoption, in a majority way. It shows high-medium adoption in the cases of DPs (1, 3, 4, 5).

Archetype IV. Clusters: (3) Metalworking and (5) Software and technology, correspond mainly to medium adoption in the cases of DPs (1, 4, 5). It indicates low or non-adoption in the cases of DPs (2, 6).

Archetype V: The Services Cluster (7) shows predominant High or Low adoption. It indicates low adoption rates for Sensors-Actuators, RFID-RTLS, Robots, VR-AR, and non-existent adoption for Additive Manufacturing, and Simulation. However, it presents high adoption rates for advanced technologies like Machine Learning, and Data Analysis, Services in the Cloud, and Cybersecurity.

## *5.1 Comparative Analysis*

Whilst the practical and theoretical applications of I4.0 continue to emerge; the maturity modeling of this concept is a growing concern for academics and practitioners [13]. Recent literature extensively covers the topic of I4.0 maturity modeling. These

**Table 8** Results of the closed questions (types I and II)

| | DP (1-7) | I4.0 Technologies (Based on Table 5.3) | Industrial Clusters (Based on Table 5.1) | | | | | | |
|---|---|---|---|---|---|---|---|---|---|
| | | | 1 | 2 | 3 | 4 | 5 | 6 | 7 |
| DESIGN – PRINCIPLES | DP(1) | (1) Sensors – Actuators | | | | | | | |
| | | (2) Communication – Networking | | | | | | | |
| | | (3) RFID – RTLS | | | | | | | |
| | | (5) Cloud | | | | | | | |
| | | (7) Cybersecurity | | | | | | | |
| | DP(2) | (2) Communication – Networking | | | | | | | |
| | | (3) RFID – RTLS | | | | | | | |
| | | (5) Cloud | | | | | | | |
| | | (6) Machine learning & Data analysis | | | | | | | |
| | | (8) Simulation | | | | | | | |
| | | (9) Additive Manufacturing | | | | | | | |
| | | (10) VR – AR | | | | | | | |
| | | (11) Mobile Technologies | | | | | | | |
| | | (12) Robotics: Adaptive – Industrial | | | | | | | |
| | DP(3) | (2) Communication – Networking | | | | | | | |
| | | (7) Cybersecurity | | | | | | | |
| | DP(4) | (1) Sensors – Actuators | | | | | | | |
| | | (3) RFID – RTLS | | | | | | | |
| | | (6) Machine learning & Data analysis | | | | | | | |
| | DP(5) | (1) Sensors – Actuators | | | | | | | |
| | | (2) Communication – Networking | | | | | | | |
| | | (3) RFID – RTLS | | | | | | | |
| | | (4) Embedded Systems | | | | | | | |
| | | (6) Machine learning & Data analysis | | | | | | | |
| | DP(6) | (8) Simulation | | | | | | | |
| | | (10) VR – AR | | | | | | | |
| | DP(7) | (5) Cloud | | | | | | | |
| | | (6) Machine learning & Data analysis | | | | | | | |

studies provide an up-to-date overview of the current developments of this concept and reflect a rising interest in I4.0 maturity modeling within academia [14–17].

The maturity models (MMs) of I4.0 highlight the significance of a strategic approach to development, where design principles systematize measurable and practicable steps for further advancement [7]. For instance, Schuh et al. [18] identify dimensions like resources, information systems, organizational structure, and culture, while Schumacher et al. [8] mention strategy, leadership, culture, people, and governance. The Acatech Industrie 4.0 Maturity Index [18] divides the structure of a company into four structural areas or dimensions (I4.0 capabilities): 1. Resources; 2. Information systems; 3. Organizational structure; and 4. Culture. This model encompasses six maturity stages: 1. Computerization; 2. Connectivity; 3. Visibility; 4. Transparency; 5. Predictive capability; and 6. Adaptability.

The work of Modrák and Šoltysová [19] proposes a research methodological framework divided into six steps: 1. Analysis of existing I4.0 MMs; 2. Development of the MM questionnaire; 3. Application of the questionnaire; 4. Identifying key requirements of SMEs based on the questionnaire results; 5. Development of the I4.0 Readiness MM; and 6. Development of a generic organizational model for mass customized manufacturing. Moreover, this work presents a structured approach to conducting self-assessment for I4.0 implementation and proposes the generic I4.0 Readiness MM as the roadmap to understand the current state in the mass customization environment.

Simetinger and Zhang [20] presented a comparative analysis of 17 I4.0 MMs, considering five main dimensions and 19 sub-dimensions. In addition, the work shows the similarities among different models and the overlapping relationship of the related dimension. The study only indicates which dimensions are most frequently used in I4.0 MMs.

Finally, the current paper proposes the development of an Industrial Cluster Maturity Assessment Tool (ICMAT) to determine the degree of adoption or maturity stages of basic and advanced I4.0-related technologies in 83 companies from seven clusters. The ICMAT is based on seven DPs related to the core and smart technologies of I4.0 and consists of a questionnaire with three types of questions, where the first two types use a 5-point scale indicating five levels of adoption or maturity.

## 5.2 *Limitations and Future Work*

One major limitation of this study is the small number of industrial clusters that responded to the questionnaire. To ensure a more comprehensive understanding of how companies adopt I4.0 technologies, it is necessary to involve more industry clusters.

The COVID-19 pandemic's disruptions demonstrated that the integration of advanced I4.0 technologies enabled long-term resilience and sustainable competitive advantage [21]. These technologies have become a strategic tool, facilitating streamlined processes through smart factories [22], and expediting the digital transformation of supply chain operations in major industries [22]. On one hand, I4.0 technologies played a crucial role in companies' responses to COVID-19, facing their most demanding test in 2020.

However, the crisis also presented challenges for the future of digital operations. A survey conducted in 2020 revealed consequences related to the adoption of these technologies, such as serving as a wake-up call for companies that have not yet initiated their digital transformation [22].

## 6 Conclusions

A research framework was proposed based on two dimensions, each with three sub-dimensions. These dimensions were connected to seven Design Principles (DPs). The DPs were then linked to the Industrial Cluster Maturity Assessment Tool (ICMAT), a questionnaire with three types of questions related to I4.0 technologies and characteristics of companies and industrial clusters. The questionnaire had a 5-point adoption scale to measure the level of adoption and maturity of I4.0 technologies. The findings from the questionnaire of 83 industrial firms in seven different clusters were utilized to identify five adoption archetypes of I4.0 technologies as follows:

Archetype I: The Energy Cluster (2) had a majority with very high to high adoption, mainly in advanced technologies of Machine Learning and Data Analysis. However, it showed low or no adoption of simulation and VR-AR technologies.
Archetype II: The Food Cluster (1) demonstrated very high and high to medium adoption, while showing low or no adoption of Additive Manufacturing, Simulation, Embedded Systems, and VR-AR technologies.
Archetype III: The Chemical (4) and Metallurgical (6) clusters exhibited similar patterns of high to medium adoption, predominantly.
Archetype IV: The Metalworking (3) and Software and Technology (5) clusters were mainly characterized by medium adoption.
Archetype V: The Services Cluster (7) primarily displayed High or Low adoption. It had low adoption rates for Sensors-Actuators, RFID-RTLS, Robots, VR-AR, Additive Manufacturing, and Simulation. On the other hand, it exhibited high adoption rates for advanced technologies like Services in the Cloud, Cybersecurity, Machine Learning and Data Analysis.

In conclusion, when we analyze the data, we find that the Energy Cluster (2) stands out with the most significant adoption rates for Machine Learning and Data Analysis, closely followed by the Food Cluster (1), where we observe high, very high, and medium adoption levels. Meanwhile, the Chemical (4) and Metallurgical (6) clusters exhibit similar trends with a combination of high and medium adoption. In contrast, the Metalworking (3) and Software and Technology (5) clusters demonstrate a medium adoption pattern. Finally, we must consider the Services Cluster (7), which demonstrates predominantly high adoption rates, especially concerning advanced technologies like Machine Learning and Data Analysis.

These results highlight the importance of adopting cutting-edge technologies to promote digital transformation in different industrial clusters in Argentina.

## References


1. Chong G, Zhiying J, Ding D (2020) The emerging business models. World Scientific Publishing Company
2. Behrens V, Trunschkey M (2020) Industry 4.0 Related Innovation and Firm Growth. ZEW—Centre for European Economic Research. Discussion Paper. No. 20-070
3. Resman M, Turk M, Herakovič N (2021) Methodology for planning smart factory. Procedia CIRP 97:401–406
4. Grabowska S, Saniuk S (2023) Business models for industry 4.0. concepts and challenges in SME organizations. Chapter 3. Routledge Studies in Innovation, Organizations and Technology. Routledge, Taylor & Francis Group, pp 62–63
5. Kruse R, Mostaghim S, Borgelt C, Braune C, Steinbrecher M (2023) Computational intelligence. In: A methodological introduction, 3rd edn. Springer. ISBN 978-3-030-42227-1. https://doi.org/10.1007/978-3-030-42227-1
6. Srinivas VS, Patibandla ML, Narayana VL, Rao BT (2023) Industry maintenance optimization using AI (Chapter 8). In: Computational intelligence applications for software engineering problems. Apple Academic Press, Inc. Co-published with CRC Press (Taylor & Francis)
7. Dikhanbayeva D, Shaikholla S, Suleiman Z, Turkyilmaz A (2020) Assessment of industry 4.0 maturity models by design principles. Sustainability 12(23):9927
8. Schumacher A, Erol S, Sihn W (2016) A maturity model for assessing industry 4.0 readiness and maturity of manufacturing enterprises. Procedia CIRP 52:161–166. https://doi.org/10.1016/j.procir.2016.07.040
9. Alcácer V, Rodrigues C, Carvalho H et al (2021) Tracking the maturity of industry 4.0: the perspective of a real scenario. Int J Adv Manuf Technol 116:2161–2181. https://doi.org/10.1007/s00170-021-07550-0
10. Quiroga OD (2022) Adoption of advanced technologies in industrial clusters. A Study in Latin American Industries. IFAC-PapersOnLine 55(10):1846–1851. ISSN 2405-8963. https://doi.org/10.1016/j.ifacol.2022.09.667
11. Yin R (2018) Case study research and applications: design and methods, 6th edn. SAGE Publications
12. Ustundag A, Cevikcan E (2018) Industry 4.0: managing the digital transformation. In: Springer Series in Advanced Manufacturing. Springer, Cham. https://doi.org/10.1007/978-3-319-57870-5
13. Elibal K, Özceylan E (2022) Comparing industry 4.0 maturity models in the perspective of TQM principles using fuzzy MCDM methods. Technol Forecast Soc Change 175:121379. ISSN 0040-1625. https://doi.org/10.1016/j.techfore.2021.121379
14. Mittal S, Khan MA, Romero D, Wuest T (2018) A critical review of smart manufacturing & industry 4.0 maturity models: implications for small and mediumsized enterprises (SMEs). J Manuf Syst 49:194–214
15. Bertolini M, Esposito G, Neroni M, Romagnoli G (2019) Maturity models in industrial internet: a review. Procedia Manuf 39:1854–1863
16. Schmitt P, Schmitt J, Engelmann B (2019) Evaluation of proceedings for SMEs to conduct I4.0 projects. Procedia CIRP 86:257–263
17. Hizam-Hanafiah M, Soomro MA, Abdullah NL (2020) Industry 4.0 readiness models: a systematic literature review of model dimensions. Information 11(7):364
18. Schuh G, Anderl R, Gausemeier J, ten Hompel M, Wahlster W (2020) Industrie 4.0 maturity index. In: Managing the digital transformation of companies (Acatech Study) Herbert Utz Verlag, (Munich). https://en.acatech.de/publication/industrie-4-0-maturity-index-update-2020/ Accessed 25 July 2023
19. Modrák V, Šoltysová Z (2020) Development of an organizational maturity model in terms of mass customization. In: Matt D, Modrák V, Zsifkovits H (eds) Industry 4.0 for SMEs. Palgrave Macmillan, Cham. https://doi.org/10.1007/978-3-030-25425-4_8
20. Simetinger F, Zhang Z (2020) Deriving secondary traits of industry 4.0: a comparative analysis of significant maturity models. Syst Res Behav Sci 37(4):663–678

21. Mancini M, Marteletti G, Patel A, Requeno L, Ye T (2021) Industry 4.0 adoption with the right focus. McKinsey and Company. https://www.mckinsey.com/business-functions/operations/our-insights/operations-blog/industry-40-adoption-with-the-right-focus
22. Agrawal M, Dutta S, Kelly R, Millan I (2021) COVID-19: an inflection point for industry 4.0. McKinsey and Company. https://www.mckinsey.com/business-functions/operations/our-insights/covid-19-an-inflection-point-for-industry-40

# Computational Intelligence in Project Planning and Monitoring

# Combining EDA and Simulated Annealing Strategies in Project Scheduling Construction

**Pedro Yobanis Piñero Pérez, Iliana Pérez Pupo, Sadeq Saeed Mahdi, Julio Madera Quintana, and Luis Alvarado Acuña**

**Abstract** Project planning can be treated as an optimization problem focused on organizing a set of tasks while respecting a set of precedent constraints and the limited use of renewable and non-renewable resources. The resulting schedule must have the properties of being executed with the least possible time and cost and with a balanced quality of the solution. In this research, a set of new distribution estimation algorithms is presented to solve this problem. Furthermore, the behavior of different evolutionary algorithms in the construction of schedules is compared. The experimental results show the viability of evolutionary algorithms for the agile construction of chronograms and their potential use in BIM environments. In the experimentation, a set of 150 instances collected in 15 databases from the PSPLib library were used. The sensitivity of the behavior of the algorithms is evaluated in the following scenarios: variation in the number of execution modes, in the number of tasks, in the number of renewable resources, and in the number of non-renewable resources, demonstrating the feasibility of the solutions.

**Keywords** Agile project scheduling · Computational intelligence · Estimation destribution algorithms · Optimization · Project management

---

P. Y. Piñero Pérez (✉) · I. P. Pupo · S. S. Mahdi
Artificial Intelligence for a Sustainable Development Group, IADES, Havana, Cuba
e-mail: pppyob@gmail.com; iades@businessredmine.com

I. P. Pupo
e-mail: iliperezpupo@gmail.com; iades@businessredmine.com

S. S. Mahdi
e-mail: businessredmine@gmail.com; iades@businessredmine.com

J. M. Quintana
Vicerrectoría de Informatización, Universidad de Camagüey, Camagüey, Cuba
e-mail: julio.madera@reduc.edu.cu

L. A. Acuña
Departamento de Ingeniería de La Construcción, Universidad Católica del Norte, Antofagasta, Chile
e-mail: lualvar@ucn.cl



P. Y. Piñero Pérez et al. (eds.), *Computational Intelligence in Engineering and Project Management*, Studies in Computational Intelligence 1134,
https://doi.org/10.1007/978-3-031-50495-2_6

# 1 Introduction

In recent years, there has been a growing development of Building Information Modeling (BIM) technologies [1, 2] for construction projects [3]. These technologies aim to simulate the complete execution of projects in virtual environments. Furthermore, the agility of project development imposes the need to frequently adjust schedules. In [4], it is proposed that the main factor that affects the inappropriate use of resources is inefficient planning. In this context, the use of metaheuristics that facilitate the rapid updating of schedules in the face of changes in projects is essential.

Among the project management schools, the Guide to the Project Management Body of Knowledge (PMBOK) stands out [5], the International Project Management Association (IPMA) and ISO 21500 (International Organization for Standardization) [6, 7]. The guides and standards proposed by these schools do not propose specific techniques based on optimization models for the schedule construction processes. In general, all schools agree on the need to identify and plan tasks logically to achieve the project objectives. This scheduling problem is known as Project Scheduling Problem (PSP).

The first approaches to conceptualizing and solving PSP problems were developed in the 1950s [8, 9]. In the literature consulted, planning problems can be classified by the objectives they pursue, the types of resources they use, and the different modes of execution [10–12] en.

Resource-Constrained Project Scheduling Problem (RCPSP) [13, 14]: It is characterized by the presence of limited resources for the development of a single project, where each task can be executed in a single way.

Resource-Constrained Multi-Project Scheduling Problem (RCMPSP) [15, 16]: Characterized by the need to harmonize the planning of multiple projects, where each task can be executed in a single way and in a context where the projects share the limited use of resources.

Multi-Mode Resource-Constrained Project Scheduling Problem (MMRCPSP) [17, 18]: This problem is characterized by the need to plan the tasks of a project, considering that they can be executed in different ways or modes in environments with limited resources.

Multi-Mode Resource-Constrained Multi-Project Scheduling Problem (MMRCMPSP) [19, 20]: It constitutes a generalization of the previous ones, multiple projects share resources that are presented in a restricted way, and tasks can be executed in multiple ways or modes.

Based on the analysis of numerous works reported in the bibliography and international project management standards, this research identifies the following opportunities for improvement and lines of research in solving project planning problems:

Generally, the MMRCPSP problem, present in the construction of schedules during project planning, is treated by standards in the absence of approaches that promote the use of optimization techniques or metaheuristics [5, 16, 21–23].

Project planning problems are generally presented as optimization problems in the presence of correlated variables. But in the searches carried out, few works were found that applied EDA to solve these problems [24–27]. There are no related works reported in EDA that use information on the Constraintsin the learning process of the probabilistic model [21].

The treatment of the problem of the balance between cost and time, or the influence of risks on planning processes and the effectiveness of the solutions proposed, is still insufficient [28, 29].

Emerging technologies such as BIM require the development of new research and algorithms for effective construction of schedules and planning in general [2, 3, 30, 31]. In this scenario, several engineering specialties converge, such as weak currents, civil engineering, and hydraulic engineering, among others, and each specialty can execute its tasks in multiple modes.

In this research, the authors propose to address project planning problems by taking advantage of the advantages of Distribution Estimation Algorithms associated with the treatment of the correlation of variables. Furthermore, compare different evolutionary algorithms for solving project planning problems. The work is structured in the following sections: the second section presents a brief theoretical framework where the fundamental aspects and the main schools and standards of project management are described, emphasizing the planning process. Section 3 proposes new EDAs that incorporate the constraints of the planning problem within the probabilistic model. Section 4 presents the comparison results of the proposed algorithms and other evolutionary algorithms for solving planning problems. Finally, conclusions and future work are presented.

## 2 Characterization and Modeling of the Problem, MMRCMPSP

As a precedent to this research, another chapter of this same book presents a study of the state of the art of solving project planning problems.

A summary of the study carried out indicated that the greatest amount of research to solve this problem focuses on the use of Genetic Algorithms (GA) [25, 32]. Although, works are also presented that combine GA with other local search methods, [33–36]. But the intrinsic nature of traditional genetic algorithms does not facilitate work with problems where there are correlated variables [37–39]. This element was one of the factors that motivated the emergence of Algorithms with Estimation of Distributions [39, 40].

The authors of this research identify that EDA, by its nature, can solve project planning problems with the presence of correlated variables, but in the searches carried out, few works were found that apply EDA in project planning [24, 41–43]. Furthermore, compared to other metaheuristics, there are few works that use EDA to solve problems with constraints [44]. This whole situation constitutes one of the motivations for this research, focused on the application of Algorithms with Distribution Estimation in the solution of project planning problems with restrictions.

The present research focuses on solving the Project Planning Problem with limited resources and tasks that are executed in multiple ways (MMRCPSP) because the solution to this problem can help address solutions for any of the other three planning problems. The RCPSP Problem refers to tasks of a single project that can be executed in a single way and, therefore, is a particular case of the MMRCPSP. The RCMPSP problem occurs in the management of project programs with shared resources. It is modeled as a single macro project, trying to combine the tasks of all the projects into a single schedule and also considering that each task can be executed in a single mode. Therefore, it could also be modeled from the solution of the MMRCPSP problem. For its part, the MMRCMPSP problem can also be transformed into the MMRCPSP problem, considering, as in the previous case, that it is a single macro-project where all the tasks of the subprojects must coexist in a single schedule while maintaining its multimode characteristic. Two fundamental concepts in the project planning process that are involved in the solution are.

Waiting slack is the period of time between the completion of a task and the start of its successor. This slack is introduced because the start of a task depends on the completion of all of its predecessors, and therefore a wait time is created between the task and its predecessors that have already finished.

Free slack is the period of time between the completion of a task and the scheduled date to make a project cut. This slack is planned as part of the reserve time for risk management processes. During project execution, this slack allows for necessary adjustments to the schedules in the event of unforeseen delays.

On the other hand, the MMRCPSP is a problem with the presence of correlated variables and is characterized by the following elements.

There are precedence Constraintsbetween tasks that establish a correlation between the variables that represent the start date of a task and the end date of the predecessor tasks based on logical dependencies between tasks.

In the case of planning problems with multiple modes, correlations between execution modes are evident. For example, in construction projects, if a technology is assumed in the early stages of the project, this will affect the rest of the tasks. In this environment, another example that generates dependencies between tasks and modes is caused by the complex interrelationships between the specialties that converge in a construction project.

The selection of the execution mode determines the amount of renewable and non-renewable resources to be consumed and the duration of the task. In the case of projects with a high dependence on human resources, for example, software projects, there may be variability depending on the human resources used and the execution modes, with an impact on time and quality. Also, there are psychological and personality factors that impact the ways of doing things, generating dependencies between the human resources that make up the team and the ways of execution.

There are Constraintson the availability of renewable resources assigned for each day of the project and establish logistical or physical dependencies between tasks based on the use of critical resources, which are those resources that are highly demanded and are presented in small quantities, causing waiting slack.

Finding a feasible solution for an MMRCPSP is an NP-Complete problem [45, 46]. Therefore, in the proposal for this research, this problem is addressed through the use of metaheuristics. The following section presents the mathematical modeling of this problem and its computational solution using different algorithms.

## 2.1 *Modeling of the Optimization Problem Associated with the Problem, MMRCPSP*

Next, the notation used in modeling the problem is introduced:

A project contains a set of tasks $j = \{1, \ldots, n\}$.

Each task $j$ has a set of predecessor tasks $H_j$ that represent precedence constraints such that, in a feasible solution to the problem, no task can begin until the tasks preceding it have finished.

There are $m_j$ modes to do the tasks $j$. $M_j = \{1, \ldots .m_j\}$ represents the set of these modes.

$t = \{1, \ldots, T\}$ It is the set of moments in which different tasks can be started or finished; generically, each of these moments will be called T "day". It is an upper bound on the project duration time that is equal to the longest possible duration according to the task processing modes.

$d_{jm}$ represents the duration of the "j" task in the given mode "m", the unit of this is number of days.

There are two categories of resources: renewable and non-renewable.

$K_v$ Represents all renewable resources. These are resources that can be used more than once and for different tasks in the context of the project, for example, personnel and equipment.

$K_{nv}$ Represents the set of non-renewable resources that are characterized because they can only be used once during the life of the project. Examples of this are expendable materials such as paper, fuel, etc.

$k \in K_v$, $R_k$ Represents the maximum amount of the resource available each day.

$C_k$ Represents the unit cost per unit of k-resource used., $k \in K_v \cup K_{nv}$.

Each j- task, depending on the execution mode "m", will consume $r_{jmk}$ units of resource with type $k$.

Let's $j = 1$ and $z = n$ they are two fictitious tasks. The first represents the beginning of the project, and the second its end. These two tasks have a single execution mode, their duration is 0, and they do not consume any type of resource; therefore, the start day of the n-task represents the duration of the project.

In this research, the MMRCPSP problem is presented as a multi-objective optimization problem that combines the objective $f_1$ duration of the project and the objective $f_2$ project cost associated with the execution of the tasks, applying a bottom-up estimate. We want to minimize that combines both objectives as presented in the Eq. 1.

The model variables are:

$$x_{jmt} = \begin{cases} 1, \ j - \text{tasks start day } t \text{with mode } m \\ 0, \ \text{any other case} \end{cases}$$

From the definition of the previous variable.

$\sum_{m=1}^{m_j} \sum_{t=1}^{T} t x_{jmt}$ represents start day of j-tasks and $\sum_{m=1}^{m_j} \sum_{t=1}^{T} d_{jm} x_{jmt}$, duration.$\sum_{m=1}^{m_j} \sum_{t=1}^{T} t x_{jmt} + \sum_{m=1}^{m_j} \sum_{t=1}^{T} d_{jm} x_{jmt}$ represents the first day on which the j- task has been completed and the tasks that follow it can begin

$$\min F = (f_1, f_2) \tag{1}$$

$$f_1 = \sum_{t=1}^{T} t x_{n1t} \tag{2}$$

$$f_2 = \sum_{j=1}^{n} \sum_{m=1}^{m_j} \sum_{t=1}^{T} \left[ \sum_{k \in K_v} r_{jmk} \cdot C_k \cdot d_{jm} + \sum_{k \in K_{nv}} r_{jmk} \cdot C_k \right] x_{jmt} \tag{3}$$

$f_1, f_2$ they represent the value of the duration and the cost of the project, respectively. Each solution must be subject to the following constraints:

$$\sum_{m=1}^{m_j} \sum_{t=1}^{T} x_{jmt} = 1, \quad \forall j = [1, \ldots, n] \tag{4}$$

Equation (4) represents the constraint that guarantees that each task is started on a day in a single mode.

$$\sum_{m=1}^{m_j} \sum_{t=1}^{T} t x_{jmt} \geq \sum_{m=1}^{m_i} \sum_{t=1}^{T} t x_{imt} + \sum_{m=1}^{m_i} \sum_{t=1}^{T} d_{im} x_{imt}, \quad i \in H_j \tag{5}$$

Equation (5) Represents the constraint that ensures that precedence relationships between tasks are not violated

$$\sum_{j=1}^{n} \sum_{m=1}^{m_j} \sum_{t=1}^{T} r_{jmk} x_{jmt} \leq R_k, k \in K_{nv} \tag{6}$$

Equation (6) represents the restriction that guarantees that the use of non-renewable resources as expendable material resources is not exceeded, considering the mode of execution of the project tasks.

$$\sum_{j=1}^{n}\sum_{m=1}^{m_j}\sum_{t=max\{\bar{t}-d_{jm}+1,0\}}^{\bar{t}} r_{jmk}x_{jmt} \leq R_{k\bar{t}}, \quad \forall\bar{t}=1,\ldots,T, k \in K_v \tag{7}$$

Equation (7) represents the constraint that guarantees the availability of renewable resources for each day of the project, such as human resources and equipmen.

Due to the complexity of the multi-objective problem, the most common thing to do is to consider the weights $\alpha f_1 + (1-\alpha) f_2$, $\alpha \in [0, 1]$ as objective functions and solve the resulting optimization problems. In this thesis, it would be min $\alpha f_1 + (1-\alpha) f_2$ subject to the Constraintsrepresented in Eqs. (4)–(7). The numerical results illustrate its solution for suitable values of the parameter $\alpha$.

In the relationships between tasks, different slack times are defined, which are known as waiting slack and free slack.

It is identified that the MMRCPSP is a problem in the presence of correlated variables and is characterized by the following elements.

There are precedence Constraintsbetween tasks that establish a correlation between the variables that represent the start date of a task and the end date of the predecessor tasks based on logical dependencies between tasks.

In the case of planning problems with multiple modes, correlations between execution modes are evident. For example, in construction projects, if a technology is assumed in the early stages of the project, this will affect the rest of the tasks. In this environment, another example that generates dependencies between tasks and modes is caused by the complex interrelationships between the specialties that converge in a construction project.

The selection of the execution mode determines the amount of renewable and non-renewable resources to be consumed and the duration of the task. In the case of projects with a high dependence on human resources, for example, software projects, there may be variability depending on the human resources used and the execution modes, with an impact on time and quality. Also, there are psychological and personality factors that impact the ways of doing things, generating dependencies between the human resources that make up the team and the ways of execution.

There are Constraints on the availability of renewable resources assigned for each day of the project and establish logistical or physical depen dencies between tasks based on the use of critical resources, which are those resources that are highly demanded and are presented in small quantities, causing waiting slack.

## 2.2 Computational Design of the Individual to Solve the Optimization Problem

In the modeling of the computational solution to respond to the optimization problem posed, each individual represents a possible solution and is characterized by.

Be formed by a sequence of tasks, where each task represents a data structure composed of the following elements: the start day of the j-task (start date), the mode in which it should be executed (m) and a set of complementary attributes.

The execution mode determines the duration of the task, the number of renewable resources, and the number of non-renewable resources. These complementary attributes are included in the representation of the solution with the objective of facilitating the calculation of costs and the evaluation of the satisfaction of the constraints.

Have a boolean field that indicates whether the solution it represents is feasible or not. This is a complementary field used to facilitate the search process.

Save information about the reserves of renewable and non-renewable resources, which can be used in the process of improving solutions.

Save the learning strategy with which it was generated:

No reinforcement: It refers to the fact that when considering learning improvement alternatives, only the individuals of the current iteration are considered.

Reinforcement: When considering learning improvement alternatives, the individuals of the current iteration and previous iterations are considered.

Save the alternative selection of the task-mode combination during the solution improvement process described later in Algorithm 1.

Also save your evaluation regarding the different objectives of the optimization problem.

Have a boolean field that indicates whether it has been evaluated or not. This field is used to avoid unnecessarily evaluating individuals who have already been evaluated.

Save information about the changes it undergoes during the improvement process. The new task-mode combinations are saved, an element that helps identify the best combinations in subsequent iterations of the search process.

An example of the structure of an individual in json format is presented belowm see Output structure 1

**Output structure 1. An example of the representation of a solution in json format**

{:evaluation=>
{:fitness=>{:time=>21.0, :cost=>538.0, :cpinv=>1.2063, :renewable=>0.0,
:nonrenewable=>0.0},
:rfitness=>{:time=>0.0048, :cost=>0.0069, :cpinv=>0.0069, :renewable=>0.0,
:nonrenewable=>0.0, :global=>0.0058},
:status=>:evaluated, :is_feasible=>true,
:alternative=>[0.0, 0.0, 0.0, 0.0, 0.0, 0.0, 0.0],
:changes=>[],
:eval=>0.0029,
reserve=>{
:renewable=>{
0=>{:resource_renewable1=>29, :resource_renewable2=>23},
1=>{:resource_renewable1=>15, :resource_renewable2=>12},
2=>{:resource_renewable1=>15, :resource_renewable2=>12},
3=>{:resource_renewable1=>11, :resource_renewable2=>11},
4=>{:resource_renewable1=>11, :resource_renewable2=>11},
5=>{:resource_renewable1=>11, :resource_renewable2=>11},
6=>{:resource_renewable1=>17, :resource_renewable2=>11},
7=>{:resource_renewable1=>9, :resource_renewable2=>3},
8=>{:resource_renewable1=>9, :resource_renewable2=>3},
9=>{:resource_renewable1=>16, :resource_renewable2=>1},
10=>{:resource_renewable1=>16, :resource_renewable2=>1},
11=>{:resource_renewable1=>16, :resource_renewable2=>1},
12=>{:resource_renewable1=>16, :resource_renewable2=>1},
13=>{:resource_renewable1=>22, :resource_renewable2=>16}, ...
:nonrenewable=>{:resource_nonrenewable1=>26, :resource_nonrenewable2=>48}},
:generation_method=>{:unique=>1, :random_method=>3, :parents_dependency=>14}},

:job_1=>{:start_day=>0, :mode=>1, :due_date=>0, :resources=>{:duration=>0,
:resource_renewable1=>0, :resource_renewable2=>0, :resource_nonrenewable1=>0,
:resource_nonrenewable2=>0}},
:job_2=>{:start_day=>1, :mode=>3, :due_date=>6, :resources=>{:duration=>6,
:resource_renewable1=>1, :resource_renewable2=>1,
:resource_nonrenewable1=>10, :resource_nonrenewable2=>0}},
:job_3=>{:start_day=>1, :mode=>1, :due_date=>2, :resources=>{:duration=>2,
:resource_renewable1=>7, :resource_renewable2=>8, :resource_nonrenewable1=>7,
:resource_nonrenewable2=>0}},
:job_4=>{:start_day=>1, :mode=>3, :due_date=>6, :resources=>{:duration=>6,
:resource_renewable1=>6, :resource_renewable2=>2, :resource_nonrenewable1=>0,
:resource_nonrenewable2=>6}},
:job_5=>{:start_day=>3, :mode=>1, :due_date=>5, :resources=>{:duration=>3,
:resource_renewable1=>7, :resource_renewable2=>4,
:resource_nonrenewable1=>10, :resource_nonrenewable2=>0}},
:job_11=>{:start_day=>3, :mode=>2, :due_date=>4, :resources=>{:duration=>2,
:resource_renewable1=>4, :resource_renewable2=>5, :resource_nonrenewable1=>0,
:resource_nonrenewable2=>7}},
:job_7=>{:start_day=>6, :mode=>1, :due_date=>12, :resources=>{:duration=>7,
:resource_renewable1=>5, :resource_renewable2=>9, :resource_nonrenewable1=>0,
:resource_nonrenewable2=>8}},
:job_6=>{:start_day=>7, :mode=>1, :due_date=>8, :resources=>{:duration=>2,
:resource_renewable1=>10, :resource_renewable2=>5,
:resource_nonrenewable1=>0, :resource_nonrenewable2=>7}},
:job_8=>{:start_day=>7, :mode=>3, :due_date=>15, :resources=>{:duration=>9, :resource_renewable1=>1, :resource_renewable2=>2, :resource_nonrenewable1=>0, :resource_nonrenewable2=>2}}

## 3 Distribution Estimation Algorithms for Solving the Planning Problem

This section presents three EDA algorithms for solving scheduling problems. The proposed algorithms are characterized by the following elements.

These algorithms are inpspired in Constraint Learning Estimation of Distribution Algorithms proposed in [47].

The proposed algorithms are adjusted for the solution of Scheduling Problems of the MMRCPSP type. In the proposal for this research, time and cost objectives are established, and two fictitious tasks located at the beginning and end of the schedule are defined.

In the construction process of the new individuals, the proposed algorithms combine the potential of EDA with a local search algorithm based on simulated annealing.

See Algorithm 1.

**Algorithm 1. A generic algorithm that combines EDA with local search, which we will call EDA_BRA1**

*objectives*: objectives of the problem to be optimized, example time, cost
*max_iteration*s: maximum number of iterations
*error_threshold:* Optional parameter for threshold-based stop condition

constraints: constraints of the problem in question.

*indicators = {g: max_iterations, threshold: error_threshold}*
*learn_centrality = Identify relevant information from constraints.*
*popul = initial_population(constraints)*
*popul = evaluation(popul, constraints, objectives)*
*while not stop_condition(popul, indicators)*
    *sel_size = calculate_selection_size (popul)*
    *S = schedule_selection(popul, sel_size)*
    *g = indicators[:g]*
    $p_e^S(x, g)$ *= distribution_learn(popul, S, constraints, learn_centrality)*
    *new_popul = schedule_sow_elitism(popul, S)*
    *new_popul = schedule_new_solutions(new_popul,* $p_e^S(x, g)$*, constraints)*
    *new_popul = improve_solutions(new_popul,* $p_e^S(x, g)$*)*
    *popul = evaluation (new_popul, constraints, objectives)*
    *indicators[:g] += 1*
*end while*
*return popul*

The input to the algorithm that represents the constraints of the problem in question refers to two types of constraints: precedence constraints between tasks and constraints regarding the availability of renewable and non-renewable resources. As part of the restrictions, a group of execution modes for each task is defined, and for each mode, the duration and the amount of renewable and non-renewable resources involved in performing this task in this mode are defined.

In step 1, a data structure called indicators is built, where the information that will be used in defining the stop condition is stored.

Subsequently, in step 2 of the proposal, the identification and learning of other relevant information from the restrictions are established. In this step, different explicit indicators are calculated based on the restrictions posed by the problem, which are then used in the generation processes of new individuals. In particular, we proceed to discover the weight of the nodes or tasks by constructing a PERT diagram [48] oriented to nodes, and an analysis of the graph that makes up the precedence relationships between tasks is carried out. In this way, the degree of relevance or centrality of each task is learned from the calculation of the interior and exterior degrees of each node.

In step 3, the initial population is generated randomly, but applying the following strategies that mitigate the possibility of generating non-feasible solutions:

- In the generation of the initial population, explicit knowledge about the precedence of the tasks is used and the random generation of the start date of a task is conditioned so that no task has a date earlier than its predecessors.
- In the initial generation, a preorder is established for assigning a start date to each task, taking into consideration the graph with the precedence restrictions. For each day, the possible tasks that can be executed and that compete for renewable resources are identified. Which tasks should start that day and their execution mode are randomly selected, so that the amount of renewable resources established for each day is not exceeded.

In steps 4 and 13, the generated population of individuals is evaluated; To this end, individuals are evaluated under a multi-objective approach, also taking into account compliance with the restrictions. In the MMRCPSP problem, the definition of the execution mode of a task determines the number of resources that will be used in it, an element that directly influences the costs of the solution. During the evaluation process, those individuals who have been evaluated before and have not undergone modifications do not need to be reevaluated, but the population and the quality of the individual with respect to the population are reevaluated.

Steps 6–14 are executed iteratively based on the defined stopping condition. Examples of alternatives are:

- Find the optimal schedule for the project.
- Find feasible solutions that are good enough that they can be considered by specialists as part of the solution to the problem. This variant assumes that the project

team pre-establishes time and cost thresholds and that the search ends when at least one solution is found that falls below these thresholds.

- Reach a maximum number of iterations of the algorithm.
- Achieve a number of iterations without improving the best solution obtained so far.

In step 6, the size of the selected set that will be considered in the learning of the probabilistic model and as part of the elitism process is determined. Then, in step 7, we proceed to select the best individuals. Different alternatives can be considered, such as roulette or establishing a ranking and truncation. In the experimentation of this research specifically, the ranking method is used, and 30% of the population is selected.

The construction of the ranking of solutions is carried out considering a multi-objective approach, for example, time and cost. Prior to this step, all individuals have been evaluated with respect to each of the objectives independently. Lists are formed in ascending order for each objective, so that individuals who meet the restrictions occupy the top of each list. Then the merge sort method is used to mix the lists, verifying at each moment the non-repetition of individuals. In each iteration of the merge sort, the elements located in the first position of the lists by objectives are compared, inspired by the concept of dominant solutions of the Pareto optimal principal [49].

If there is a solution that dominates the rest, it is inserted into the mixed list. If neither solution dominates the other than both are included in the mixed list. Each time an individual is selected to go to the mixed list, he or she is eliminated from the objective list that gave rise to it.

In step 9, the "Probabilistic Model" is learned, which describes the behavior of the population of the selected individuals. In the probabilistic model, a single distribution function is not learned, but rather a set of distribution functions that cover the restrictions and are used at different times during the generation of the new individuals is estimated. Considering the specific conditions of the problem in question, it is necessary to select what type of specific algorithm from the family of Distribution Estimation Algorithms with treatment of restrictions in the probabilistic model should be used. In this research, two specific algorithms are presented: FDA_BRA6 and UMDA_BRA8. This characteristic requires that this family of algorithms have a high degree of adaptability to the specific optimization problem.

The learning of the probabilistic model is divided into two moments: the first is associated with "learning of distribution functions," while in the second moment, "function adjustment," changes are introduced in what has been learned from the restrictions. In general, in the first moment of "learning distribution functions," the learning of the following distribution functions is proposed:

1. Learning the relationships between tasks and their start dates
2. Learning the probability of occurrence of the task-execution mode combination
3. Learning the best combinations of tasks and modes that are executed on a given day is an element that supports the process of lessons learned in each project so that they can be used in the planning of the project itself or other similar projects. Those combinations are learned that maximize the number of tasks that can start

on the same day, reducing the presence of waiting slack. This concept reinforces the idea of learning solution patterns that can be repeated in different projects, a fact that occurs in both the construction and information technology sectors.

4. Learning of alternatives to improve solutions, which influences the best use of resources. This learning presupposes the existence of a step to improve the solutions based on their partial reconstruction, following a mode selection approach that minimizes the reserves of renewable and non-renewable resources present in the solution. This learning process is inspired by the heuristic that assumes that the maximum use of resources, by minimizing untapped reserves, implies a reduction in the duration of tasks and the project in general.

In the second moment, "function adjustment" changes are introduced in what has been learned from the restrictions, for example, in a project planning problem where there are precedence restrictions between tasks, the pruning of arcs of the probabilistic model that do not comply with constraints of precedence previously introduced. The probability of assignment to a *j*-task with a start date less than the end dates of the tasks that precede it must have a value of zero, and that arc is pruned.

In the proposal, we work with populations of fixed size, and elitism is established in step 10 so that the best individuals identified in the selection process go directly to the new population. Then in step 11, the population is completed, whereas many new individuals are generated as necessary until the established population size is reached, using the different learned distribution functions that make up the probabilistic model (see Algorithm 2).

In step 12, the proposed algorithm includes a solution adjustment or improvement step. In this step, the local search for solutions is enhanced through simulate annealing strategies. In this way, the adjustment of the solutions is intended to obtain solutions that meet the restrictions of the problem.

Subsequently, in step 13, the new population is evaluated, and in step 14, the iteration counter is incremented. Finally, the algorithm returns the final set of generated solutions.

**Algorithm 2. Population generation: schedule_new_solutions (new_popul, method, $\mathbf{p_e^S(x, g)}$, constraints)**

Inputs:

*new_popul:* new population to generated.

*method:* Method of generating individuals with or without reinforcement

$p_e^S(x, g)$: learned probabilistic model.

*constraints:* constraints of the optimization problem

1. *renewable = constraints[:resource_renewable]*
2. *nonrenewable = constraints[:resource_nonrenewable]*
3. *amount = population_size - new_popul.length + 1*
4. *amount.times do*
5. *individual = generate_solution(method,* $p_e^S(x, g)$*, constraints)*
6. *new_popul.store(individual)*
7. *end*
8. *return new_popul*

The process of generating new individuals can be carried out using the following methods:

- Generation of new individuals (pre-candidates) considering the information learned in the probabilistic model.
- Introduction of new solutions in the vicinity of the pre-candidates that improve the original solutions, promoting hybridization with local search methods (for example: simulated annealing, hill climbing, etc.). This step is optional. If carried out, it should focus on maximizing the number of tasks that can start on the same day, reducing the presence of waiting slack or due to renewable resource restrictions. In this step, the knowledge associated with learning $p_e^S(x, t)$ the best combinations of task-mode sets can be used.

In Algorithm 2, step 5 is essential because a feasible solution is generated. See Algorithm 3. The essence of this step lies in the fact that the generation of new individuals is carried out day by day, considering the set of tasks that are carried out. Can execute each day, ensuring compliance with precedence restrictions. At this point, the learned probabilistic model is applied, and a task and its execution mode are selected. On this day, the process is repeated as long as tasks can be included without violating the restrictions associated with renewable resources.

**Algorithm 3. Generation of solutions generate_solution (method, $p_e^S(x, g)$, constraints)**

Inputs:

*method: algorithm type UMDA or FDA*

$p_e^S(x, g)$: probabilistic learned model.

*constraints:* constraints in the optimization problems.

```
new_individual = {}
curr_day = 1
waiting_jobs = all_jobs
while not waiting_jobs.empty
  available_jobs = detect_available_jobs(curr_day, waiting_jobs)
  one_day_sol = more_probable_combination(curr_day, available_jobs,
 pe^S(x,g))
  new_individual = aggregate_sol(new_individual, one_day_sol)
  waiting_jobs = update_waiting_jobs(one_day_sol, waiting_jobs)
  curr_day = update_current_day(new_individual)
end while
calculate_reserve(new_individual, constraints)
return new_individual
```

### 3.1 Algorithm for Improving Individuals Based on Local Search Strategies

In step 12 of the algorithm, local search methods are incorporated that allow the solutions to be adjusted. In this way, the potential of the global search of the EDA is combined with the advantages of the local search provided by simulate annealing (see Algorithm 4).

**Algorithm 4. improve_solutions (popul, $p_e^S(x, g)$)**

Inputs:

*popul:* population to be improve.

$p_e^S(x, g)$: probabilistic learned model.

1. *idx = 0*
2. *while idx < popul.length do*
3.   *individual = popul[idx]*
4.   *if not individual.isfeasible or individual.has_reserve then*
5.     *fixed_individual = improve_individual(individual,* $p_e^S(x, g)$*)*
6.     *popul[idx] = fixed_individual*
7.   *end if*
8.   *idx = idx + 1*
9. *end while*
10. *return popul*

Another novelty of the proposal is that the best local solutions found throughout the search process are stored as learned knowledge. At the beginning of the algorithm, this memory is empty, but with each iteration of the algorithm, it increases. In essence, this process combines local search methods with elements learned and represented in the "Probabilistic Model" obtained in step 11 of Algorithm 1.

In Algorithm 4, the past population is used as a parameter, and those individuals that do not represent feasible solutions or that are feasible but have reservations about the use of renewable or non-renewable resources are detected. The reserves of a solution are made up of the set of resources not planned for the project in this solution. An improvement opportunity is a combination $task_j - mode_m$, that $task_j$ could be executed in $mode_m$ without violating the constraints on the use of resources. In step 5 of this algorithm, these repairable solutions are improved using a local search method inspired by simulate annealing, such as the one presented in Algorithm 5.

The most important method of this algorithm is responsible for evaluating a solution, "calculate_profit(sol)," which aims to evaluate the quality of a solution by combining the evaluation of the objectives proposed in the optimization problem with the efficient use of resources. The particular solution. Between steps 4 and 16 of this algorithm, a cycle is executed with a predetermined number of iterations in which new solutions are generated in the neighborhood of the current solution that you want to improve ("neighborhood (current_solution, $p_e^S(x, g)$)". If a better solution is obtained, this is stored as the best solution obtained so far "best_solution". The "neighborhood (current_solution),$p_e^S(x, g)$)" method uses the information learned $p_e^S(x, g)$ to construct new individuals, focusing on the analysis of resource reserves and violations of constraints. The tasks to change on a certain day are selected with one of the following alternatives:

- Minimum path, an alternative that between two tasks gives greater weight to the one that has the shortest path, in the PERT graph that represents the solution, until the final task.
- Maximum path, an alternative that between two tasks gives greater weight to the one that has the longest path to reach the final task of the project.
- Alternative task-mode combination that maximizes reserve consumption.
- Alternative task-mode combination that minimizes reserve consumption.
- Alternative of the selection of those tasks that maximize the external degree.
- Alternative selection of those tasks that minimize the external degree.

To generate a new task-mode combination, the information learned in the probabilistic model $p_e^S(x, g)$ is used, in particular in the database of the best combinations of task-mode sets. A partial reconstruction of the solutions is carried out based on the date (day) on which the improvement was generated. The applied improvement alternative is remembered in the solution itself and used during the learning of the next iteration of the algorithm.

**Algorithm 5. improve_individual (sol, $\mathbf{p_e^S(x,\ g)}$)**

Inputs:

*sol:* individual to be improve.

*MAX_GENERATION:* constant previously defined

```
1.  sol_profit = calculate_profit(sol)
2.  best_solution = current_solution = {solution: sol, profit: sol_profit}
3.  epoch = 1, temp = 1
4.  while epoch <= MAX_GENERATION
5.    new_solution = neighbourhood(current_solution, p_e^S(x, g))
6.    new_profit = calculate_profit(new_solution)
7.    if new_profit < current_solution[:profit] then
8.      current_solution = {solution: new_solution, profit: new_profit}
9.      if current_solution[:profit] < best_solution[:profit] then
10.             best_solution = {solution: current_solution[:solution],
                                 profit: current_solution[:profit]}
11.   else
12.    factor = exp((current_solution[:profit]- new_profit) / temp)
13.    if factor < rand then
               current_solution = {solution: new_solution, profit: new_profit}
14.   end if line 7
15.   epoch += 1, temp = (MAX_GENERATION - epoch) / MAX_GENERATION
16. end
   return best_solution[:solution]
```

## 3.2 *FDA_BRA6 Algorithm for Solving the MMRCPSP Problem*

This subsection presents a variant of the EDA_BRA1 algorithm that takes essential elements from the FDA algorithm and will be called FDA_BRA6. This new algorithm tries to solve the problem of high complexity in the solution of the MMRCPSP problem associated with the dependency between variables by applying factorization that simplifies the search for solutions based on the approach of the restrictions of the problem in question. See below for the pseudocode in Algorithm 6.

**Algorithm 6. FDA_BRA6 algorithm pseudocode**

Inputs:

objectives: objectives of the problem to be optimized, example time, cost

*max_iterations:* maximum number of iterations.

*error_threshold:* Optional parameter for threshold-based stop conditions.

*constraints:* constraints of the optimization problem.

1. *indicators = {g: max_iterations, threshold: error_threshold}*
2. *learn_centrality = identify relevant information from constraints*
3. *popul = initial_population(constraints)*
4. *popul = evaluation(popul, constraints, objectives)*
5. *while not stop_condition(popul, indicators)*
6. *sel_size = calculate_selection_size (popul)*
7. *S = schedule_selection(popul, sel_size)*
8. *g = indicators[:g]*
9. $p_e^S(x, g)$ = *distribution_FDA_learn(popul, S, constraints, learn_centrality)*
10. *new_popul = schedule_sow_elitism(popul, S)*
11. *new_popul = schedule_new_solutions(popul, new_popul,* $p_e^S(x, g)$, *constraints)*
12. *new_popul = improve_solutions(new_popul,* $p_e^S(x, g)$*)*
13. *popul = evaluation(new_popul, constraints, objectives)*
1. *indicators[:g] += 1*
14. *end while*
15. *return popul*

Most of the steps in this algorithm are similar to those explained in the previous subsection for the EDA_BRA1 algorithm; For this reason, only the steps that are different between the two algorithms will be explained in detail. In particular, step 9 differs because it depends heavily on the probabilistic model.

In step 9, the probabilistic model is learned from the factorization defined based on the problem in question. The learned probabilistic model is formed by the conjunction of different probabilistic distribution functions. In the definition of the factorization, two possible scenarios are proposed, as explained below.

Factorization based on the logical precedence restrictions between project tasks, considering each factorization formed by the variables that identify the tasks and their direct predecessors, see Eq. 8.

$$p^S(x_1, x_{2,}, \ldots, x_n, g) = \prod_{i=1}^{n} p(x_i|\text{parents}(x_i)) \tag{8}$$

Another possible factorization is considering the restrictions associated with the use of renewable and non-renewable resources that belong to the project's shared resource fund. These restrictions cause groups of tasks that could be executed in parallel to compete for resources and can be grouped into small sets, giving rise to logistical dependency relationships that are not explicit in the problem statement. These small subsets of tasks can be grouped as factorization and simplify the learning process of the probabilistic model, see Eq. 9.

$$p^S(x_1, x_{2,}, \ldots, x_n, g) = \prod_{i=1}^{n} p(x_i|\text{tasks competion}(x_i)) \tag{9}$$

Either of the two possible factorizations, will establish small groupings of tasks that will be used in the estimation of the distribution functions that are part of the probabilistic model. The learning process is detailed in Algorithm 7 and responds to the following elements:

1. Learning the relationships between tasks and their start dates, given the factorization defined in the previous steps. The conditional probability of the occurrence of tasks in days is determined, given the occurrence dates of their predecessors. For example, the probability of assignment to a task with a start date less than the end dates of the tasks that precede it has a value of zero, and that arc is pruned.
2. Learning probability of occurrence task-execution mode, given the factorization defined in the previous steps, the conditional probability of occurrence of each task-mode combination is calculated.
3. We also learn the best combinations of tasks and modes that are executed on a given day and the distribution function that describes the alternatives for improving solutions, which influences the best use of resources. These are explained in step 9 of the EDA_BRA1 algorithm.

**Algorithm 7. Algorithm Learning the probabilistic model distribution_FDA_ learn (popul, sel_popul, constraints, learn_centrality, g)**

Inputs:

*popul*: current population.

*sel_popul*: selected population.

*constraints*: constraints for the optimization problem.

*learn_centrality*: relevant information for constraints.

$g$: iteration

1. $p^S(x, g) = \{\}$
2. $p^S(x, g) \Leftarrow$ *learning probabilistic model of relationships between tasks-start dates* $(p_f^S(x_1, x_{2,}, \ldots, x_n, g))$ *and tasks-modes* $(p_m^S(x_1, x_{2,}, \ldots, x_n, g))$

   *case :relationship_method_learn*

   *when :parents*

   *Learning considering factorization by precedence relations*

$$p_f^S(x_1, x_{2,}, \ldots, x_n, g) = \prod_{i=1}^{n} p_f(x_i | parents(x_i))$$

$$p_m^S(x_1, x_{2,}, \ldots, x_n, g) = \prod_{i=1}^{n} p_m(x_i | parents(x_i))$$

   *when :siblings*

   *Learning considering factorization due to competition for resources*

$$p_f^S(x_1, x_{2,}, \ldots, x_n, g) = \prod_{i=1}^{n} p_f(x_i | task\ in\ competition(x_i))$$

$$p_m^S(x_1, x_{2,}, \ldots, x_n, g) = \prod_{i=1}^{n} p_m(x_i | task\ in\ competition(x_i))$$

   *end*

3. $p_e^S(x, g) \Leftarrow$ *Pruning arches that violate precedence restrictions* $p^S(x, g)$

   *lets* $p_f^S(x_1, x_{2,}, \ldots, x_n, g)$.

   $p_e^S(x, g) \Leftarrow$ *Learning the best combinations of task sets-modes that are executed on the same day as lessons learned inspired by case-based systems*

   $p_e^S(x, g) \Leftarrow$ *Learning about alternative solutions' improvement, which influences the best use of resources*

4. *return* $p_e^S(x, g)$

The computational complexity of FDA_BRA6 is based on FDA analyzes [50], it is required to determine the complexity of the initialization step and the internal iteration steps. That is, initialization, selection, learning and generation of new solutions.

- Initialization: The initialization step of FDA_BRA6 consists of assigning the values to all the solutions of the initial population, which has the complexity $O(N \cdot jobs)$, where N is the size of the population and jobs is the number of tasks that make up the solution.
- Selection: Although FDA_BRA6 can use different selection methods, currently it uses ranking selection, which has a time complexity of $O(Slog S)$, where S is the size of the selected solutions.
- Learning: the complexity of the learning step is given by Eq. (3).

Where S represents the number of solutions selected, l represents the number of possible factorizations $\gamma = |tasks| + |modes|$ (cardinality of the modes and tasks of the problem in question), $|c_i|$ represents cardinality of the i-esima factorization, $c_i$ represents the set of variables that form the ith factorization.

Therefore, the complexity of the learning step is $O\left(l \cdot \gamma^{|\hat{c}|} \cdot S\right)$, where $\left|\hat{c}\right|$ is the average size of the factorizations.

Generation: The complexity of the generation step depends on the total number of solutions produced and the number of tasks. So, the complexity of this step is $O(N \cdot jobs)$.

From the above analysis, the computational complexity of FDA_BRA6 is $O(g \cdot l \cdot \gamma^{|\hat{c}|} \cdot S)$, which basically depends on the factorizations and their average size.

### *3.3 UMDA_BRA8 Algorithm for Solving the MMRCPSP Problem*

In the case of the UMDA_BRA8 algorithm, it is defined as shown in Algorithm 8. Most of the steps in this algorithm are similar to those explained in the previous subsection for the EDA_BRA1 algorithm, for this reason, only the steps that are different between the two algorithms will be explained in detail.

Particularly, step 9 is one of those that characterizes each of the algorithms in the family of algorithms with the treatment of restrictions in the probabilistic model. In the case of UMDA_BRA8, the distribution_UMDA_learning method is implemented as shown in Algorithm 8, where the univariate marginal distribution of the combinations is learned: tasks-start dates, tasks-modes and improvement alternatives.

The method of generating the new solutions in schedule_new_solutions in step 11 of the algorithm is implemented similarly in the EDA_BRA1 algorithm, but considering the learned probabilistic model and the univariate marginal approach.

**Algorithm 8. Pseudocode of UMDA_BRA8 Algorithm**

Input:

objectives of the problem to be optimized, for example, time and cost;

*max_iterations:* maximum number of iterations.

*error_threshold:* Optional parameter for a threshold-based stop condition.

constraints: restrictions on the problem in question

1. *indicators = {g: max_iterations, threshold: error_threshold}*
2. *learn_centrality =Identify relevant information from restrictions*
3. *popul = initial_population(constraints)*
4. *popul = evaluation(popul, constraints, objectives)*
5. *while not stop_condition(popul, indicators)*
6. *sel_size = calculate_selection_size (popul)*
7. *S = schedule_selection(popul, sel_size)*
8. *g = indicators[:g]*
9. $p_e^S(x,g)$ *= distribution_UMDA_learn(popul, S, constraints, learn_centrality)*
10. *new_popul = schedule_sow_elitism(popul, S)*
11. *new_popul = schedule_new_solutions(popul, new_popul,* $p_e^S(x,g)$*, constraints)*
12. *new_popul = Improve_solutions(new_popul,* $p_e^S(x,g)$*)*
13. *popul = evaluation(new_popul, constraints, objectives)*
14. *indicators[:g] += 1*
15. *end*
16. *return popul*

In step 3 of Algorithm 9, zero is assigned to the probabilities of the value options that do not comply with the precedence restrictions of the problem in question.

**Algorithm 9. Algorithm Learning the probabilistic model distribution_ UMDA_learn (popul, sel_popul, constraints, learn_centrality, g)**

Inputs:

*popul*: current population

*sel_popul*: selected population.

*constraints*: constraints in the optimization problem.

*learn_centrality*: relevant information from constraints.

$g$: iteration.

1. $p^s(x, g) = \{\}$

   $p^s(x, g) \Leftarrow$ *Learning probabilistic model of relationships between tasks-start dates*

   $p_f^S$ *mismo día* $(x_1, x_{2,}, \ldots, x_n, g)$ *and tasks-modes* $p_m^S(x_1, x_{2,}, \ldots, x_n, g)$

   $p_f^S(x_1, x_{2,}, \ldots, x_n, g) = \prod_{i=1}^{n} p_f(x_i)$ , $p_m^S(x_1, x_{2,}, \ldots, x_n, g) = \prod_{i=1}^{n} p_m(x_i)$

   $p_e^S(x, g) \Leftarrow$ *Assign zero to the probabilities of non-compliant situations precedence constraints on*

   $p^s(x, t)$ *dado* $p_f^S(x_1, x_{2,}, \ldots, x_n, g)$

   $p_e^S(x, g) \Leftarrow$ *learning the best combinations of task sets-modes that are executed on the same day*

   $p_e^S(x, g) \Leftarrow$ *Learning about alternative solutions improvement, which influences in the best use of resources.*

2. *return* $p_e^S(x, g)$

Using the same variables and notation from the complexity analysis of FDA_BRA6, the computational complexity of UMDA_BRA8 can be calculated as $O(g \cdot jobs \cdot S)$.

### 3.4 Scope and Limitations of the Proposed Algorithms

The complexity of the EDA_BRA1 algorithm family depends fundamentally on the complexity of learning the probabilistic model and the objective function. The proposal made seeks to reduce the computational cost of the search process, following two fundamental ideas:

- Reduce the need for evaluation whenever possible, inspired by a partial evaluation approach.
- Reduce the cost of evaluating algorithms by using simpler probability models that allow estimating the interrelationship between variables.

In the first case, elitism is exploited and in the evaluation processes, those individuals who have not suffered changes from one generation to the next are not re-evaluated. In the second case, it is about being able to model complex problems such as the MMRCPSP from simpler models that simplify the computational complexity.

Table 1 summarizes the complexity of the UMDA_BRA8 and FDA_BRA6 algorithms.

The complexity of FDA_BRA6 is greater than that of UMDA_BRA8, however, FDA_BRA6, by its nature, can better model the restrictions of the problem in question from the probabilistic model itself. A novel element is that the algorithms do not learn a single function but rather a set of probabilistic distribution functions that describe the dependency relationships between variables, the task-start date relationships, and the task-modes relationships.

In the proposal, the FDA_BRA6 factorization is associated with the dependency constraints between the tasks. The simplest case of factorization is that based on logical precedence restrictions between project tasks. Each factorization is made up of the variables that identify the tasks and their direct predecessors.

The UMDA_BRA8 is inspired by the UMDA [51]. The UMDA, despite assuming that all variables are independent, has shown excellent performance in many problems where the variables are interrelated [52]; by its nature, the UMDA_BRA8 algorithm should have similar behavior.

These algorithms are designed to be applied to problems with the presence of constraints. In their current design they do not take into account the existence of task priorities, an element that can be addressed in future research.

**Table 1** Comparison of the complexity of algorithms

| Algorithm | Complexity |
|---|---|
| FDA_BRA6 | $O(g \cdot l \cdot \gamma^{\lvert \hat{c} \rvert} \cdot S)$ |
| UMDA_BRA8 | $O(g \cdot jobs \cdot S)$ |

## 4 Results Analysis

The algorithms are compared, considering the variable effectiveness of project planning in environments with limited resources and tasks that are performed in multiple modes. The algorithms are compared considering the following dimensions: effectiveness in the face of variations in the number of resources, behavior in the face of variations in the number of modes, stability of the algorithms in the face of variations in the number of tasks, and overall performance of the algorithms.

In the first three dimensions, the following set of indicators will be used equally:

- Average duration time: it is desired to minimize the average duration time of the project considering 20 runs. It is denoted by Average_Time_Duration, see Eq. (1).
- Minimum duration of the project: you want to minimize this indicator and it is denoted by Minimum_Time_Duration. This indicator appears in the most frequently consulted bibliography. Considering this fact, it will be used in this research for comparison with different works.
- Number of times the optimal value is found: we want to maximize this indicator and the optimal value is known for each of the database instances used in the experimentation. It is denoted by Quantity_Times_Optimal.

In the evaluation of the dimension "Overall performance of the algorithms" the indicators are used: Average_Time_Duration and Quantity_Times_Optimal.

"Execution time" indicator, which represents the execution time of the algorithms in seconds.

The following comparisons are made using the databases of the PSPLib library [53], as it's shown in the following:

Test 1. Test with databases with 16 tasks, three modes, two renewable resources and two non-renewable resources (databases: c5_12, c5_32).

Test 2. Analysis of the influence of variations in non-renewable resources. It is tested with databases of 16 tasks in three modes, with two renewable resources (11 instances of database n0 from n0_38_1 to n0_48_1) and with three non-renewable resources (databases n3_12 and n3_32).

Test 3. Analysis of the influence of variations in the amount of renewable resources. It is tested with databases of 16 tasks, three modes for renewable resources (databases r4_12 and r4_32), and five renewable resources (databases r5_12 and r5_32).

Test 4. Analysis of the dimension "Effectiveness of algorithms in the face of variations in the number of modes". It is tested with databases of 16 tasks, two non-renewable resources, two renewable resources, tasks with four execution modes (databases: m4_12 and m4_32), and tasks with five execution modes (databases: m5_12 and m5_32).

Test 5. Analysis of the dimension "Effectiveness of algorithms in the face of variations in the number of tasks". It is tested with databases of 30 tasks, three modes, two non-renewable resources and two renewable resources (databases j30_12 and j30_32). Analysis of the global performance of the proposed algorithms on the different PSPLib databases selected for experimentation with respect to the indicators Average_Time_Duration and Quantity_Optimum_Times.

As the first step of the experimentation for each indicator and in each database, an exploratory analysis is applied to each one, which is complemented with a normality test using the Shapiro-Wilk tests [54–56] and Kolmogorov-Smirnov [57, 58]. In cases where the results conform to a normal distribution, the algorithms are compared using the T-Student test for two related samples. In the event that the normality of the data is not tested, the non-parametric tests are applied in the comparison: Friedman for multiple related samples and Wilcoxon for two related samples. In all cases, we work with a significance level of 0.05 and confidence intervals of 99%. As a result of the comparisons in each database, the algorithms are organized into groups in such a way that.

The algorithms in the same group do not present significant differences between them. While the algorithms in different groups present significantly different results. The algorithms in "Group A" report better results than the algorithms in "Group B," such that "Group A results" > "Group B results" > "Group C results," and so on.

### *4.1 Parameters of the Algorithms Used in the Experimentation*

Table 2 presents the different algorithms used in the experimentation, as well as the parameters used in the proposed algorithms.

Table legend:

- Algorithms proposed by the author in the research.
- NR parameter refers to the fact that only the individuals of the current iteration are considered when learning the improvement alternatives.
- Parameter R indicates that when learning the improvement alternatives, the individuals from the current iteration and previous iterations are considered.
- The configuration regarding the optimization objectives is represented using the notation/T for time while/MO for the cost and time objectives.

Both the proposed algorithms of the UMDA_BRA8 and FDA_BRA6 families and the traditional UMDA and FDA were run with the following parameters:

- In each of the runs, fixed-size populations of 150 individuals are used.
- A maximum number of iterations equal to 100 was set as a stopping condition.
- In the selection processes of the implemented EDAs, truncation and selection by ranking of 30% of the population are applied.
- In the generation of the new population, elitism is applied and 70% of the population is generated from the learned distributions.
- To mitigate the bias in the experimentation, 20 runs of each algorithm are carried out in each instance of the databases, making a total of 36,240 runs.

**Table 2** Set of algorithms used in the experimentation

| | Opotimization approach | | Estimation distribution (Improve strategy) | | |
|---|---|---|---|---|---|
| Algorithms | Time | Multiobjetive | NR | R | Complete name or reference |
| FDA_BRA6/T.NR * | X | | X | | FDA_BRA6_just_ time_ no_ reinforcement |
| FDA_BRA6/T.R * | X | | | X | FDA_BRA6_just_ time_ reinforcement |
| FDA_BRA6/ MO.NR * | | X | X | | FDA_BRA6_ multiobjective_no_ reinforcement |
| FDA_BRA6/ MO.R * | | X | | X | FDA_BRA6_ multiobjective_ reinforcement |
| UMDA_BRA8/ T.NR * | X | | X | | UMDA_BRA8_just_ time_no_ reinforcement |
| UMDA_BRA8/ T.R * | X | | | X | UMDA_BRA8_just_ time_reinforcement |
| UMDA_BRA8/ MO.NR * | | X | X | | UMDA_BRA8_ multiobjective_no_ reinforcement |
| UMDA_BRA8/ MO.R * | | X | | X | UMDA_BRA8_ multiobjective_ reinforcement |
| TRAD_FDA/T | X | | | | TRAD_FDA_just_ time_none |
| TRAD_FDA/MO | | X | | | TRAD_FDA_ multiobjective_none |
| TRAD_UMDA/T | X | | | | TRAD_UMDA_just_ time_none |
| TRAD_UMDA/ MO | | X | | | TRAD_UMDA_ multiobjective _none |
| B&B | X | | | | [59] |
| $GA_{Coe\text{-}Van}$ | X | | | | [60] |
| $GA_{hart}$ | X | | | | [61] |
| CP-SAT | X | | | | [62] |
| MILP | X | | | | [63] |
| hGMEDA | X | | | | [64] |
| Robust hGMEDA | X | | | | [27] |
| SPEA2 | X | | | | [65] |

- All runs were carried out using an Intel Core i3 Microprocessor with 1.7 GHz, 4 cores and 8 GB of RAM.

The reported data of the B&B, GACoe-Van, GAhart, CP-SAT, MILP, hGMEDA, Robust hGMEDA and SPEA2 algorithms were taken from the referenced articles and from the PSPLib repository [53]. The results of these algorithms are not available in all the databases used in the experimentation, therefore, in the comparisons, only the algorithms with results reported in said instances participate in each database instance.

## 4.2 *Test 1. With Databases with 16 Tasks, Three Modes, Two Renewable Resources and Two Non-renewable Resources*

### 4.2.1 Analysis of the "Average_Time_Duration" Indicator

Table 3 shows the results of comparisons of the algorithms with respect to the "Average_Time_Duration" indicator. In the case of the c15_12 database, the results obey a normal distribution. It is demonstrated that the proposed algorithms based on UMDA_BRA8 located in "Group A" report significantly better results than the traditional algorithms and the proposed FDA_BRA6 variants. In turn, the algorithms based on FDA_BRA6 "Group B" reported better results than traditional algorithms that do not address the constraints in the probabilistic model.

There are no statistically significant differences between the proposed algorithms regarding the optimization approach or the learning of the probabilities of the individual generation strategies.

In the case of the c15_32 database, it has been identified that the results do not follow a normal distribution. The best results correspond to the proposed algorithms based on UMDA_BRA8 "Group A", without distinction between its different variants and those based on FDA_BRA6, but in the "just_time" variant.

In "Group B," and with significant differences from the first group, the FDA_BRA6/MO algorithms were located, representing FDA_BRA6 with a multi-objective optimization approach. The proposed algorithms obtain better results than traditional algorithms.

### 4.2.2 Analysis of the "Minimum_Time_Duration" Indicator

It is identified that the results obtained in the two databases follow a normal distribution, so parametric tests are applied. It is shown that there are no significant differences between the best results reported in the literature [59], which is denoted in the comparison as BestPSPLib, with the algorithms based on UMDA_BRA8 and FDA_BRA6 located in "Group A".

**Table 3** Comparison of algorithms, "Average_Time_Duration" indicator, c15_12 and c15_32 databases

| c15_12/algorithms T-test two related samples | sig. | c15_32/algorithms comparing all algorithms | sig. |
|---|---|---|---|
| (Group A) | | Friedman test (Group A) | |
| UMDA_BRA8/T.NR - UMDA_BRA8/T.R | 1.000<br>0.057<br>**0.042** | UMDA_BRA8/T.NR - UMDA_BRA8/T.R | 0.064 |
| UMDA_BRA8/T.NR - UMDA_BRA8/MO.NR | | – FDA_BRA6/T.R - UMDA_BRA8/MO.R | |
| UMDA_BRA8/T.NR - UMDA_BRA8/MO.R | 0.790 | – UMDA_BRA8/MO.NR - FDA_BRA6/T.NR | |
| (Group B) | 0.708<br>0.537<br>0.494 | Wilcoxon test | |
| UMDA_BRA8/MO.R - FDA_BRA6/MO.NR | **0.000** | UMDA_BRA8/T.NR - FDA_BRA6/MO.R | **0.018** |
| UMDA_BRA8/MO.R - FDA_BRA6/T.R | | (Group B) | 0.753 |
| UMDA_BRA8/MO.R - FDA_BRA6/MO.R | 0.076<br>0.076 | FDA_BRA6/MO.R - FDA_BRA6/MO.NR | **0.005** |
| UMDA_BRA8/MO.R - FDA_BRA6/T.NR | **0.000** | FDA_BRA6/MO.R - TRAD_FDA/T | 0.721 |
| UMDA_BRA8/MO.R - TRAD_UMDA/T | | (Group C) | **0.036** |
| | | TRAD_FDA/T - TRAD_UMDA/T | |
| (Group C) | | TRAD_FDA/T - TRAD_FDA/MO | **0.005** |
| TRAD_UMDA/T - TRAD_FDA/T | | (Group D) | |
| TRAD_UMDA/T - TRAD_FDA/MO | | TRAD_FDA/MO - TRAD_UMDA/MO (Group E) | |
| TRAD_UMDA/T - TRAD_UMDA/MO (Group D) | | | |

Regarding the behavior of the different variants of the UMDA_BRA8 and FDA_BRA6 algorithms, no significant differences were found between the different learning strategies or with respect to the optimization approaches.

The results of the T-test confirm that the proposed algorithms present statistically significant differences compared to traditional approaches without treatment of restrictions in the probabilistic model, see Table 4.

**Table 4** Comparison of proposed algorithms regarding the "Minimum_Time_Duration" indicator in the c15_12 and c15_32 databases

| c15_12/algorithms T-test two related samples | sig. | c15_32/algorithms T-test two related samples | sig. |
|---|---|---|---|
| (Group A) | | (Group A)<br>UMDA_BRA8/T.NR - UMDA_BRA8/MO.NR<br>UMDA_BRA8/T.NR - BestPSPLib<br>UMDA_BRA8/T.NR - UMDA_BRA8/T.NR | |
| BestPSPLib - UMDA_BRA8/T.NR | 0.168 | UMDA_BRA8/T.NR - FDA_BRA6/T.NR | 1.000 |
| BestPSPLib - UMDA_BRA8/T.R | 0.168 | UMDA_BRA8/T.NR - UMDA_BRA8/MO.R | 1.000 |
| BestPSPLib - FDA_BRA6/T.R | 0.193 | UMDA_BRA8/T.NR - FDA_BRA6/MO.R | 1.000 |
| BestPSPLib - FDA_BRA6/MO.NR | 0.104 | UMDA_BRA8/T.NR - FDA_BRA6/T.R | 0.343 |
| BestPSPLib - FDA_BRA6/MO.R | 0.104 | UMDA_BRA8/T.NR - FDA_BRA6/MO.NR | 0.343 |
| BestPSPLib - UMDA_BRA8/MO.NR | 0.104 | UMDA_BRA8/T.NR - UMDA_BRA8/T.R | 0.343 |
| BestPSPLib - UMDA_BRA8/MO.R | 0.104 | UMDA_BRA8/T.NR - TRAD_UMDA/T | 0.193 |
| BestPSPLib - FDA_BRA6/T.NR | **0.024** | | 0.193 |
| (Group B) | | | 0.193 |
| FDA_BRA6/T.NR - TRAD_UMDA/T | **0.012** | | **0.001** |
| (Group C) | | (Group B) | |
| TRAD_UMDA/T - TRAD_FDA/T | **0.009** | TRAD_UMDA/T - TRAD_FDA/T | 0.081 |
| (Group D) | | TRAD_UMDA/T - TRAD_FDA/MO | **0.044** |
| TRAD_FDA/T - TRAD_FDA/MO | 1.000 | (Group C) | |
| TRAD_FDA/T - TRAD_UMDA/MO | 0.678 | TRAD_FDA/MO - TRAD_UMDA/MO | 0.343 |

### 4.2.3 Analysis of the "Quantity_Times_Optimal" Indicator

Table 5 reports the results obtained regarding this indicator in the two databases; They do not obey a normal distribution. For c15_12, it is shown that there are no significant differences between the proposed Algorithms UMDA_BRA8 and FDA_BRA6 located in "Group A". Regarding the c15_32 database, it is confirmed that the Algorithms (UMDA_BRA8/T.NR, UMDA_BRA8/T.R and FDA_BRA6/T.NR) report the best results.

It can be seen that there are significant differences between the different variants of the optimization approach, with the "just_time" approach reporting better results

**Table 5** Comparison of algorithms, "Optimal_Quantity_Times" indicator in c15_12 and c15_32

| c15_12/algorithms | sig. | c15_32/algorithms | sig. |
|---|---|---|---|
| Friedman test (Group A) | | Friedman test (Group A) | |
| UMDA_BRA8/T.R - UMDA_BRA8/T.NR - UMDA_BRA8/MO.NR - UMDA_BRA8/MO.R - FDA_BRA6/MO.NR - FDA_BRA6/T.R - FDA_BRA6/MO.R | 0.051 | UMDA_BRA8/T.NR - UMDA_BRA8/T.R - FDA_BRA6/T.NR | 0.161 |
| Wilcoxon test | | Wilcoxon test | |
| UMDA_BRA8/T.R - FDA_BRA6/T.NR | **0.018** | UMDA_BRA8/T.NR - FDA_BRA6/T.R | **0.018** |
| (Group B) | | Friedman test (Group B) | |
| FDA_BRA6/T.NR - TRAD_FDA/MO | 0.066 | FDA_BRA6/T.R - UMDA_BRA8/MO.R - UMDA_BRA8/MO.NR - FDA_BRA6/MO.NR | 0.797 |
| Friedman test (Group C) | 0.368 | Friedman test (Group C) | 1.000 |
| TRAD_FDA/T - TRAD_UMDA/T - TRAD_UMDA/MO | | TRAD_FDA/T - TRAD_UMDA/T - TRAD_UMDA/MO - TRAD_FDA/MO | |

than the multi-objective approach. In both databases it is shown that the proposed Algorithms obtain better results than the traditional variants.

## 4.3 *Test 2. Analysis of the Influence of Variations in Non-renewable Resources (Databases: n0, n3_12 and n3_32)*

### 4.3.1 Analysis of the "Average_Time_Duration" Indicator

Regarding this indicator, in databases n0 and n3_32, it is identified that the results do not comply with the normal distribution, unlike n3_12, where the data do obey a normal distribution. It is observed that in the three DBs, the algorithms UMDA_BRA8/T.NR and UMDA_BRA8/T.R reports the best results, see Tables 6 and 7.

It is also observed that the Algorithms proposed in the research obtain better results than the traditional variants of FDA and UMDA. In particular, in database n3_12 it is shown that there are no significant differences between the different variants of UMDA_BRA8 and that these obtain better results than the rest of the algorithms.

In the bibliography consulted [27, 64] results are reported from the application of the Robust Algorithms hGMEDA, hGMEDA and SPEA2 in DB n0, in the instances: n041_1, n042_1, n043_1, n044_1 and n045_1.

**Table 6** Comparison of algorithms, "Average_Time_Duration" indicator, database n0

| n0 "Average_Time_Duration"/comparing all algorithms | sig. |
|---|---|
| *Wilcoxon test (Group A)* | |
| UMDA_BRA8/T.R - UMDA_BRA8/T.NR | 0.798 |
| *Wilcoxon test* | |
| UMDA_BRA8/T.R - FDA_BRA6/T.R | **0.007** |
| *Friedman test (Group B)* | |
| FDA_BRA6/T.R - UMDA_BRA8/MO.R - FDA_BRA6/T.NR - FDA_BRA6/MO.R - UMDA_BRA8/MO.NR - FDA_BRA6/MO.NR | 0.309 |
| *Wilcoxon test (Group C)* | |
| TRAD_UMDA/T - TRAD_FDA/T | **0.032** |
| *Wilcoxon test (Group D)* | |
| TRAD_FDA/T - TRAD_FDA/MO | 0.122 |
| *Wilcoxon test (Group E)* | |
| TRAD_FDA/T - TRAD_UMDA/MO | **0.008** |

### 4.3.2 Analysis of the "Minimum_Time_Duration" Indicator

Regarding the normality of the data, this indicator behaves similarly to the "Average_Time_Duration" indicator in the three databases.

It is observed that the proposed variants do not have significant differences from the best results reported in the literature [59], which is denoted in the comparison as BestPSPLib, while the proposals made in the research do report significantly better performance than the traditional variants used in experimentation. The algorithms UMDA_BRA8/T.NR, UMDA_BRA8/TR report the best results, see Tables 8 and 9.

### 4.3.3 Analysis of the "Quantity_Times_Optimal" Indicator

Regarding this indicator, it is identified that the results in the three databases do not fit the normal distribution and that the UMDA_BRA8/T.NR and UMDA_BRA8/TR algorithms obtain the best results. In particular, in BD n3_12, no significant differences are observed between the different variants of UMDA_BRA8 and FDA_BRA6, while these do obtain significantly better results than the rest of the algorithms (see Tables 10 and 11).

**Table 7** Comparison of algorithms, indicator "Average_Time_Duration", BD n3_12 and n3_32

| n3_12 "Average_Time_Duration"/algorithms T-test two related samples | sig. | n3_32 "Average_Time_Duration"/comparing all algorithms | sig. |
|---|---|---|---|
| (Group A) | | Wilcoxon test (Group A) | |
| UMDA_BRA8/T.NR - UMDA_BRA8/T.R | 0.168 | UMDA_BRA8/T.NR - UMDA_BRA8/T.R | 0.705 |
| UMDA_BRA8/T.NR - UMDA_BRA8/MO.NR | 0.154 | Wilcoxon test | |
| UMDA_BRA8/T.NR - UMDA_BRA8/MO.R | 0.052 | UMDA_BRA8/T.NR - FDA_BRA6/T.R | **0.026** |
| UMDA_BRA8/T.NR - FDA_BRA6/T.R | **0.002** | Friedman test (Group B) | |
| (Group B) | | FDA_BRA6/T.R - FDA_BRA6/T.NR | 0.503 |
| FDA_BRA6/T.R - FDA_BRA6/T.NR | 0.915 | - FDA_BRA6/MO.R - UMDA_BRA8/MO.NR | |
| FDA_BRA6/T.R - FDA_BRA6/MO.R | 0.390 | - FDA_BRA6/MO.NR - UMDA_BRA8/MO.R | |
| FDA_BRA6/T.R - FDA_BRA6/MO.NR | 0.126 | Wilcoxon test | |
| FDA_BRA6/T.R - TRAD_UMDA/T | **0.000** | FDA_BRA6/T.R - TRAD_FDA/T | **0.005** |
| TRAD_UMDA/T - TRAD_FDA/T (Group C) | **0.040** | Friedman test (Group C) | |
| (Group D) | | TRAD_FDA/T - TRAD_FDA/MO | 0.741 |
| TRAD_FDA/T - TRAD_FDA/MO | 0.299 | - TRAD_UMDA/T | |
| (Group E) | | Wilcoxon test (Group D) | |
| TRAD_FDA/T - TRAD_UMDA/MO | **0.010** | TRAD_FDA/T - TRAD_UMDA/MO | **0.007** |

## 4.4 *Test 3. Analysis of the Influence of Variations in the Number of Renewable Resources (Databases: R4_12, R4_32, R5_12 and R5_32)*

### 4.4.1 Analysis of the "Average_Time_Duration" Indicator

Regarding this indicator, it is identified that the results obtained in the four databases follow a normal distribution and that, in three of them, the algorithms are UMDA_BRA8/T. R and UMDA_BRA8/T.NR report better performance than the rest. No significant differences are detected regarding the learning strategy used. In the case

**Table 8** Comparison of algorithms, "Minimum_Time_Duration" indicator in database n0

| n0 "Minimum_Time_Duration"/algorithms | | | |
|---|---|---|---|
| Friedman test (Group A) | sig. | Wilcoxon test | sig. |
| BestPSPLib - FDA_BRA6/T.NR - UMDA_BRA8/T.NR - UMDA_BRA8/T.R | 0.057 | BestPSPLib - UMDA_BRA8/MO.NR | 0.066 |
| Wilcoxon test | | BestPSPLib - TRAD_UMDA/T | **0.003** |
| BestPSPLib - FDA_BRA6/MO.R | 0.066 | TRAD_UMDA/T - TRAD_FDA/T (Group B) | **0.007** |
| BestPSPLib - UMDA_BRA8/MO.R | 0.066 | Wilcoxon test (Group C) | |
| BestPSPLib - FDA_BRA6/MO.NR | 0.068 | TRAD_FDA/T - TRAD_FDA/MO | 0.317 |
| BestPSPLib - FDA_BRA6/T.R | 0.066 | TRAD_FDA/T - TRAD_UMDA/MO | 0.317 |

**Table 9** Comparison of algorithms, "Minimum_Time_Duration" indicator in DBs n3_12 and n3_32

| n3_12 "Minimum_Time_Duration"/algorithms T-test two related samples | sig. | n3_32 "Minimum_Time_Duration"/algorithms | sig. |
|---|---|---|---|
| (Group A) | | Friedman test (Group A) | |
| BestPSPLib - UMDA_BRA8/MO.NR | 0.104 | FDA_BRA6/T.NR - FDA_BRA6/T.R | 1.000 |
| BestPSPLib - UMDA_BRA8/T.NR | 0.052 | - FDA_BRA6/MO.NR - FDA_BRA6/MO.R | |
| BestPSPLib - UMDA_BRA8/T.R | 0.052 | - UMDA_BRA8/T.NR - UMDA_BRA8/T.R | |
| BestPSPLib - FDA_BRA6/T.NR | 0.051 | - UMDA_BRA8/MO.NR - UMDA_BRA8/MO.R | |
| BestPSPLib – FDA_BRA6/MO.R | 0.066 | - BestPSPLib | |
| BestPSPLib - FDA_BRA6/T.R | **0.029** | Wilcoxon test | |
| FDA_BRA6/T.R - TRAD_UMDA/T (Group B) | **0.016** | FDA_BRA6/T.NR - TRAD_UMDA/T | 0.005 |
| TRAD_UMDA/T - TRAD_FDA/MO (Group C) | **0.003** | Friedman test (Group B) | |
| (Group D) | | TRAD_UMDA/T - TRAD_FDA/T - TRAD_FDA/MO | 0.289 |
| TRAD_FDA/MO - TRAD_FDA/T | 0.168 | Wilcoxon test (Group C) | |
| TRAD_FDA/MO - TRAD_UMDA/MO | 0.138 | TRAD_UMDA/T - TRAD_UMDA/MO | 0.018 |

**Table 10** Comparison of algorithms, "Optimal_Quantity_Times" indicator at n0

| n0/algorithms | sig. |
|---|---|
| *Wilcoxon test (Group A)* | |
| UMDA_BRA8/T.NR - UMDA_BRA8/T.R | 0.197 |
| UMDA_BRA8/T.NR - FDA_BRA6/T.R | **0.027** |
| *Friedman test (Group B)* | |
| FDA_BRA6/T.R - FDA_BRA6/MO.R - FDA_BRA6/T.NR - FDA_BRA6/MO.NR - UMDA_BRA8/MO.R - UMDA_BRA8/MO.NR | 0.721 |
| TRAD_FDA/T - TRAD_FDA/MO - TRAD_UMDA/T - TRAD_UMDA/MO | 1.000 |

**Table 11** Comparison of algorithms "Optimal_Quantity_Times" indicator in n3_12 and n3_32

| n3_12/algorithms | sig. | n3_32/algorithms | sig. |
|---|---|---|---|
| Friedman test (Group A) | | Wilcoxon test (Group A) | |
| UMDA_BRA8/T.NR - UMDA_BRA8/T.R | 0.076 | UMDA_BRA8/T.NR - UMDA_BRA8/T.R | 0.480 |
| - UMDA_BRA8/MO.NR - FDA_BRA6/T.NR | | UMDA_BRA8/T.NR - FDA_BRA6/T.R | **0.026** |
| - UMDA_BRA8/MO.R - FDA_BRA6/MO.R | | Friedman test (Group B) | |
| - FDA_BRA6/T.R - FDA_BRA6/MO.NR | | FDA_BRA6/T.R - FDA_BRA6/T.NR | 0.800 |
| Wilcoxon test | | - FDA_BRA6/MO.R - UMDA_BRA8/MO.R | |
| UMDA_BRA8/T.NR - TRAD_UMDA/T | **0.043** | - UMDA_BRA8/MO.NR - FDA_BRA6/MO.NR | |
| Friedman test (Group B) | | Wilcoxon test | |
| TRAD_UMDA/T - TRAD_UMDA/MO | 0.112 | FDA_BRA6/T.R - TRAD_UMDA/MO | **0.005** |
| - TRAD_FDA/MO - TRAD_FDA/T | | Friedman test Group C) | |
| | | TRAD_UMDA/MO - TRAD_FDA/T | 1.000 |
| | | - TRAD_FDA/MO - TRAD_UMDA/T | |

of the different variants of the FDA_BRA6 algorithm, no significant differences are observed with respect to the optimization approach. In all the experiments, all the variants of the algorithms proposed in the research obtained better results than the traditional variants of FDA and UMDA (see Tables 12 and 13).

**Table 12** Comparison of algorithms, "Average_Time_Duration" indicator, databases r4_12 and r4_32

| r4_12 "Average_Time_Duration"/ comparing all algorithms T-test two related samples | sig. | r4_32 "Average_Time_Duration"/ comparing all algorithms T-test two related samples | sig. |
|---|---|---|---|
| (Group A) | | (Group A) | |
| UMDA_BRA8/T.NR - UMDA_BRA8/T.R | **0.039** | UMDA_BRA8/T.R - UMDA_BRA8/T.NR | 0.276 |
| (Group B) | | UMDA_BRA8/T.R - FDA_BRA6/T.R | **0.011** |
| UMDA_BRA8/T.R - UMDA_BRA8/MO.NR | 0.154 | (Group B) | |
| UMDA_BRA8/T.R - UMDA_BRA8/MO.R | 0.177 | FDA_BRA6/T.R - FDA_BRA6/T.NR | 0.264 |
| UMDA_BRA8/T.R - FDA_BRA6/T.NR | **0.028** | FDA_BRA6/T.R - UMDA_BRA8/MO.NR | 0.346 |
| (Group C) | | FDA_BRA6/T.R - UMDA_BRA8/MO.R | 0.222 |
| FDA_BRA6/T.NR - FDA_BRA6/T.R | 0.293 | FDA_BRA6/T.R - FDA_BRA6/MO.R | **0.030** |
| FDA_BRA6/T.NR - FDA_BRA6/MO.R | 0.153 | (Group C) | |
| FDA_BRA6/T.NR - FDA_BRA6/MO.NR | **0.018** | FDA_BRA6/MO.R - FDA_BRA6/MO.NR | 0.370 |
| (Group D) | | FDA_BRA6/MO.R - TRAD_FDA/T | **0.000** |
| FDA_BRA6/MO.NR - TRAD_UMDA/T | **0.000** | (Group D) | |
| TRAD_UMDA/T - TRAD_FDA/T (Group E) | **0.010** | TRAD_FDA/T - TRAD_UMDA/T | 0.458 |
| TRAD_FDA/T - TRAD_FDA/MO (Group F) | 0.067 | TRAD_FDA/T - TRAD_FDA/MO (Group E) | **0.001** |
| TRAD_FDA/T - TRAD_UMDA/MO (Group G) | **0.002** | TRAD_FDA/MO - TRAD_UMDA/MO (Group F) | **0.000** |

### 4.4.2 Analysis of the "Minimum_Time_Duration" Indicator

The results regarding this indicator have a behavior according to the normal distribution, similar to the "Average_Time_Duration" indicator. There are no statistically significant differences between the proposed algorithms and the best results reported in the literature [59], which are denoted in the comparison as BestPSPLib, except in the r4_12 database. The results of the T-Student test confirm that the proposed algorithms achieve better performance than the traditional implementations of UMDA and FDA in the four databases, see Tables 14 and 15.

**Table 13** Comparison of algorithms, "Average_Time_Duration", BD r5_12 and r5_32

| r5_12 "Average_Time_Duration"/ comparing all algorithms T-test two related samples | sig. | r5_32 "Average_Time_Duration"/ comparing all algorithms T-test two related samples | Sig. |
|---|---|---|---|
| (Group A) | | (Group A) | |
| UMDA_BRA8/T.NR - UMDA_BRA8/T.R | 0.726 | UMDA_BRA8/T.NR - UMDA_BRA8/T.R | 0.630 |
| UMDA_BRA8/T.NR - UMDA_BRA8/MO.R | **0.028** | UMDA_BRA8/T.NR - FDA_BRA6/T.NR | **0.022** |
| (Group B) | | (Group B) | |
| UMDA_BRA8/MO.R - UMDA_BRA8/MO.NR | 0.081 | FDA_BRA6/T.NR - FDA_BRA6/T.R | 0.364 |
| UMDA_BRA8/MO.R - FDA_BRA6/T.NR | 0.480 | FDA_BRA6/T.NR - UMDA_BRA8/MO.R | 0.211 |
| UMDA_BRA8/MO.R - FDA_BRA6/MO.R | 0.052 | FDA_BRA6/T.NR - FDA_BRA6/MO.NR | 0.216 |
| UMDA_BRA8/MO.R - FDA_BRA6/T.R | **0.041** | FDA_BRA6/T.NR - FDA_BRA6/MO.R | 0.052 |
| (Group C) | | FDA_BRA6/T.NR - UMDA_BRA8/MO.NR | **0.015** |
| FDA_BRA6/T.R - FDA_BRA6/MO.NR | 0.847 | (Group C) | |
| FDA_BRA6/T.R - TRAD_FDA/T | **0.000** | UMDA_BRA8/MO.NR - TRAD_FDA/T | **0.000** |
| (Group D) | | (Group D) | |
| TRAD_FDA/T - TRAD_UMDA/T | 0.729 | TRAD_FDA/T - TRAD_UMDA/T | 0.479 |
| TRAD_FDA/T - TRAD_FDA/MO | 0.088 | TRAD_FDA/T - TRAD_FDA/MO (Group E) | **0.002** |
| TRAD_FDA/T - TRAD_UMDA/MO (Group E) | **0.003** | TRAD_FDA/MO - TRAD_UMDA/MO (Group F) | **0.000** |

### 4.4.3 Analysis of the "Quantity_Times_Optimal" Indicator

Regarding the indicator "Quantity_Optimum_Times" it is identified that, in the four databases, the results do not follow a behavior according to the normal distribution. It is shown that the proposed algorithms are UMDA_BRA8/T.NR and UMDA_BRA8/T. R reported good results in the four databases. However, in most databases, no statistically significant differences were found between the UMDA_BRA8 and FDA_BRA6 variants. There are also no significant differences between the different learning strategies. Regarding the optimization approach, in 50% of the databases, significantly better results are achieved from the "just_time" approach over the "multi-objective" approach. The results explained above can be seen in Tables 16 and 17.

**Table 14** Comparison of algorithms, "Minimum_Time_Duration" in BD r4_12 and r4_32

| r4_12/algorithms T-test two related samples | sig. | r4_32/algorithms T-test two related samples | sig. |
|---|---|---|---|
| (Group A) | | (Group A) | |
| BestPSPLib - UMDA_BRA8/T.NR | **0.025** | FDA_BRA6/T.R - FDA_BRA6/MO.NR | 1.000 |
| (Group B) | | FDA_BRA6/T.R - FDA_BRA6/MO.R | 1.000 |
| UMDA_BRA8/T.NR - UMDA_BRA8/T.R | 0.343 | FDA_BRA6/T.R - UMDA_BRA8/T.R | 1.000 |
| UMDA_BRA8/T.NR - FDA_BRA6/T.NR | 0.168 | FDA_BRA6/T.R - UMDA_BRA8/MO.NR | 1.000 |
| UMDA_BRA8/T.NR - UMDA_BRA8/MO.NR | 0.168 | FDA_BRA6/T.R - BestPSPLib | 1.000 |
| UMDA_BRA8/T.NR - UMDA_BRA8/MO.R | 0.168 | FDA_BRA6/T.R - UMDA_BRA8/T.NR | 0.343 |
| UMDA_BRA8/T.NR - FDA_BRA6/T.R | 0.081 | FDA_BRA6/T.R - FDA_BRA6/T.NR | 0.168 |
| UMDA_BRA8/T.NR - FDA_BRA6/MO.NR | 0.081 | FDA_BRA6/T.R - UMDA_BRA8/MO.R | 0.193 |
| UMDA_BRA8/T.NR - FDA_BRA6/MO.R | **0.037** | FDA_BRA6/T.R - TRAD_UMDA/T | **0.000** |
| (Group C) | | (Group B) | |
| FDA_BRA6/MO.R - TRAD_UMDA/T | **0.002** | TRAD_UMDA/T - TRAD_FDA/T | **0.019** |
| (Group D) | | (Group C) | |
| TRAD_UMDA/T - TRAD_FDA/T | **0.003** | TRAD_FDA/T - TRAD_FDA/MO | 0.168 |
| (Group E) | | TRAD_FDA/T - TRAD_UMDA/MO | 0.140 |
| TRAD_FDA/T - TRAD_FDA/MO | 0.443 | | |
| TRAD_FDA/T - TRAD_UMDA/MO | 0.104 | | |

## 4.5 *Test 4. Validation of the Dependent Variable in the Dimension "Effectiveness of the Algorithms in the Face of Variations in the Number of Modes*

The influence on the behavior of the algorithms in the face of variations in the number of execution modes in the experimentation environment is analyzed. In particular, comparisons are made on the databases m4_12, m4_32, m5_12 and m5_32.

**Table 15** Comparison of algorithms, "Minimum_Time_Duration" in BD r5_12 and r5_32

| r5_12 "Minimum_Time_Duration"/ algorithms T-test two related samples | sig. | r5_32 "Minimum_Time_Duration"/ algorithms T-test two related samples | sig. |
|---|---|---|---|
| (Group A) | | (Group A) | |
| BestPSPLib - UMDA_BRA8/T.NR | 0.168 | BestPSPLib - FDA_BRA6/MO.NR | 0.343 |
| BestPSPLib - UMDA_BRA8/T.R | 0.168 | BestPSPLib - UMDA_BRA8/T.NR | 0.343 |
| BestPSPLib - FDA_BRA6/MO.R | 0.138 | BestPSPLib - FDA_BRA6/T.NR | 0.343 |
| BestPSPLib - UMDA_BRA8/MO.NR | 0.138 | BestPSPLib - FDA_BRA6/MO.R | 0.343 |
| BestPSPLib - FDA_BRA6/T.NR | 0.081 | BestPSPLib - UMDA_BRA8/T.R | 0.343 |
| BestPSPLib - FDA_BRA6/T.R | 0.081 | BestPSPLib - UMDA_BRA8/MO.NR | 0.343 |
| BestPSPLib - UMDA_BRA8/MO.R | 0.168 | BestPSPLib - UMDA_BRA8/MO.R | 0.343 |
| BestPSPLib - FDA_BRA6/MO.NR | 0.066 | BestPSPLib - FDA_BRA6/T.R | 0.343 |
| BestPSPLib - TRAD_UMDA/T | **0.002** | BestPSPLib - TRAD_UMDA/T | **0.001** |
| (Group B) | | (Group B) | |
| TRAD_UMDA/T - TRAD_FDA/MO | 0.217 | TRAD_UMDA/T - TRAD_FDA/T | **0.012** |
| TRAD_UMDA/T - TRAD_FDA/T | 0.053 | (Group C) | |
| TRAD_UMDA/T - TRAD_UMDA/ MO (Group C) | **0.037** | TRAD_FDA/T - TRAD_FDA/MO | 1.000 |
| | | TRAD_FDA/T - TRAD_UMDA/MO | 0.223 |

### 4.5.1 Analysis of the "Average_Time_Duration" Indicator

The results obtained regarding this indicator in the databases m4_12 and m4_32 show that they do not comply with a normal distribution, unlike m5_12 and m5_32, which do fit this distribution. Tables 18 and 19 show that the UMDA_BRA8/T.NR and UMDA_BRA8/TR algorithms report the best results.

It is also demonstrated that the algorithms proposed in the research, with their different variants, report results significantly better than the implementations of traditional FDA and UMDA algorithms. There are no statistically significant differences between the variants of the learning strategies. In the case of m5_12, no significant differences are detected between the UMDA_BRA8 and FDA_BRA6 variants proposed in the research. There are also no statistically significant differences in the results obtained following the optimization approaches (just_time and multi-objective).

### 4.5.2 Analysis of the "Minimum_Time_Duration" Indicator

The results regarding this indicator in the m4_12 and m4_32 databases do not comply with a normal distribution, unlike m5_12 and m5_32, which do fit this distribution. There are no statistically significant differences between the best results reported in the literature [59], which is denoted in the comparison as BestPSPLib, and

**Table 16** Comparison of Algorithms, indicator "Quantity_Times_Optimal" in BD r4_12 and r4_32

| r4_12/algorithms | sig. | r4_32/algorithms | sig. |
|---|---|---|---|
| Friedman test (Group A) | | | |
| UMDA_BRA8/T.NR - UMDA_BRA8/T.R<br>- UMDA_BRA8/MO.NR - UMDA_BRA8/MO.R<br>- FDA_BRA6/T.NR | 0.183 | Wilcoxon test (Group A)<br>UMDA_BRA8/T.NR - UMDA_BRA8/T.R<br>UMDA_BRA8/T.NR - FDA_BRA6/T.R | 0.102<br>**0.018** |
| Wilcoxon test<br>UMDA_BRA8/T.NR - FDA_BRA6/T.R<br>UMDA_BRA8/T.NR - FDA_BRA6/MO.R | 0.068<br>**0.042** | Friedman test (Group B)<br>FDA_BRA6/T.R - FDA_BRA6/T.NR<br>- UMDA_BRA8/MO.NR - UMDA_BRA8/MO.R<br>- FDA_BRA6/MO.NR - FDA_BRA6/MO.R | 0.901 |
| Friedman test (Group B)<br>FDA_BRA6/MO.R - FDA_BRA6/MO.NR | 0.581 | | |
| Wilcoxon test<br>FDA_BRA6/MO.R - TRAD_UMDA/MO | **0.042** | Wilcoxon test<br>FDA_BRA6/T.R - TRAD_UMDA/MO | **0.018** |
| Friedman test (Group C)<br>TRAD_UMDA/MO - TRAD_UMDA/T<br>- TRAD_FDA/MO - TRAD_FDA/T | 1.000 | Friedman test (Group C)<br>TRAD_UMDA/MO - TRAD_UMDA/T<br>- TRAD_FDA/MO - TRAD_FDA/T | 1.000 |

the algorithms proposed in the research. There are also no significant differences between the different optimization approaches or between the different learning strategies. The proposed algorithms achieve significantly better results than traditional implementations of the UMDA and FDA algorithms; see Tables 20 and 21.

### 4.5.3 Analysis of the "Quantity_Times_Optimal" Indicator

Regarding the "Quantity_Times_Optimal" indicator, it is identified that, in the four databases analyzed, the results do not comply with the normal distribution. Tables 22 and 23 show that the UMDA_BRA8 variants report better results than the rest of the algorithms, except in the m5_12 database, where no statistically significant differences are found between the different variants of UMDA_BRA8 and FDA_BRA6.

Something similar occurs with the different learning strategies, where no statistically significant differences are detected between the variants of algorithms UMDA_BRA8 and FDA_BRA6 in three of the four databases. The proposed algorithms report better results than traditional FDA and UMDA implementations; the latter never found the optimum in any of the 40 instances analyzed.

**Table 17** Comparison of Algorithms, indicator "Quantity_Times_Optimal" in r5_12 and r5_32

| r5_12/algorithms | sig. | r5_32/algorithms | sig. |
|---|---|---|---|
| Friedman test (Group A) UMDA_BRA8/T.NR - UMDA_BRA8/T.R - FDA_BRA6/MO.R - UMDA_BRA8/MO.R - UMDA_BRA8/MO.NR - FDA_BRA6/MO.NR | 0.066 | Friedman test (Group A) UMDA_BRA8/T.NR - UMDA_BRA8/T.R - FDA_BRA6/T.NR | 0.054 |
| Wilcoxon test UMDA_BRA8/T.NR - FDA_BRA6/T.NR | 0.027 | Wilcoxon test UMDA_BRA8/T.NR - FDA_BRA6/T.R | 0.027 |
| (Group B) FDA_BRA6/T.NR - FDA_BRA6/T.R FDA_BRA6/T.NR - TRAD_UMDA/MO | 0.285 0.042 | Friedman test (Group B) FDA_BRA6/T.R - FDA_BRA6/MO.NR - FDA_BRA6/MO.R - UMDA_BRA8/MO.R - UMDA_BRA8/MO.NR | 0.056 |
| Friedman test (Group C) TRAD_UMDA/MO - TRAD_UMDA/T - TRAD_FDA/MO - TRAD_FDA/T | 1.000 | Wilcoxon test UMDA_BRA8/MO.NR - TRAD_UMDA/MO | 0.007 |
| | | Friedman test (Group C) TRAD_UMDA/MO - TRAD_UMDA/T - TRAD_FDA/MO - TRAD_FDA/T | 1.000 |

## 4.6 Test 5. Validation of the Dependent Variable in the Dimension "Effectiveness of the Algorithms in the Face of Variations in the Number of Tasks"

The influence of variations in the number of tasks in the project on the behavior of the algorithms is analyzed. Particularly, it is tested with the databases j30_12 and j30_32, with 30 tasks, three modes, two non-renewable resources and two renewable resources on the same indicators used in the previous tests.

### 4.6.1 Analysis of the "Average_Time_Duration" Indicator

Regarding this indicator, in the j30_12 database, the results obey a normal distribution, while the results of the experimentation in the j30_32 database do not comply with this distribution. In general, algorithms based on UMDA_BRA8 report significantly better results than the proposed variants based on FDA_BRA6. It is also shown that the proposed algorithms report significantly better results than the traditional UMDA and FDA algorithms (see Table 24).

**Table 18** Comparison of algorithms, indicator "Average_Time_Duration" in BD: m4_12 and m4_32

| m4_12 "Average_Time_Duration"/ algorithms | sig. | m4_32 "Average_Time_Duration"/ algorithms | sig. |
|---|---|---|---|
| Friedman test (Group A) UMDA_BRA8/T.NR - UMDA_BRA8/T.R - UMDA_BRA8/MO.NR - UMDA_BRA8/MO.R | 0.061 | Wilcoxon test (Group A) UMDA_BRA8/T.NR - UMDA_BRA8/T.R UMDA_BRA8/T.NR - FDA_BRA6/T.R | 0.400 |
| Wilcoxon test UMDA_BRA8/T.NR - FDA_BRA6/MO.R | **0.011** | Friedman test (Group B) FDA_BRA6/T.R - UMDA_BRA8/MO.NR - UMDA_BRA8/MO.R - FDA_BRA6/T.NR - FDA_BRA6/MO.R | **0.008** |
| Friedman test (Group B) FDA_BRA6/MO.R - FDA_BRA6/T.NR - FDA_BRA6/T.R - FDA_BRA6/MO.NR | 0.709 | Wilcoxon test FDA_BRA6/T.R - FDA_BRA6/T.NR (Group C) | 0.069 |
| Wilcoxon test FDA_BRA6/MO.R - TRAD_UMDA/T (Group C) | **0.005** | FDA_BRA6/T.NR - TRAD_FDA/T (Group D) | **0.038** **0.005** |
| TRAD_UMDA/T - TRAD_FDA/T (Group D) | **0.025** | TRAD_FDA/T - TRAD_UMDA/T | 0.202 |
| TRAD_FDA/T - TRAD_FDA/MO (Group E) | **0.027** | | |
| TRAD_FDA/MO - TRAD_UMDA/MO (Group F) | **0.005** | TRAD_UMDA/T - TRAD_FDA/MO | 0.086 |

With the increase in the number of tasks, no significant differences are seen in the behavior of the different variants of the algorithms, either considering the learning strategies or with respect to changes in the optimization approaches.

### 4.6.2 Analysis of the "Minimum_Time_Duration" Indicator

It should be noted that in the j30_12 and j30_32 databases, there are a greater number of algorithms published with data available for comparison [66] that us e the Minimum_Time_Duration indicator. Therefore, this subsection takes advantage of this fact to compare the proposed algorithms with others reported in the bibliography.

*Comparison of Algorithms UMDA_BRA8 and FDA_BRA6 Configurations*

The results regarding this indicator in the j30_12 database comply with the normal distribution, so parametric tests are applied, while the results from the j30_32 database do not obey this distribution (see Table 25).

**Table 19** Comparison of algorithms, indicator "Average_Time_Duration" in BD m5_12 and m5_32

| m5_12 "Average_Time_Duration"/ algorithms T-test two related samples | sig. | m5_32 "Average_Time_Duration"/ algorithms T-test two related samples | sig. |
|---|---|---|---|
| (Group A)<br>UMDA_BRA8/T.NR - UMDA_BRA8/MO.R<br>UMDA_BRA8/T.NR - UMDA_BRA8/T.R<br>UMDA_BRA8/T.NR - UMDA_BRA8/MO.NR | | (Group A)<br>UMDA_BRA8/T.NR - UMDA_BRA8/T.R<br>UMDA_BRA8/T.NR - UMDA_BRA8/MO.R | |
| UMDA_BRA8/T.NR - FDA_BRA6/T.R | | (Group B)<br>UMDA_BRA8/MO.R - UMDA_BRA8/MO.NR<br>UMDA_BRA8/MO.R - FDA_BRA6/T.NR | 0.809<br>**0.026** |
| UMDA_BRA8/T.NR - FDA_BRA6/T.NR | 0.735<br>0.309 | (Group C)<br>FDA_BRA6/T.NR - FDA_BRA6/T.R<br>FDA_BRA6/T.NR - FDA_BRA6/MO.NR<br>FDA_BRA6/T.NR - FDA_BRA6/MO.R<br>FDA_BRA6/T.NR - TRAD_FDA/T | 0.853<br>**0.031**<br>0.867<br>0.134<br>0.104<br>**0.000**<br>0.778<br>**0.000**<br>**0.000** |
| UMDA_BRA8/T.NR - FDA_BRA6/MO.R | 0.453<br>0.354 | | |
| UMDA_BRA8/T.NR - FDA_BRA6/MO.NR | 0.152<br>0.126 | | |
| UMDA_BRA8/T.NR - TRAD_UMDA/T | 0.063<br>**0.000** | | |
| (Group B) | 0.079<br>**0.037** | | |
| TRAD_UMDA/T - TRAD_FDA/T | | | |
| TRAD_UMDA/T - TRAD_FDA/MO | **0.000** | | |
| (Group C)<br>TRAD_FDA/MO-TRAD_UMDA/MO (Group D) | | (Group D)<br>TRAD_FDA/T - TRAD_UMDA/T<br>TRAD_FDA/T - TRAD_FDA/MO<br>(Group E)<br>TRAD_FDA/MO - TRAD_UMDA/MO (Group F) | |

In both databases, no significant differences are reported between the different variants of algorithms proposed in this research. The best results correspond to the UMDA_BRA8/T.NR and FDA_BRA6/T.NR variants, respectively. These variants will be used in the comparison in the next subsection.

**Table 20** Comparison of Algorithms, indicator "Minimum_Time_Duration" in BD m4_12 and m4_32

| m4_12 "Minimum_Time_ Duration"/algorithms | sig. | m4_32 "Minimum_Time_ Duration"/algorithms | sig. |
|---|---|---|---|
| Friedman test (Group A)<br>BestPSPLib - UMDA_BRA8/T.NR - UMDA_BRA8/T.R - FDA_ BRA6/MO.R - UMDA_BRA8/MO.NR - UMDA_BRA8/MO.R - FDA_BRA6/T.NR - FDA_BRA6/ T.R - FDA_BRA6/MO.NR | 0.126 | Friedman test (Group A)<br>BestPSPLib - FDA_BRA6/T.NR - UMDA_BRA8/T.NR - FDA_ BRA6/MO.NR - UMDA_BRA8/T.R - UMDA_ BRA8/MO.NR - UMDA_BRA8/MO.R - FDA_ BRA6/T.R | 0.056 |
| Wilcoxon test<br>BestPSPLib - TRAD_UMDA/T (Group B)<br>TRAD_UMDA/T - TRAD_FDA/T | **0.007**<br>**0.016** | Wilcoxon test<br>BestPSPLib - TRAD_FDA/T (Group B)<br>TRAD_FDA/T - TRAD_ UMDA/T | **0.005**<br>**0.005** |
| Wilcoxon test (Group C)<br>TRAD_FDA/T - TRAD_FDA/MO<br>TRAD_FDA/T - TRAD_UMDA/ MO | 0.102<br>0.102 | Wilcoxon test (Group C)<br>TRAD_UMDA/T - TRAD_ FDA/MO | 0.157 |
| | | Wilcoxon test<br>TRAD_UMDA/T - TRAD_ UMDA/MO (Group D) | **0.046** |

*Comparison of the best configurations of the proposed algorithms with others reported in the bibliography*

In this subsection, the best results of the proposed algorithms UMDA_BRA8/T.NR and FDA_BRA6/T.NR are compared with others reported in the literature. In the case of the j30_12 database, the best results reported in the bibliography and that are available correspond to the algorithms GAhart [61], B&B [59], CP-SAT [62], GACoe-Van [60] and MILP [63]. In this database, the best results reported in PSPLib are distributed among the following works: the B&B algorithm [59] in the instances j3012_3, j3012_9 and j3012_10, while in the rest of the instances, the best results are reported by GAhart [61]. Consequently, the best results in PSPLib are denoted in the comparison as BestPSPLib. Normality tests are performed, and it is shown that all algorithms comply with the normal distribution. The parametric T-Student test is applied in pairs for related samples, and the results are obtained in Table 25.

In the case of the j30_32 database, the best results reported in the bibliography and that are available correspond to the algorithms: B&B [59], CP-SAT [62], GACoe-Van [60] and MILP [63]. Normality tests are carried out on this database, and it is shown that the normal distribution is not met, so non-parametric tests are applied, and the results of Table 26 are obtained.

Table 26 reveals that in the j30_12 database, the UMDA_BRA8/T.NR algorithm has no significant differences with the best results reported in the literature and is

**Table 21** Comparison of Algorithms, indicator "Minimum_Time_Duration" in BD m5_12 and m5_32

| m5_12 "Minimum_Time_Duration"/ algorithms T-test two related samples | sig. | m5_32 "Minimum_Time_Duration"/ algorithms T-test two related samples | sig. |
|---|---|---|---|
| (Group A) | | (Group A) | |
| FDA_BRA6/T.NR - UMDA_BRA8/T.NR | 1.000 | BestPSPLib - UMDA_BRA8/T.NR | 0.343 |
| FDA_BRA6/T.NR - UMDA_BRA8/T.R | 1.000 | BestPSPLib - UMDA_BRA8/T.R | 0.223 |
| FDA_BRA6/T.NR - UMDA_BRA8/MO.NR | 1.000 | BestPSPLib - UMDA_BRA8/MO.NR | 0.343 |
| FDA_BRA6/T.NR - UMDA_BRA8/MO.R | 1.000 | BestPSPLib - FDA_BRA6/T.R | 0.138 |
| FDA_BRA6/T.NR - BestPSPLib | 1.000 | BestPSPLib - FDA_BRA6/MO.NR | 0.138 |
| FDA_BRA6/T.NR - FDA_BRA6/T.R | 0.343 | BestPSPLib - FDA_BRA6/MO.R | 0.177 |
| FDA_BRA6/T.NR - FDA_BRA6/MO.NR | 0.343 | BestPSPLib - UMDA_BRA8/MO.R | 0.138 |
| FDA_BRA6/T.NR - FDA_BRA6/MO.R | 0.168 | BestPSPLib - TRAD_UMDA/T | **0.000** |
| FDA_BRA6/T.NR - TRAD_UMDA/T | **0.000** | (Group B) | |
| (Group B) | | TRAD_UMDA/T - TRAD_FDA/T | **0.001** |
| TRAD_UMDA/T - TRAD_FDA/MO | **0.001** | (Group C) | |
| (Group C) | | TRAD_FDA/T - TRAD_FDA/MO | 0.443 |
| TRAD_FDA/MO - TRAD_FDA/T | 0.343 | TRAD_FDA/T - TRAD_UMDA/MO | **0.007** |
| TRAD_FDA/MO - TRAD_UMDA/MO | 0.555 | (Group D) | |

also located in Group A. On the other hand, the proposed algorithm FDA_BRA6/T.NR is located in Group B. In the j30_32 database, the UMDA_BRA8/T.NR and FDA_BRA6/T.NR variants do not report significant differences with the best results from the literature.

On the other hand, in both databases, the proposed algorithms UMDA_BRA8/T.NR and FDA_BRA6/T.NR report significantly better results than the traditional variants based on UMDA and FDA.

### 4.6.3 Analysis of the "Quantity_Times_Optimal" Indicator

Regarding the "Quantity_Times_Optimal" indicator, it is identified that, in both databases, the results do not comply with the normal distribution. In the case of j30_12, it is shown that the UMDA_BRA8 variants report significantly better results than the rest of the algorithms, unlike the j30_32 database, where the algorithms based on UMDA_BRA8 and FDA_BRA6 do not present statistically significant differences between them (see Table 27).

**Table 22** Comparison of Algorithms, indicator "Quantity_Times_Optimal" in m4_12 and m4_32

| m4_12/algorithms | sig. | m4_32/algorithms | sig. |
|---|---|---|---|
| Friedman test (Group A)<br>UMDA_BRA8/T.NR - UMDA_BRA8/T.R<br>- UMDA_BRA8/MO.R | | | |
| Wilcoxon test<br>UMDA_BRA8/T.NR - UMDA_BRA8/MO.NR | 0.113 | Wilcoxon test (Group A)<br>UMDA_BRA8/T.NR - UMDA_BRA8/T.R<br>UMDA_BRA8/T.NR - FDA_BRA6/T.R | |
| Friedman test (Group B)<br>- FDA_BRA6/MO.R | **0.042**<br>0.062 | Friedman test (Group B)<br>FDA_BRA6/T.R - UMDA_BRA8/MO.NR<br>- FDA_BRA6/MO.R - UMDA_BRA8/MO.R<br>- FDA_BRA6/MO.NR - FDA_BRA6/T.NR | 0.102<br>**0.018** |
| Wilcoxon test<br>UMDA_BRA8/MO.NR - FDA_BRA6/MO.NR | **0.042** | | 0.901 |
| (Group C)<br>FDA_BRA6/MO.NR - FDA_BRA6/T.NR<br>FDA_BRA6/MO.NR - TRAD_UMDA/T | 0.833<br>**0.027** | Friedman test<br>TRAD_FDA/T - TRAD_FDA/MO | |
| Friedman test (Group D)<br>TRAD_UMDA/T - TRAD_UMDA/MO<br>- TRAD_FDA/MO - TRAD_FDA/T | 0.392 | - TRAD_UMDA/T - TRAD_UMDA/MO | 1.000 |

## *4.7 Test 6. Validation of the Dependent Variable in the Dimension "Global Performance of the Algorithms"*

In this section, an analysis of the behavior of the algorithms is carried out from a global perspective in all experimental databases with respect to the indicators "Average_Time_Duration" and "Quantity_Times_Optimal". In this test, two techniques are combined: the first consists of the application of Page's L Test [54, 55, 67] and the second a graphic analysis of the behavior of the average of each indicator considering all the databases.

**Procedure for applying the Page's L Test**

1. Selection of the criteria for establishing the ranking.

**Table 23** Comparison of algorithms, indicator "Quantity_Times_Optimal" in m5_12 and m5_32

| m5_12/algorithms | sig. | m5_32/algorithms | sig. |
|---|---|---|---|
| Friedman test (Group A) UMDA_BRA8/MO.R - UMDA_BRA8/T.NR - UMDA_BRA8/MO.NR - UMDA_BRA8/T.R - FDA_BRA6/T.R - FDA_BRA6/T.NR | 0.788 | Friedman test (Group A) UMDA_BRA8/T.R - UMDA_BRA8/T.NR - UMDA_BRA8/MO.R - UMDA_BRA8/MO.NR | 0.124 |
| Wilcoxon test UMDA_BRA8/MO.R - FDA_BRA6/MO.R | **0.039** | Wilcoxon test UMDA_BRA8/T.R - FDA_BRA6/T.NR | **0.027** |
| (Group B) FDA_BRA6/MO.R - FDA_BRA6/MO.NR | 0.275 | Friedman test (Group B) FDA_BRA6/T.NR - FDA_BRA6/MO.R - FDA_BRA6/T.R - FDA_BRA6/MO.NR | 0.955 |
| Friedman test TRAD_FDA/T - TRAD_FDA/MO - TRAD_UMDA/T - TRAD_UMDA/MO | 1.000 | Wilcoxon test FDA_BRA6/T.NR - TRAD_UMDA/MO | **0.028** |
| | | Friedman test (Group C) TRAD_UMDA/MO - TRAD_UMDA/T - TRAD_FDA/MO - TRAD_FDA/T | 1.000 |

2. Assignment of the ranking values in such a way that the ranking position is assigned to each algorithm in each database according to the established criteria. To assign the ranking, the following steps are followed:
    (a) Each algorithm in each database is assigned a ranking value that corresponds to the group where it was located according to the statistical tests carried out. The algorithms that were located in the first group are assigned lower positions.
    (b) If algorithms in the same database share the same ranking position, because they do not have significant differences between them, they are all assigned the same value that corresponds to the midpoint of the positions they would occupy if they had significant differences. For example, if two algorithms are in first place in a certain database, each one is assigned a value of 1.5.
3. Ordering of the analyzed measures (algorithms) based on the sums of the rankings.
4. Application of the non-parametric test Page's L Test. In all cases, the level of statistical significance is 0.05, which means that the null hypothesis will be rejected whenever the result (in this step the Statext tool is used). The null hypothesis is defined as the equality between analyzed measures (algorithms) that can be rejected in favor of an ordered alternative.

**Table 24** Comparison of Algorithms, indicator "Average_Time_Duration" in BD j30_12 and j30_32

| j30_12 "Average_Time_Duration"/ algorithms T-test two related samples | sig. | j30_32 "Average_Time_ Duration"/algorithms | sig. |
|---|---|---|---|
| (Group A)<br>UMDA_BRA8/T.NR - UMDA_BRA8/ T.R<br>UMDA_BRA8/T.NR - - UMDA_BRA8/ MO.R<br>UMDA_BRA8/T.NR - UMDA_BRA8/ MO.NR | | Wilcoxon test (Group A)<br>UMDA_BRA8/T.R - UMDA_ BRA8/T.NR | |
| | | Wilcoxon test<br>UMDA_BRA8/T.R - FDA_ BRA6/T.NR | |
| (Group B)<br>UMDA_BRA8/MO.NR - FDA_BRA6/ T.R | 0.193<br>0.106<br>**0.003** | Friedman test (Group B)<br>FDA_BRA6/T.NR - UMDA_ BRA8/MO.NR<br>- FDA_BRA6/T.R - UMDA_ BRA8/MO.R<br>- FDA_BRA6/MO.NR - FDA_ BRA6/MO.R | 0.765<br>**0.011** |
| FDA_BRA6/T.R - FDA_BRA6/T.NR<br>FDA_BRA6/T.R - FDA_BRA6/MO.R | **0.002** | | |
| (Group C)<br>FDA_BRA6/MO.R - TRAD_UMDA/T<br>(Group D) | 0.101<br>**0.032** | Wilcoxon test<br>FDA_BRA6/T.NR - TRAD_ UMDA/T | 0.301<br>**0.005** |
| TRAD_UMDA/T - TRAD_FDA/T<br>(Group E) | **0.000**<br>**0.000** | TRAD_UMDA/T - TRAD_ FDA/T (Group C) | **0.007**<br>**0.024**<br>**0.005** |
| TRAD_FDA/T - TRAD_FDA/MO<br>(Group F) | 0.260<br>**0.002** | TRAD_FDA/T - TRAD_FDA/ MO (Group D) | |
| TRAD_FDA/T - TRAD_UMDA/MO<br>(Group G) | | TRAD_FDA/MO - TRAD_ UMDA/MO (Group E) | |

(a) Calculate using the following expression:

$$L = \sum_{j=1}^{kj} j\,\mathcal{R}_j = \mathcal{R}_1 + 2\mathcal{R}_2 + \ldots + k\mathcal{R}_k \quad (10)$$

where $\mathcal{R}_j = \sum_{j=1}^{n} r_i^j$ is the sum of the ranking of each algorithm on the $n$ BD.

Taking into account that $k = 8$ and $n = 15$, it is necessary to consider a normal approximation to the value of $L$ [67], which is given by the following expression

**Table 25** Comparación respecto al indicator "Minimum_Time_Duration" in BD j30_12 and j30_32

| j30_12/algorithms T-test two related samples | sig. | j30_32/algorithms | sig. |
|---|---|---|---|
| (Group A)<br>UMDA_BRA8/T.NR - UMDA_BRA8/T.R | | | |
| UMDA_BRA8/T.NR - UMDA_BRA8/MO.R<br>UMDA_BRA8/T.NR - UMDA_BRA8/MO.NR<br>UMDA_BRA8/T.NR - FDA_BRA6/T.NR<br>UMDA_BRA8/T.NR - FDA_BRA6/T.R<br>UMDA_BRA8/T.NR - FDA_BRA6/MO.NR<br>UMDA_BRA8/T.NR - FDA_BRA6/MO.R | 1.000<br>0.443<br>0.343<br>0.279<br>0.269<br>0.279<br>0.138 | Friedman test (Group A)<br>UMDA_BRA8/T.NR - UMDA_BRA8/T.R<br>- FDA_BRA6/T.NR - FDA_BRA6/T.R<br>- FDA_BRA6/MO.NR - FDA_BRA6/MO.R<br>- UMDA_BRA8/MO.R - UMDA_BRA8/MO.NR | 0.328 |

**Table 26** Comparison of proposed algorithms with bibliography, indicator "Minimum_Time_Duration" in DBs j30_12 and j30_32

| j30_12 "Minimum_Time_Duration"/algorithms T-test two related samples | sig. | j30_32 "Minimum_Time_Duration"/algorithms | sig. |
|---|---|---|---|
| (Group A)<br>BestPSPLib - CP-SAT<br>BestPSPLib - GACoe-Van<br>BestPSPLib - MILP<br>BestPSPLib - UMDA_BRA8/T.NR | 1.000<br>1.000<br>1.000<br>0.051 | Wilcoxon test (Group A)<br>BestPSPLib - CP-SAT<br>BestPSPLib - GACoe-Van<br>BestPSPLib - MILP<br>BestPSPLib - UMDA_BRA8/T.NR<br>BestPSPLib - FDA_BRA6/T.NR | 1.000<br>1.000<br>1.000<br>0.066<br>0.066 |
| (Group B)<br>BestPSPLib - FDA_BRA6/T.NR | **0.019** | (Group B)<br>BestPSPLib - TRAD_UMDA/T | **0.005** |
| (Group C)<br>FDA_BRA6/T.NR - TRAD_UMDA/T | **0.000** | (Group C) | **0.014** |
| (Group D)<br>TRAD_UMDA/T - TRAD_FDA/T | **0.001** | TRAD_UMDA/T - TRAD_FDA/T | |

$$Z = \left[ \frac{12L}{n\left(k^3 - k\right)} - \frac{3(k+1)}{(k-1)} \right] [n(k-1)]^{0.5} \tag{11}$$

where $k$ = number of Algorithms, $n$ = number of BD.

**Table 27** Comparison of Algorithms, indicator "Quantity_Times_Optimal" in j30_12 and j30_32

| j30_12/algorithms | sig. | j30_32/algorithms | sig. |
|---|---|---|---|
| Friedman test (Group A)<br>UMDA_BRA8/T.R - UMDA_BRA8/T.NR<br>- UMDA_BRA8/MO.R - UMDA_BRA8/MO.NR | | Friedman test (Group A)<br>UMDA_BRA8/T.NR - UMDA_BRA8/T.R<br>- FDA_BRA6/MO.NR - FDA_BRA6/T.NR<br>- FDA_BRA6/T.R - FDA_BRA6/MO.R<br>- UMDA_BRA8/MO.R | 0.074 |
| Wilcoxon test<br>UMDA_BRA8/T.R - FDA_BRA6/MO.R | 0.142 | Wilcoxon test<br>UMDA_BRA8/T.NR - UMDA_BRA8/MO.NR | **0.043** |
| Friedman test (Group B)<br>FDA_BRA6/MO.R - FDA_BRA6/T.NR<br>- FDA_BRA6/MO.NR - FDA_BRA6/T.R<br>- TRAD_UMDA/MO | **0.042**<br>0.082 | (Group B)<br>UMDA_BRA8/MO.NR - TRAD_UMDA/MO | **0.043** |
| Wilcoxon test<br>FDA_BRA6/MO.R - TRAD_UMDA/T<br>FDA_BRA6/MO.R - TRAD_FDA/MO<br>FDA_BRA6/MO.R - TRAD_FDA/T | 0.068<br>0.068<br>0.068 | Friedman test (Group C)<br>TRAD_UMDA/MO - TRAD_UMDA/T<br>- TRAD_FDA/MO - TRAD_FDA/T | 1.000 |

### 4.7.1 Analysis Regarding the Indicator "Average_Time_Duration"

In this analysis, the "Average_Time_Duration" indicator is considered a criterion when assigning the ranking to each algorithm. Then the Page's L Trend Test is applied, obtaining the following results:

$$H_0 : m_1 = m_2 = \ldots$$

$H_1 : m_1 < m_2 < \ldots m_i$ represents the measure of central tendency of the sample $i$.

$$L = 9512, \ Z = 11.419981, \ P(> Z) = 0.$$

The null hypothesis is rejected and it is shown that the data comply with the order $m_1 < m_2 < m_1 < m_2 <$.

The integrated analysis of the indicator considering the 15 databases shows that the family of Algorithms based on UMDA_BRA8 reports the best results considering the value of the means and their dispersion, thus corroborating the results of the application of the statistical tests.

It is shown that, with respect to the "Average_Time_Duration" indicator, the algorithms maintain the following ranking:

1. UMDA_BRA8/T.NR
2. UMDA_BRA8/T.R
3. UMDA_BRA8/MO.R
4. UMDA_BRA8/MO.NR
5. FDA_BRA6/T.NR
6. FDA_BRA6/T.R
7. FDA_BRA6/MO.R
8. FDA_BRA6/MO.NR
9. TRAD_UMDA/T
10. TRAD_FDA/T
11. TRAD_FDA/MO
12. TRAD_UMDA/MO

### 4.7.2 Analysis Regarding the Indicator "Quantity_Times_Optimal"

Regarding the "Quantity_Times_Optimal" indicator. Applying the Page's L Trend Test the following results are obtained:

$$H_0 : m_1 = m_2 = \ldots$$

$H_1 : m_1 < m_2 < \ldots$ where $m_i$ represents the measure of central tendency of the sample $i$.

$L = 9310$, $Z = 10.210313$, $P(> Z) = 0$. The null hypothesis is rejected and it is shown that the data comply with the order $m1 < m2 < \ldots$...

The integrated analysis of the "Quantity_Times_Optimal" indicator shows that the Algorithms have the following ranking:

1. UMDA_BRA8/T.NR
2. UMDA_BRA8/T.R
3. UMDA_BRA8/MO.R
4. UMDA_BRA8/MO.NR
5. FDA_BRA6/T.NR
6. FDA_BRA6/T.R
7. FDA_BRA6/MO.NR
8. FDA_BRA6/MO.R
9. TRAD_FDA/MO
10. TRAD_FDA/T
11. TRAD_UMD

## 5 Conclusions

An analysis of the behavior of the proposed algorithms was carried out in the c15_12 and c15_32 databases, which have 16 tasks, three modes, two renewable resources and two non-renewable resources. This step was consered a template for the sensitivity analysis of the algorithms to variations with respect to the quantity and types of resources. The proposed algorithms were compared with the best results reported in the literature and with the traditional variants of FDA and UMDA. From the experiments carried out, it was possible to demonstrate that in the databases used, no significant changes were observed in the behavior of the proposed algorithms when faced with changes in the quantity and types of resources.

In particular it is shown that:

- In the sensitivity analysis regarding the variation in the number of resources in both the case of non-renewable and renewable resources, the family of proposed algorithms (UMDA_BRA8, FDA_BRA6), with treatment of restrictions in the probabilistic model, report better results than the algorithms based on traditional FDA and UMDA implementations, a result that behaves similarly in all the indicators that measure the effectiveness of the algorithms.
- In database n0, regarding the "Average_Time_Duration" indicator, the algorithms proposed in this research achieve better results than the robust algorithms hGMEDA, hGMEDA and SPEA2 reported in the bibliography.
- In most databases, the variants of Algorithms UMDA_BRA8/T.R and UMDA_BRA8/T.NR, with respect to the indicators related to effectiveness, report superior results than those obtained by the rest of the Algorithms.
- Between the proposed algorithms, there are no significant differences in the "Minimum_Time_Duration" indicator; in the latter, it is also shown that there are no statistically significant differences with the best result reported in the bibliography. For the indicators "Average_Time_Duration" and "Quantity_Times_Optimal" the algorithms with the "just_time" optimization approach over the "multi-objective" approach.
- In general, the proposed algorithms reported better results than the traditional variants based on UMDA and FDA.
- Given the variation in the number of tasks, it was identified that the UMDA_BRA8 variants report significantly better results than the rest of the Algorithms. Regarding the optimization approach, slightly higher results were reported for "just_time" and in some cases, the "reinforcement" learning strategy reported better results.
- The Page's L Test was applied to compare the global performance of the algorithms in all the databases analyzed, and it was shown that the variants of the UMDA_BRA8 algorithm obtained better results than those of the FDA_BRA6, and that in general, the research proposals report superior results compared to traditional approaches.

- It is evident that the UMDA_BRA8 algorithm is capable of finding satisfactory solutions to problems with correlated variables, with a behavior similar to the UMDA, a result that confirms what was stated in the consulted bibliography.

In future work, it is proposed to incorporate elements associated with task prioritization into the proposed algorithms. Additionally, include partial evaluation techniques or other probabilistic model construction methods to reduce the computational cost without affecting the effectiveness of the algorithms.

## References

1. ISO (2018) ISO 19650-1:2018. Organization and digitization of information about Buildings and Civil Engineering Works, Including Building Information Modelling (BIM)—Information management using building information modelling. Part 1: concepts and principles. In: ISO. https://www.iso.org/standard/68078.html. Accessed 10 Sep 2020
2. Mesároš P, Mandičák T, Behúnová A (2020) Use of BIM technology and impact on productivity in construction project management. Wireless Netw. https://doi.org/10.1007/s11276-020-02302-6
3. Piñero Ramírez PE, Pérez Pupo I, Piñero Pérez PY, Marquez Ruiz Y, Fustiel Alvarez and (2022) A software ecosystem for project management in BIM environments assisted by artificial intelligent techniques. In: Piñero Pérez PY, Bello Pérez RE, Kacprzyk J (eds) Artificial intelligence in project management and making decisions. Springer International Publishing, Cham, pp 191–212
4. Dave B, Buda A, Nurminen A, Främling K (2018) A framework for integrating BIM and IoT through open standards. Autom Constr 95:35–45. https://doi.org/10.1016/j.autcon.2018.07.022
5. Project Management Institute (2021) A guide to the project management body of knowledge, 7th edn. Project Management Institute Inc., Newtown Square, Pennsylvania
6. ISO (2012) ISO 21500: guidance on project management. In: ISO. https://www.iso.org/cms/render/live/en/sites/isoorg/contents/data/standard/05/00/50003.html. Accessed 16 Jan 2021
7. Stellingwerf R, Zandhuis A (2013) ISO 21500 Guidance on project management—A pocket guide, 1st edn. Van Haren Publishing, Zaltbommel
8. Kelley JE (1961) Critical-path planning and scheduling: mathematical basis. Oper Res 9:296–320. https://doi.org/10.1287/opre.9.3.296
9. Malcolm DG, Roseboom JH, Clark CE, Fazar W (1959) Application of a technique for research and development program evaluation. Oper Res 7:646–669. https://doi.org/10.1287/opre.7.5.646
10. Hartmann S, Briskorn D (2010) A survey of variants and extensions of the resource-constrained project scheduling problem. Eur J Oper Res 207:1–14. https://doi.org/10.1016/j.ejor.2009.11.005
11. 1Issa SB, Tu and (2020) A survey in the resource-constrained project and multi-project scheduling problems. 105267/j.jpm 117–138. https://doi.org/10.5267/j.jpm.2019.11.001
12. Hartmann S, Briskorn D (2021) An updated survey of variants and extensions of the resource-constrained project scheduling problem. Eur J Oper Res. https://doi.org/10.1016/j.ejor.2021.05.004
13. Habibi F, Barzinpour F, Sadjadi S (2018) Resource-constrained project scheduling problem: review of past and recent developments. J Project Manag 3:55–88. https://doi.org/10.5267/j.jpm.2018.1.005
14. Ding H, Zhuang C, Liu J (2023) Extensions of the resource-constrained project scheduling problem. Autom Constr 153:104958. https://doi.org/10.1016/j.autcon.2023.104958

15. Yassine AA, Mostafa O, Browning TR (2017) Scheduling multiple, resource-constrained, iterative, product development projects with genetic algorithms. Comput Ind Eng 107:39–56. https://doi.org/10.1016/j.cie.2017.03.001
16. Gómez Sánchez M, Lalla-Ruiz E, Fernández Gil A, Castro C, Voß S (2023) Resource-constrained multi-project scheduling problem: a survey. Eur J Oper Res 309:958–976. https://doi.org/10.1016/j.ejor.2022.09.033
17. Kolisch R, Sprecher A (1997) PSPLIB—A project scheduling problem library: OR software—ORSEP operations research software exchange program. Eur J Oper Res 96:205–216. https://doi.org/10.1016/S0377-2217(96)00170-1
18. Dang Q-V, van Diessen T, Martagan T, Adan I (2021) A matheuristic for parallel machine scheduling with tool replacements. Eur J Oper Res 291:640–660. https://doi.org/10.1016/j.ejor.2020.09.050
19. Cheng C-B, Lo C-Y, Chu C-P (2019) Solving multi-mode resource-constrained multi-project scheduling problem with combinatorial auction mechanisms. Int J Inf Manag Sci 30:143–167. https://doi.org/10.6186/IJIMS.20190630(2).0004
20. Chen JC, Lee H-Y, Hsieh W-H, Chen T-L (2022) Applying hybrid genetic algorithm to multi-mode resource constrained multi-project scheduling problems. J Chin Inst Eng 45:42–53. https://doi.org/10.1080/02533839.2021.1983461
21. Mahdi G, Quintana J, Pérez P, Al-subhi S (2020) Estimation of distribution algorithm for solving the multi-mode resource constrained project scheduling problem. EAI Endorsed Trans Energy Web 7. https://doi.org/10.4108/eai.13-7-2018.164111
22. Piñero Pérez P, Armas Guerrero J, Pérez Pupo I, Menéndez Rizo J (2019) Suite de Gestión de Proyectos 19.05. UCI, Universidad de las Ciencias Informáticas
23. Villavicencio NE (2016) Modelo integrado para la mejora de la productividad in organizaciones orientadas a proyectos de tecnologías de la información. Tesis para optar al grado de: Máster in Diseño, Gestión and Dirección de de Proyectos, Fundación Universitaria Iberoamericana
24. Ayodele M (2018) Effective and efficient estimation of distribution algorithms for permutation and scheduling problems. Ph.D. thesis, Robert Gordon University
25. Ayodele M, McCall J, Regnier-Coudert O (2017) Estimation of distribution algorithms for the multi-mode resource constrained project scheduling problem. In: 2017 IEEE congress on evolutionary computation (CEC). IEEE, Donostia, San Sebastián, Spain, pp 1579–1586
26. Gao S, de Silva CW (2018) Estimation distribution algorithms on constrained optimization problems. Appl Math Comput 339:323–345. https://doi.org/10.1016/j.amc.2018.07.037
27. Tian J, Hao X, Gen M (2019) A hybrid multi-objective EDA for robust resource constraint project scheduling with uncertainty. Comput Ind Eng 130:317–326. https://doi.org/10.1016/j.cie.2019.02.039
28. Medrano BE (2013) Planificación de múltiples proyectos de desarrollo de software utilizando métodos metaheurísticos. Master Thesis, Universidad de La Habana
29. Pupo IP, Vacacela RG, Pérez PP, Mahdi GSS, Peña M (2020) EXPERIENCIAS in EL USO DE TÉCNICAS DE SOFTCOMPUTING in LAEVALUACIÓN DE PROYECTOS DE SOFTWARE. Investigación Oper 41:108–119
30. Christophe C, Daan A, Ingrid A (2018) BIM and ISO 19650 from a project management perspective
31. Loyola M (2019) Encuesta Nacional BIM 2019: Informe de resultados. University of Chile, Chile
32. Mahdi GSS, Al-subhi SH, Vacacela RG, Pupo IP, Quintana JM, Pérez PP (2019) ALGORITMO DE ESTIMACIÓN DE DISTRIBUCIÓN CON TRATAMIENTO DE RESTRICCIONES in EL MODELO PROBABILÍSTICO in PROBLEMAS DE SCHEDULING. Investigación Oper 40:674–686
33. Beşikci U, Bilge Ü, Ulusoy G (2015) Multi-mode resource constrained multi-project scheduling and resource portfolio problem. Eur J Oper Res 240:22–31. https://doi.org/10.1016/j.ejor.2014.06.025

34. Kuehn M, Zahid T, Voelker M, Zhou Z, Rose O (2016) Investigation of genetic operators and priority heuristics for simulation based optimization of multi-mode resource constrained multi-project scheduling problems (MMRCMPSP). In: Thorsten C, Frank H, Michael M, Oliver R (eds) ECMS 2016 proceedings. ECMS, pp 481–487
35. Afshar MR, Shahhosseini V, Sebt MH (2019) A genetic algorithm with a new local search method for solving the multimode resource-constrained project scheduling problem. Int J Constru Manag 1–9. https://doi.org/10.1080/15623599.2019.1623992
36. Goncharov EN, Leonov VV (2017) Genetic algorithm for the resource-constrained project scheduling problem. Autom Remote Control 78:1101–1114. https://doi.org/10.1134/S0005117917060108
37. Miquélez Echegaray MT (2010) Avances in Algoritmos de Estimación de Distribuciones. Alternativas in el Aprendizaje and Representación de Problemas. Ph.D. thesis, Universidad del País Vasco
38. Kronsjö L, Shumsheruddin D (1992) Advances in parallel algorithms. Wiley, United States
39. Mühlenbein H, Mahnig T, Rodriguez AO (1999) Schemata, distributions and graphical models in evolutionary optimization. J Heuristics 5:215–247. https://doi.org/10.1023/A:1009689913453
40. Mühlenbein H, Paaß G (1996) From recombination of genes to the estimation of distributions I. Binary parameters. In: Voigt H-M, Ebeling W, Rechenberg I, Schwefel H-P (eds) Parallel problem solving from nature—PPSN IV. Springer, Berlin Heidelberg, Berlin, Heidelberg, pp 178–187
41. Ayodele M, McCall J, Regnier-Coudert O (2016) BPGA-EDA for the multi-mode resource constrained project scheduling problem. In: 2016 IEEE congress on evolutionary computation (CEC). Vancouver, Canada, pp 3417–3424
42. Soliman OS, Elgendi EAR (2014) A Hybrid estimation of distribution algorithm with random walk local search for multi-mode resource-constrained project scheduling problems. IJCTT 8:57–64. https://doi.org/10.14445/22312803/IJCTT-V8P111
43. Wang L, Fang C (2012) A hybrid estimation of distribution algorithm for solving the resource-constrained project scheduling problem. Expert Syst Appl 39:2451–2460. https://doi.org/10.1016/j.eswa.2011.08.095
44. Coello CAC (2019) Constraint-handling techniques used with evolutionary algorithms. In: Proceedings of the genetic and evolutionary computation conference companion. ACM, Prague Czech Republic, pp 485–506
45. Balouka N, Cohen I (2019) A robust optimization approach for the multi-mode resource-constrained project scheduling problem. Eur J Oper Res. https://doi.org/10.1016/j.ejor.2019.09.052
46. Tirkolaee EB, Goli A, Hematian M, Sangaiah AK, Han T (2019) Multi-objective multi-mode resource constrained project scheduling problem using Pareto-based algorithms. Computing 101:547–570. https://doi.org/10.1007/s00607-018-00693-1
47. Mahdi G, Piñero P, García R, Madera J, Al-subhi S, Peréz I (2020) Nuevo algoritmo de estimación de distribuciones con diferentes métodos de aprendizaje aplicado a planificación de proyectos. Investig Oper 41:945–960
48. Applbaum RL, Anatol K (1971) Pert: a tool for communication research planning. J Commun 21:368–380. https://doi.org/10.1111/j.1460-2466.1971.tb02936.x
49. Emmerich MTM, Deutz AH (2018) A tutorial on multiobjective optimization: fundamentals and evolutionary methods. Nat Comput 17:585–609. https://doi.org/10.1007/s11047-018-9685-y
50. Mühlenbein H, Mahnig T (1999) FDA-A scalable evolutionary algorithm for the optimization of additively decomposed functions. Evol Comput 7:353–376. https://doi.org/10.1162/evco.1999.7.4.353
51. Mühlenbein H (1997) The equation for response to selection and its use for prediction. Evol Comput 5:303–346. https://doi.org/10.1162/evco.1997.5.3.303
52. Soto M (2003) A single connected factorized distribution algorithm and its cost of evaluation. Ph.D. thesis. Universidad de La Habana

53. (2017) Project Scheduling Problem Library—PSPLib. http://www.om-db.wi.tum.de/psplib. Accessed 13 Dec 2018
54. Bartz-Beielstein T, Doerr C, Bossek J, Chandrasekaran S, Eftimov T, Fischbach A, Kerschke P, Lopez-Ibanez M, Malan KM, Moore JH, Naujoks B, Orzechowski P, Volz V, Wagner M, Weise T (2020) Benchmarking in optimization: best practice and open issues. arXiv:200703488
55. Carrasco J, García S, Rueda MM, Das S, Herrera F (2020) Recent trends in the use of statistical tests for comparing swarm and evolutionary computing algorithms: practical guidelines and a critical review. Swarm Evol Comput 54:100665. https://doi.org/10.1016/j.swevo.2020.100665
56. Shapiro SS, Wilk MB (1965) An analysis of variance test for normality (complete samples). Biometrika 52:591–611. https://doi.org/10.2307/2333709
57. Lilliefors HW (1967) On the Kolmogorov-Smirnov test for normality with mean and variance unknown. J Am Stat Assoc 62:399–402. https://doi.org/10.1080/01621459.1967.10482916
58. Sheskin DJ (2020) Handbook of parametric and nonparametric statistical procedures, 5th edn. Chapman and Hall/CRC, Florida, USA
59. Sprecher A, Drexl A (1998) Multi-mode resource-constrained project scheduling by a simple, general and powerful sequencing algorithm. Eur J Oper Res 107:431–450. https://doi.org/10.1016/S0377-2217(97)00348-2
60. Coelho J, Vanhoucke M (2011) Multi-mode resource-constrained project scheduling using RCPSP and SAT solvers. Eur J Oper Res 213:73–82. https://doi.org/10.1016/j.ejor.2011.03.019
61. Hartmann S (2001) Project scheduling with multiple modes: a genetic algorithm. Ann Oper Res 102:111–135. https://doi.org/10.1023/A:1010902015091
62. Schnell A, Hartl RF (2017) On the generalization of constraint programming and boolean satisfiability solving techniques to schedule a resource-constrained project consisting of multi-mode jobs. Oper Res Perspect 4:1–11. https://doi.org/10.1016/j.orp.2017.01.002
63. Araujo JAS, Santos HG, Gendron B, Jena SD, Brito SS, Souza DS (2020) Strong bounds for resource constrained project scheduling: preprocessing and cutting planes. Comput Oper Res 113:104782. https://doi.org/10.1016/j.cor.2019.104782
64. Hao X, Sun L, Gen M (2018) Solving robust resource constrained scheduling problem by multi-objective optimization method based on hybridization of EDA and GA. Procedia Manuf 17:1187–1194. https://doi.org/10.1016/j.promfg.2018.10.001
65. Zitzler E, Laumanns M, Thiele L (2001) SPEA2: improving the strength pareto evolutionary algorithm. Eidgenössische Technische Hochschule Zürich (ETH), Institut für Technische Informatik und Kommunikationsnetze (TIK) 103. https://doi.org/10.3929/ETHZ-A-004284029
66. Vanhoucke M (2021) Best known results for the MMRCPSP. http://solutionsupdate.ugent.be/dataset/j30mm. Accessed 13 May 2021
67. Derrac J, García S, Hui S, Suganthan PN, Herrera F (2014) Analyzing convergence performance of evolutionary algorithms: a statistical approach. Inf Sci 289:41–58. https://doi.org/10.1016/j.ins.2014.06.009

# Platform as Service for Data Analysis Suppoted by Computational Intelligence Techniques

**Yosvany Márquez Ruíz, Iliana Pérez Pupo, Pedro Yobanis Piñero Pérez, Luis Alvarado Acuña, and Roberto García Vacacela**

**Abstract** In this work, a cloud-based software architecture for data analysis called BRIntelligent. Its platform allows researchers to perform data analysis by taking advantage of the availability of cloud resources. The proposed architecture constitutes a platform as services that could be extended by groups of researchers based on the development of algorithms and tools for data analysis. With its implementation, it is expected that the end customers of the platform will be able to have access through a web browser to algorithms and data analysis and visualization tools without requiring advanced programming knowledge or installing any specific software. The proposal is structured to consider different architectural views for software development and is made up of three groups of fundamental components. The first component is a data analysis project management system whose functionalities combine computational intelligence techniques and best practices for project management. The second component constitutes a system for the management of algorithm runs, making use of different groups of technological resources. The third component is a set of software assets and services that facilitate data analysis. The proposal has been partially

Y. M. Ruíz (✉) · I. Pérez Pupo · P. Y. Piñero Pérez
Artificial Intelligence for Sustainable Development Group, IADES, La Habana, Cuba
e-mail: ymarquez@gmail.com; ymarquezr@uci.cu

I. Pérez Pupo
e-mail: iades@businessredmine.com

P. Y. Piñero Pérez
e-mail: iades@businessredmine.com

L. A. Acuña
Departamento de Ingeniería de La Construcción, Universidad Católica del Norte, Antofagasta, Chile
e-mail: lualvar@ucn.cl

R. G. Vacacela
Facultad de Especialidades Empresariales, Universidad Católica Santiago de Guayaquil, Guayaquil, Ecuador
e-mail: roberto.garcia@cu.ucsg.edu.ec

 
P. Y. Piñero Pérez et al. (eds.), *Computational Intelligence in Engineering and Project Management*, Studies in Computational Intelligence 1134,
https://doi.org/10.1007/978-3-031-50495-2_7

implemented through the customization of the BusinessRedmine software ecosystem using a free software development strategy based on open core.

**Keywords** Artificial intelligence · BRIntelligent · Computation intelligence techniques · Data analysis · Platform as services · Project management

## 1 Introduction

Society has undergone a gradual transformation towards computerization, which has led to the emergence of a wide variety of information systems with applications in various scenarios, and has allowed the availability of large amounts of data. As a consequence, new possibilities have been opened for intelligent data analysis, where interdisciplinary approaches that involve statistical techniques, mathematics, artificial intelligence, pattern recognition, data mining and database technologies are combined. This continuous evolution has created a fertile environment for the discovery of relevant and non-trivial knowledge from big data.

Data analysis has become a crucial activity for many companies and organizations today. However, to carry out this task effectively, appropriate tools and systems are required that allow the processing, storage and visualization of large amounts of data and act as a unified analysis platform with a high degree of integration.

Implementing this type of effective data analytics platform can be challenging for businesses due to various factors such as lack of specialized skills and knowledge, lack of data quality, difficulties in accessing data, associated costs, integration with existing systems, inefficient use of computing resources, and the inability of tools to process large volumes of data efficiently, fundamentally wasting the use of parallel processing and distributed storage capabilities [1–3]. However, if these problems are addressed appropriately, companies can obtain significant benefits, and in this sense, software architecture becomes a proposition of great significance and importance.

Many universities including cuban universities, have been involved in the adoption of advanced technologies and data analysis platforms to improve research and education. However, the lack of resources and access to digital technologies can make it difficult to implement these tools in practice. Despite this, some academic institutions have managed to establish collaborations with foreign universities and international organizations to gain access to cutting-edge digital technologies and resources. However, in the Cuban environment, it has been possible to identify some factors that affect access to analysis tools and high-performance computing resources, such as:

- There are data centers in universities and institutions dedicated to scientific development, with little use of the installed technological capacity. Furthermore, there is a low level of integration between different installed platforms, a phenomenon present in Cuba and other countries [4, 5].

- In addition, there is a significant amount of research and projects that have proposed algorithms for data analysis [1–3, 6–8]. But the research results are not integrated into an ecosystem or platform, this is an elemental situation that affects the application of this research for decision-making.

From this situation, the need to develop a platform for data analysis is identified, considering the following requirements:

- Ease to design workflows visually through an intuitive graphical user interface, with a wide variety of components for data ingestion, processing and distribution to different systems.
- High ability to prioritize the use of processing and storage resources. Allowing it to be scalable and designed to handle large amounts of data effectively, using distributed processing and parallel storage techniques.
- High degree of compatibility with multiple data sources and systems to perform more complete and accurate analyses, guaranteeing integration with other tools and platforms to improve the quality of data analysis.
- Data visualization that allows users to see the results of their analysis clearly and effectively, providing graphs, tables, maps and other data visualization tools
- Guarantee the security of processed and stored data, protecting it from unauthorized access, cyberattacks and other security risks. Complying with the security and privacy requirements established by the regulations and laws applicable in the jurisdiction in which the platform is used [11].
- Ability to monitor data analysis processes that allow monitoring their progress, detecting possible errors and maximizing the efficiency of data analysis processes. In addition, it facilitates monitoring the performance of the platform, detecting and solving problems in a timely manner, and optimizing the use of resources.
- Facilities for the treatment of imprecision and uncertainty in decision-making by combining a set of computational intelligence techniques and facilitating decision-making in environments under uncertainty.

In order to summarize the problem, we formulate the following research questions:

1. What are the best practices and technologies to ensure the appropriate use of resources, accessibility and availability for accessing the data analysis platform more efficiently and effectively?
2. What architectural features are necessary for more effective performance and integration?
3. How can we include computational intelligence techniques to facilitate decision-making in highly uncertain environments?

In the context of this research, we understand computational intelligence as "computational intelligence". Many authors use the terms computational intelligence and soft computing interchangeably to refer to the same set of techniques as a branch of artificial intelligence. In this sense, computational intelligence is presented as the branch of AI that encompasses various techniques aimed at simulating human tolerance in decision-making processes in environments with uncertainty and imprecision.

Among the precursor techniques of computational intelligence are evolutionary algorithms, artificial neural networks, fuzzy set theory and fuzzy systems. They are also part of the computational intelligence techniques: rough set, linguistic data summary, conversational systems, fuzzy cognitive maps, collective intelligence techniques, and neutrosophic theory, among other extensions.

To solve the problem of this research, we set ourselves the following objective: to propose a reference architecture for the design and development of a data analysis platform based on the best practices and technologies available to achieve an effective solution to the current problem. In this work, a detailed exploration of the fundamental elements that must be considered when proposing a reference architecture for this type of platform will be carried out with the aim of maximizing its effectiveness and efficiency. To achieve this purpose, the work is divided into several main sections. In the first section, the main elements that make up the proposed reference architecture will be presented. The third section will focus on the analysis of the results obtained from a proof of concept of the platform in a scenario focused on decision-making support for environments oriented to project management. Finally, the conclusions of the study will be presented.

## 2 Platform Architecture Proposal as Services for Data Analysis

In response to the research questions asked, a state-of-the-art study was developed that allowed us to determine some of the best practices and technologies that can be used to ensure the design and development of an effective data analysis platform:

- Select a deployment environment that guarantees greater accessibility, flexibility, low costs and a continuous update of the components that make up a data analysis platform, which can improve the efficiency and productivity of companies or institutions and allow them to take informed and data-based decisions.
- Design a scalable and modular architecture that allows the data analysis platform to adapt to changes in requirements and increases in data volume. This involves considering a microservices' architecture, where each microservice can be developed, tested, deployed, and scaled independently.
- Choose the appropriate technology for data storage, depending on the amount and type of data to be analyzed. Some of the most popular data storage technologies include SQL and NoSQL databases, distributed file systems, and caching systems.
- Use efficient data analysis tools, essential for the processing and analysis of large volumes of data, whose integration capacity must be taken advantage of. There are many data analysis tools available, such as Apache Hadoop, Spark, and Flink, that can help process and analyze large volumes of data.

- Analysis of the potential of computational intelligence techniques for the treatment of uncertainty and imprecision in decision-making. To advance in the development of algorithms that facilitate decision-making in project management environments under uncertainty.
- Ensure data quality to allow analyzes to be accurate and reliable. To ensure data quality, it is important to establish data cleaning and normalization processes as well as protocols for data entry.
- Implement essential security measures for the data analysis platform. It is important to implement authentication, authorization and encryption protocols to ensure data privacy and security.
- Establish monitoring and maintenance processes to ensure that the data analytics platform is functioning correctly. This may include implementing monitoring and early warning tools, as well as performing load testing and stress testing.
- Encourage collaboration and communication, essential aspects for the success of any data analysis project. It is important to foster collaboration and communication between development teams and business teams, as well as establish processes for project management and problem resolution.

On the other hand, it is identified that the appropriate use of web technologies guarantees high levels of mobility and accessibility safely [9, 10].

Considering all the previous elements, a reference architecture is proposed, as shown in Fig. 1.

Additionally, the following principles are established:

- The design of this platform focuses on working in the cloud and taking advantage of container-based virtualization technologies [12–16].
- Several architectural patterns are used based on Microservices, Serverless, Containers, Automatic Scaling and events.
- Containers should be designed following the principles of low coupling and high cohesion, reducing dependencies.
- There is an information system to facilitate communication between the technologies that are part of the platform and the clients.
- An asset repository is incorporated as an important element of the platform and includes algorithms for integration, preprocessing, data analysis and visualization.

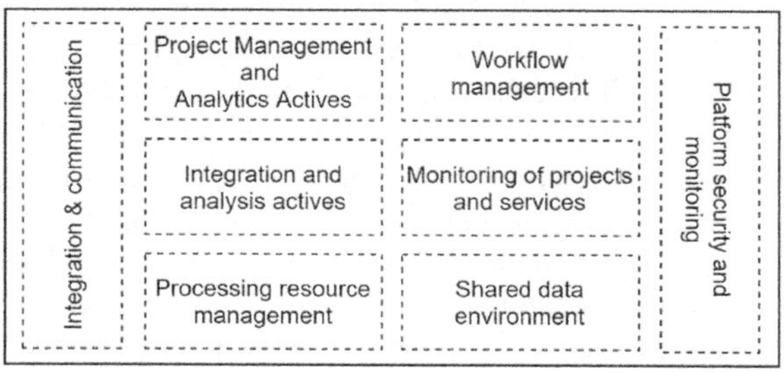


**Fig. 1** Elements that make up the data analysis platform

- An interface is provided to define and execute data processing workflows, these workflows can include multiple steps such as extracting data from various sources, transforming the data into a format suitable for analysis, and loading the data in a data warehouse or data lake (shared data environment). Once the data has been processed and is available in the data warehouse, users can use data analysis tools to explore, visualize, and extract valuable insights from the data.
- Shared data environment ensures data storage and retrieval of analysis results. In addition, the use of cache is guaranteed to reduce latency and speed up information processing.
- The security layer must guarantee secure access to analysis projects, compartmentalization of information and protection of the intellectual property of analysis results by clients.

To guarantee a better understanding of the architecture given its complexity, its presentation is broken down into views. Each view constitutes a projection of the architecture with respect to different dimensions of systems analysis and design. The following subsections present these views.

## *2.1 Systems View of the Architecture*

In this view, the information systems that make up the platform and that may be located in different layers of the proposal are analyzed. One of the most important elements of this view is the repository of processing assets that represent algorithms, although there will be others that are used for integration, preprocessing and visualization of results. Project management and analysis assets:

Asset 1. It is a set of software assets that constitute algorithms for integration, preprocessing, data analysis and visualization. We will call the assets that belong to the repository "algorithm assets" from now on. This repository is the essence of the system and is practically described in the rest of the views. Fig. 2 summarizes the steps for its construction. Fig. 2: Processes involved in the construction and management of the algorithm repository.

Assets 2. Subsystem for managing data analysis projects that allows orchestration and communication between the different actors on the platform. Includes a system for managing requests and communicating with clients. It receives the input from the algorithms in JSON format and is capable of representing these inputs in some form. It saves the information regarding each project associated with the algorithm parameters in databases. Likewise, it manages the algorithm queue and distributes algorithm execution requests. Furthermore, it can send jobs to two types of active task executors.

Assets 3. Active executor container managers that publish the algorithms in the containers as services. In the first version, all algorithm containers will be running, waiting to be executed. In higher versions, it will manage the

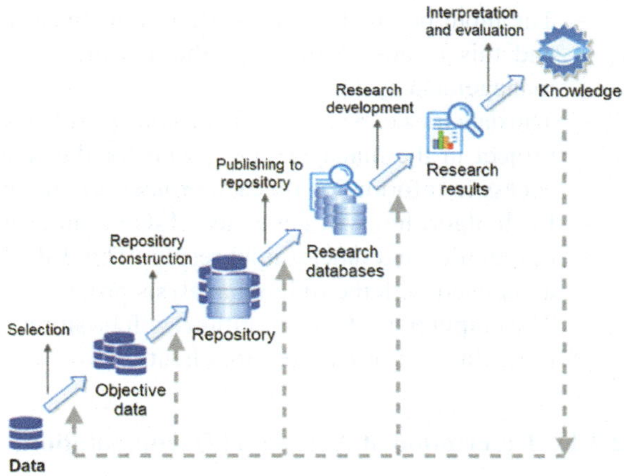


**Fig. 2** Processes involved in the construction and management of the algorithm repository

containers on demand by turning them on or off, following the principles of serverless-based architectures [17].

Asset 4. The subsystem, for communication with the HPC server system, constitutes an independent asset that allows the platform to include the possibility of using HPC servers to execute data analysis tasks [4].

## *2.2 Architecture Integration View*

As part of the integration strategy, the following elements are proposed:

- It is established that each "algorithm asset" in the repository will be encapsulated as a container and may contain one or more algorithms and tools for integration, preprocessing, data analysis and visualization of results.
- It is the responsibility of the developers of each "algorithm asset" to update it.
- There is an information system that manages data analysis projects and user analysis requests. Where they plan the sequence of analysis tasks they want to execute, and they are orchestrated through workflows defined by users.
- Each "algorithm asset" is registered on the platform and publishes a set of metadata that includes its identification, version, input and output channels (communication bus), and also the entire set of messages and events that it will use to facilitate the exploitation of these.
- A whiteboard pattern centered on the shared data environment is used for communication between the client and the data analysis environment. The information from each analysis project in this environment can only be accessed by members of the project.
- For each "algorithm asset," the following elements must be defined:

- The input parameters of the algorithms must be established in JSON format, and this is information that the algorithm developers must provide to the management system.
- The data files to be analyzed must be uploaded by clients through their analysis project in the management system for the shared data environment. Secure access to information on each project is maintained.
- Each algorithm will generate a JSON output indicating the addresses of the output files and other output results. This JSON will be stored in the database associated with the running analysis project.
- The output files of the algorithms will be stored in the shared data environment to facilitate secure access for clients.

### 2.2.1 Description of Assets and Documentation

The specification of software assets is one of the most important elements to achieve their integration and correct use. Two groups of data are proposed for each asset: a functional description and an extra-functional description.

*Functional description*

The technical implementation characteristics that define the behavior and communication mode of the asset must be defined. The functional description is classified into two types, syntactic information and semantic information. The elements that are taken into account for the documentation are:

- Identification of the asset and its description.
- Input parameters, functionalities and methods.
- Notifications and events: notifications that occur when an event happens.

*Extra-functional description of assets*

Extra-functional information is the second part of the documentation of an asset. This includes information on assembly, packaging, marketing and installation requirements.

*Assembly information*

This section is related to the assembly or packaging of an asset called a component. The elements that are taken into account for the documentation of this section are the following:

- Operating System (OS): name of the OS on which it was developed, including the distribution if it is from the GNU/Linux family and the version of the distribution.
- Recommended technological architecture.
- Language: programming language.
- File: name of the file containing all the source code.
- Platform: name of the development IDE on which it was developed.
- Library: libraries used, whether graphical or implemented, and their version.

- Database manager: name and version.

*Quality information*

Specifies the quality factors established by the supplying entity for the evaluation of the asset. The quality criteria that the asset meets are named, such as functionality, reliability, usability, maintainability and portability.

*Market documentation*

As part of this documentation, you must specify:

- Description of the position that the asset has in the market, identification of real clients and potential clients.
- Developer: data necessary to contact the manufacturer of an asset, including the physical address where they work, their telephone number, and their email address.
- Support: data necessary to contact the personnel responsible for supporting the asset.
- Price: price of the asset in the market in the event of its future commercialization.
- License: if the asset is dependent on licenses, name them. Also specify the expiration date of the license(s).
- Certificate: name of the certificate assigned to the asset as a guarantee of having been certified by a quality or safety evaluation or certification entity.
- Expiration date of the use of the asset if specified in a contract.
- URL for the sale or download of the asset.
- Notoriety of successes that the asset has had since the first time it was reused. This information includes the number of completed and canceled installations. They are classified into: positive, negative and neutral ratings.

*Asset quality attributes*

These authors call an attribute "a quality property to which a metric can be assigned, where a metric is a procedure that examines a component and produces a simple piece of data, a symbol (e.g., Excellent, Yes, No) or a number." Next, the quality attributes are organized into different tables [18].

- The evaluation of the assets associated with their functionality, efficiency, effectiveness and reliability is described in Table 1.
- Another group of criteria refers to usability and complexity; see Table 2.
- Criteria are also included for recoverability, portability of each "algorithm asset," and platform scalability. See Table 3.

## 2.3 Architecture Data View

This view includes the set of SQL and NoSQL database technologies that will be used in the platform to manage the data used in the analysis projects, as well as

**Table 1** Set of quality attributes associated with functionality, efficiency and maintainability

| | Subcriteria | Attributes |
|---|---|---|
| Functionality | Suitability | Interface coverage |
| | | Too many interfaces |
| | | Deployment coverage |
| | Interoperability | Data compatibility |
| | Accordance | Compliance with standards |
| | | certifications |
| Efficiency | Temporary consumption | Response time |
| | | Emission capacity |
| | | Reception capacity |
| | Resource consumption | Memory requirements |
| | | Disk utilization |
| Maintainability | Adaptability | Modification index |
| | Extensibility | Ability to be extended |
| Reliability | Maturity | Volatility |
| | | Evolutionality |
| | | Eliminated bugs |
| | Performance | Platform on which it works |
| | | Permissible transactions. Prosecution |
| | Availability | Mean time between failures (MTBF) |
| | | Mean time restored (MTTR) |
| | | Availability |
| | Security | Data encryption |
| | | Average false rejection value (FRR) |
| | | False acceptance average value (FAR) |
| | | SQL injection vulnerability (SIV) |
| | Reliability | Average number of previously tested components |

elements used as cache that will improve the response times of the data services. Prosecution.

Assets 5. PostgreSQL-based database manager subsystem that allows management of analysis projects and algorithm outputs.

Assets 6. Shared data environment subsystem that facilitates:

a. Loading input files for analysis if necessary.
b. The publication of output files of the runs if necessary.
c. The publication of datasets and data repositories to support scientific research.

**Table 2** Set of quality attributes associated with usability

| | Subcriteria | Attributes |
|---|---|---|
| Usability | Learning capacity | Period to use correctly |
| | | Setup period |
| | | Period to manage correctly |
| | | Period to master work |
| | Understandability | User documentation |
| | | Help system |
| | | Computational documentation |
| | | Training and qualification |
| | Operability | Effort to operate |
| | | Effort to configure |
| | | Effort to manage |
| | Complexity | Interfaces offered |
| | | External interfaces used |
| | | Complexity index |

**Table 3** Set of attributes associated with portability and recoverability

| | Subcriteria | Attributes |
|---|---|---|
| Portability | Replaceability | Backwards compatibility |
| | Installation capacity | Easily installable |
| Portability | Persistent | Persistent |
| | Transactional | Transactional |
| | Error handling | Detect errors |
| | | Detect and report errors |
| | | Deal with errors |
| Scalability | Scalability | Ability to be scalable from the point of view of technological resources and parallelization |

## *2.4 Architecture Security View*

This view guarantees the safe use of the solutions and the monitoring of access to the platform services. The assets of this layer allow compliance with the security requirements established by Cuban regulatory entities.

Assets 7. Includes the assembly of an LDAP server with an interface for management from the WEB to manage the users of the solution.

Assets 8. Rupcatcha is an asset that allows integration with captcha management and enhances secure access to information by each of the participants in the project.

Assets 9. Logstash's security is configured to capture logs associated with platform security.

Assets 10. Elastic search security is aimed at integration with log stash, which allows the structuring of security logs on the platform for their management.

Assets 11. Grafana's security is configured for alarm management and visualization of platform security logs, classifying logs associated with application errors, user access to the platform, and user behavior within the platform.

Assets 12. The firewall is active to control access to platform components.

## 2.5 View of Architecture Technologies

It includes the analysis of the basic technological infrastructure for the operation of the platform.

Asset 13. Assets for the assembly of a high-availability platform. Integration of the HAProxy and keep-alive tools that enhance high availability and performance of the main functionalities of the platform.

Asset 14. Related to the application web server layer and balancer. Table 4 summarizes an example of the technological infrastructure required for the operation of the fundamental components of the platform.

Table 4 summarizes an example of the technological infrastructure required for the operation of the fundamental components of the platform.

Servers for high-performance computing (HPC) are not included in this technological proposal.

## 2.6 Architecture Deployment View

This view includes elements of the versioning policy and licensing of platform components. A development strategy based on Open-Core is considered; the core of the solution is open under an open-source license; licenses based on GNU-GPL 2.0 are used. The use of the platform's services is established through a subscription-based model. The mechanism of use and/or marketing of the platform services in the SaaS modality will be evaluated by the project in coordination with all the Cuban entities involved, both developers and sponsors. As part of the architecture, a mixed hybrid model is established as a development strategy, where the products are based on a combination of open-source projects developed publicly by international communities (Redmine, PostgreSQL, and R, among others) and some functionalities are developed independently. Closed doors. The inclusion of new active algorithms includes those that meet the following characteristics:

**Table 4** Mid-range view of technological requirements

| Layer | Assets | CPU | RAM | CPU |
|---|---|---|---|---|
| Project management layer | Information systems projects analysis and communication | 4 | 8 | 60 |
| | HAProxy | 4 | 8 | 40 |
| Shared data layer | PostgreSQL | 4 | 8 | 60 |
| | Memcached | 2 | 4 | 20 |
| | Cluster NFS | 4 | 4 | 150 |
| Security layer | LDAP + Admin | 4 | 4 | 40 |
| | Grafana | 4 | 4 | 40 |
| | Logstash | 4 | 4 | 30 |
| | Elasticsearch | 4 | 8 | 80 |
| Asset repository layer | Each "active algorithm" | 4 | 4 | 20 |
| | 3D visualization | 4 | 4 | 20 |
| | Grafana, maps | 4 | 6 | 80 |
| Totals | | 16 | 26 | 260 |

- Algorithm assets developed under open-source technologies.
- Licensing under permissive licenses or under GNU-GPL 2.0 and higher licenses.
- Within the platform for the construction of each asset, the principle of high cohesion and low coupling was promoted.

## 2.7 *View of Processes for Managing the Algorithm Repository*

The principles that govern the design of management processes are the following:

1. Focused on quality: everyone involved in the development and contribution of assets to the platform is responsible for their quality and evolution.
2. Systemic: all processes are integrated with each other.
3. Proactivity: It is encouraged that all those interested in using the platform promote the development of new "algorithm assets".

### 2.7.1 Entities that Participate in the Methodology

Provider entity: They are those who develop the software assets either as a design or implementation to be reused and must guarantee an adequate specification of their assets. In addition, their products comply with a set of quality factors (defined in this work) depending on the software asset to be treated before publishing them to the repository:

- Fill out the meta-data requested by the repository for the publication of the assets
- Once you have published an asset, update the meta-data every time it undergoes a change.

Client entity: These are the entities that access the repository for the reuse of some asset and must report on:

- The assets you are reusing
- Bugs presented by the assets they reuse
- Products that include any asset that was reused.

Evaluation entity: is responsible for the evaluation and certification of software assets and must:

- Evaluate the assets using the quality factors described in the proposal and according to the type of asset,
- Deliver a report accepting the assets and authorizing their inclusion in the asset directory.

Publishing entity: is responsible for managing the repository and must:

- Guarantee a repository with full availability (in operating time and storage capacity) for the publication of the software assets of a given institution.
- Establish security and access policies for the repository.
- Ensure the publication of all software assets for which you have received the publication notification, the quality certification certificate, and whose asset suppliers have signed the code of ethics.
- Monitor or audit the assets to verify any inconsistencies that the metadata could present.

### 2.7.2 Processes Involved in the Methodology

This section shows the processes involved in the exploitation of software asset repositories that represent the algorithms, the entities involved and the artifacts for the packaging of each asset [6].

*Process 1 Contribute to the Network*

This is the process through which the different institutions that participate in the development process share, through the repository available on the network, the algorithms and tools for the analysis of the data generated.

In this process, other subprocesses are defined; three of them are evaluation, publication, and update, carried out by the asset provider entity. There are also adaptation and integration, which are carried out by the client entity in solving the problems.

*Subprocess 1.1: Assets Evaluation*

This activity is carried out internally by the development entity as part of the tests carried out before the publication of a software asset, according to the quality factors defined by the entity that will be taken into account for the evaluation of an asset.

In turn, it is also carried out by a third party that has been established as a quality, certification or reviewer team in charge of certifying the asset for publication. The evaluating entity is responsible for verifying that the data associated with the asset description is real and that the asset meets the quality factors.

*Subprocess 1.2: Assets Evaluation*

Publishing is a process carried out by the developer entity that consists of establishing software assets in containers for publication. Two important elements that any type of container or information store must guarantee are the reliability and authenticity of the data.

The first element is the reliability of the components, which can be guaranteed in three fundamental steps:

1. Save the components in a shared space where supplier entities can access them to upload their reuse assets.
2. Review of the assets by the publishing entity for their classification as publishable or non-publishable, which will depend on the correctness and consistency of the meta-data information.
3. Publication of the assets by the publishing entity responsible for the repository. Clearly identify the reference that should be used to access each specific asset.

A second element is the security and protection of the assets, which can be guaranteed through some user authentication service to restrict access to people who are not related to both the publishing and client entities. Before publishing a component, it must be packaged together with a set of documents that guarantees its best reuse:

- Functionalities and services that offer examples of use scenarios.
- Technical specification of the active inputs and outputs that it provides and includes a user manual for its use, which interfaces the component should call in case it requires any input interface due to dependencies on libraries or another service.
- Required extra-functional information.
- Quality attributes associated with time or other indicators associated with its use.
- Installation manual: way to integrate into the platform, considering that it is integrated into a container.
- A log of the most common problems and questions. This document is optional. It is where the solutions that the component development team gave to the problems that arose during operation with the component by the different users who reused it are recorded. Likewise, the answers to the most frequently asked questions.

*Subprocess 1.2: Assets Updating*:

It is the responsibility of the publishing entity to correctly comply with this process. He is the one who must update the metadata of the components that he has published in the repository, that is, any variation in the technical and/or general characteristics of the components must be recorded, such as versioning, the environments of his clients where it has been integrated, functionalities added or removed, among other data. Only by having the correctness of the component's metadata as support does its degree of maturity become evident.

*Process 2 Consume from the Network*

Process that allows the consumption or reuse of assets from the repository. The consumption process is evidenced in the meta-data of the projects in execution by specifying the name of the asset being reused and the references to the assets used. In this process, search and selection are defined as sub-processes, both carried out by the client entity.

*Subprocess 2.1: Assets Search*

Search is one of the fundamental processes to facilitate the location and use of repository assets. It is carried out by the client entity, the search for assets consists of reviewing among the directory of available services which one or which ones best fit the specifications of the environment where it will be integrated, that is, which one satisfies the requirements imposed by both the client and the architecture of the application.

To do this, the client uses the meta-data recorded for each asset, where they can find a summary or description of the functionalities or services offered by each asset. It is generally done through a mediator (a component search program in a repository) or some other search algorithm implemented by the asset provider to offer "asset search" as an added service (see Fig. 3) search levels can be classified:

- Basic: only the description or synthesis recorded by assets is reviewed.
- Media: the meta-data associated with the asset is reviewed in addition to its description.
- Advanced: the technical specification document of the asset is reviewed, where the meta-data values are included.

*Subprocess 2.2: Selection*

In this activity, the client selects assets that fit a set of given specifications

1. Technical tests to verify ideal compliance with the functionalities and compatibility in the use of the services provided by the platform. This includes an analysis of the client architecture and the service provided by the platform asset.
2. Quality tests to determine the degree of satisfaction of the asset with the quality requirements of interest to the client. Depending on the degree of satisfaction with the evaluation, they are selected to be integrated into the client system.

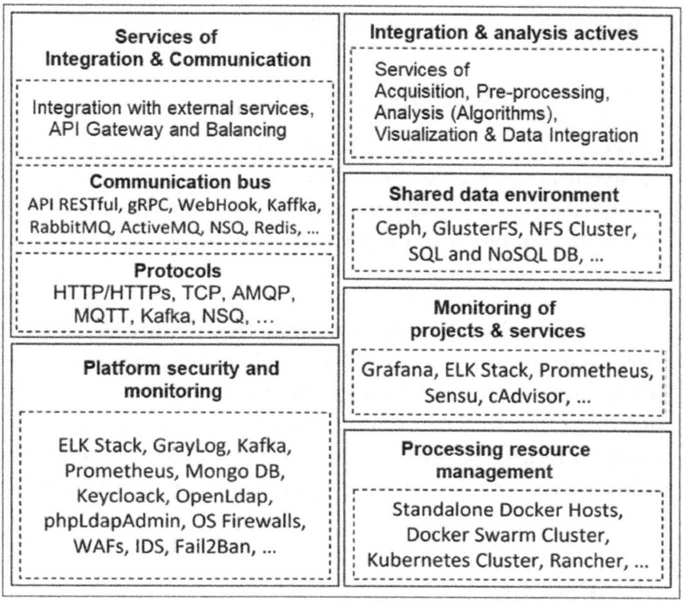


**Fig. 3** General view of the prototype for decision-making in project management

*Process 3: Manage repository*

In the proposed methodology, repository management is a process carried out by the publishing entity to guarantee, on the one hand, the proper use of the reuse environment. The management of the repository is supported by a code of ethics and guidelines for security and access control to the repository. In addition, stimulation or reward guidelines are established by virtue of increasing the degree of reuse, favoring the most contributors, and grouping the least in the list of those isolated from the reuse environment.

*Sub-process 3.1: Define security and access policies for the repository*

The publishing entity must guarantee certain levels of confidentiality, i.e., establish certain policies that reduce the intrusion of unauthorized personnel or third parties into the system. To do this, it is necessary to establish policies and restrictions that limit access to the component repository to only authorized personnel. Furthermore, for the security of the information registered in the repository, a redundancy system for both data and network must be established.

*Sub-process 3.2: Define a Code of Ethics*

It is also necessary to establish guidelines for discipline and commitment among participating entities in a software asset reuse environment. To do this, it is convenient to define in a code of ethics the ethical and moral statutes of the institution, and the signature is proof of the commitment that the participating entities have to comply with them.

*Sub-processes 3.3: Design Monitoring System*

This sub-process consists of reviewing the information recorded in the meta-data to detect inconsistencies, identifying who are the entities that have the greatest participation in the reuse environment and who have the least, and which are the most used software assets or the entities with the greatest contribution to the network.

*Thread 3.4: Establish a reward system*

This is another of the actions that the publishing entity must carry out within the "Repository Management" process to stimulate a better exchange in an environment of reuse of software assets. It would be appropriate to establish a reward system that, as a consequence of the stimulation itself, increases the publication of assets and components and, in turn, strengthens reusability. The following figure represents a summary of the concepts and way of working of the repository management process.

## 3 Analysis of Results

This section presents a proof of concept of the proposed architecture for a decision support scenario in project-oriented organizations. The proof of concept is carried out in a real environment dedicated to decision-making in project management and constitutes a working prototype of the platform proposal and its architecture (see Fig. 3).

The list of current platform services organized in containers is listed below.

Security layer:

- Logstash: security log management
- Elasticsearch: document indexing
- Grafana: graph information on security dashboards
- Osixia-OpenLdap: identity management
- Osixia-phpLdapAdmin: openldap visual interface.

High availability layer:

- Nginx: load balancer
- HAproxy: reverse proxy request management.

Shared data environment:

- PostgreSQL: database manager.
- PGAdmin4: postgres database administration.
- MariaDB: database manager.
- Memcached: cache management, improve performance.

- NFS: to share data volumes.

Analysis project management layer:

- Analytics Project Manager: two instances based on BusinessRedmine.
- HPC Manager: currently under construction.
- Jenkings: automated task manager.

Data analysis layer, repositories:

- graphana: graph analysis information on dashboards.
- IFC 3D viewer: for 3D visualization of IFC files.
- ProdAnalysis: library of data analysis algorithms (ruby programming language) Below are different views of parts of the platform that exemplify the assembly of the architecture.

## *3.1 Analysis Project Layer*

This layer includes information systems for project management and modeling for data analysis. It facilitates the management of data analysis requests and communication with the platform's clients from the web with a responsive design that facilitates mobility.

## *3.2 Algorithm Repository Layer*

In this layer, there will be as many containers as "algorithm assets" included in the platform, with the possibility of being extended following contracts and collaborative strategies by researchers from different institutions. Each container may contain algorithms developed on different platforms and languages. Below is an example of a container for the development of algorithms in the Ruby language.

### 3.2.1 "Assets Algorithm" ProdAnalysis Container and Ways to Use It

This service facilitates the design of experiments for data analysis following the CRISP-DM methodology and establishes a non-mandatory workflow because interested parties can access algorithms at different stages independently of the flow.

This "algorithm asset" consists of a container made up of two fundamental components, an algorithm manager and the "prodanalysis" plugin. This plugin incorporates the necessary interfaces for algorithm selection, parameter management, and execution task management. And visualization of the results. The other component of this "active algorithm" is the library of algorithms that it includes based on the Ruby language.

A set of algorithms implemented in this library [1, 3, 6–8] are listed below:

- DataPartitioner: which facilitates the creation of partitions for the design of experiments. PCA: which allows principal component analysis
- LPA: builds probabilistic graphs from data.
- Apriori_LDS: which allows the generation of linguistic summaries of data using association rules [1].
- PCA_LDS: which allows the generation of linguistic summaries of data by combining principal component analysis and the association rule learning algorithm [1].
- RoughSet_LDS: which allows the generation of linguistic summaries of data by applying rough set techniques [1].
- AG_LDS: which allows the generation of linguistic data summaries by applying genetic algorithms.
- ACO_LDS: which allows the generation of linguistic data summaries applying ant colonies,
- HYBRID_LDS: which allows the generation of linguistic summaries by combining the RoughSet_LDS, PCA_LDX and LPA_LDS algorithms [1].
- Outliers_LDS: which allows the generation of linguistic summaries from Outliers [1]. SA_Cut_2d: algorithm for 2D cutting and packaging that combines the simulated annealing and GRASP strategies.
- Scheduling_AG: builds project schedules using genetic algorithms [7].
- Scheduling_UMDA: builds project schedules using univariate distribution estimation algorithms [7].
- Scheduling_FDA: builds project schedules using factored distribution estimation algorithms [7].
- FCM_Multistages: allows inference from fuzzy cognitive maps [8].
- NCM_Multistages: allows inference from neutrosophic cognitive maps [8].
- NCM_LDS: allows inference by combining cognitive maps and linguistic summaries of data [8].

This container also includes facilities for data recovery from databases as part of data preparation actions for analysis.

### 3.2.2 "Active Algorithm" Container 3D Visualization IFC Format

This service constitutes a container that allows the visualization and basic modification of IFC files to support construction projects, see Fig. 4. The IFC format, "Industry Foundation Classes", is an open specification data format developed by the IAI (International Alliance for Interoperability) and is a standard that facilitates interoperability between programs in the construction sector [19–21]. IFC classes and objects represent both a geometric and alphanumeric information model, made up of a set of more than 600 classes and continually expanding.

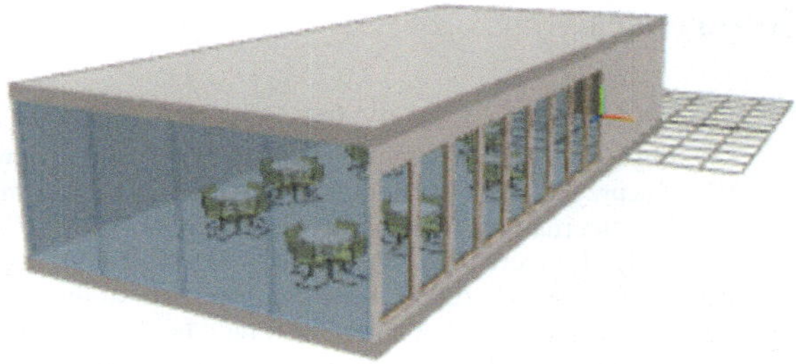

**Fig. 4** Image of the IFC-format 3D visualization service

## 4 Conclusions

In this work, an architectural proposal was presented for the development of a platform for intelligent data analysis based on free software technologies. The development of the proposal focuses on the use of technologies for virtualization, with an emphasis on service-based communication.

The possibility of extending the functionalities of the platform is proposed through the development of assets specialized in the processing, analysis and visualization of information. This approach allows those interested in extending the platform to make contributions using different technologies or languages.

The proposed architecture has been partially implemented as part of an ecosystem to support decision-making in project management. Which demonstrates its viability.

This ecosystem constitutes a prototype implementation of the platform and presents concrete examples of extension and development of new algorithms. As part of future development, it is planned:

- To extend the system of "assets algorithms" and enhance its application to help decision-making in investment projects.
- Support the development and integration of new algorithms into the algorithm library of the ProdAnalysis asset.
- Support the development of new "algorithm assets" for incorporation into the repository of data analysis assets based on other technologies and languages. Taking advantage of the potential of virtualization technologies based on containers.
- Development of "active algorithms" that allow the platform to communicate with the HPC technological infrastructure installed in different scientific institutions.
- Strengthen work with researchers and institutions through collaboration and transparency.

## References


1. Pérez Pupo I, Piñero Pérez PY, Bello Pérez RE, García Vacacela R, Villavicencio Bermúdez N (2022) Linguistic data summarization: a systematic review. In: Piñero Pérez PY, Bello Pérez RE, Kacprzyk J (eds) Artificial intelligence in project management and making decisions. Springer International Publishing, Cham, pp 3–21
2. Pérez Pupo I, Piñero Pérez PY, Al-subhi SH, García Vacacela R, Martínez Noriega HA, Villavicencio Bermúdez N (2022) New linguistic data summarization approach for prediction problems in project management applications. In: Piñero Pérez PY, Bello Pérez RE, Kacprzyk J (eds) Artificial intelligence in project management and making decisions. Springer International Publishing, Cham, pp 23–38
3. Piñero P et al (2021) Proyecto Biblioteca de algoritmos y herramientas para la ayuda al a toma de decisiones en el proceso inversionista, Universidad de las Ciencias Informáticas
4. Sherif T et al (2014) CBRAIN: a web-based, distributed computing platform for collaborative neuroimaging research. Front Neuroinform. https://doi.org/10.3389/fninf.2014.00054,Vol. 8,No.54
5. Mathew AS (2023) Green cloud computing: a literature survey. Int J Emerg Res Areas (IJERA) 3:322–328. https://doi.org/10.5281/zenodo.8210329
6. Piñero Pérez PY, Pérez Pupo I, Piñero Ramírez PE, Marquez Ruiz Y, Fustiel Alvarez Y (2022) Project management repository for decision-making researches. In: Piñero Pérez PY, Bello Pérez RE, Kacprzyk J (eds) Artificial intelligence in project management and making decisions. Springer International Publishing, Cham, pp 303–317
7. Mahdi G (2021) Algoritmos de estimación de distribuciones con tratamiento de restricciones para la construcción de cronogramas de proyectos. Tesis presentada en opción al grado científico de Doctor en Ciencias Técnicas. Departamento de Investigaciones en Gestión de Proyectos, Universidad de las Ciencias Informáticas, La Habana, Cuba
8. Hasan S (2021) Nuevas extensiones de mapas cognitivos difusos para problemas de toma de decisiones secuenciales multietapa. Tesis presentada en opción al grado científico de Doctor en Ciencias Técnicas. Departamento de Investigaciones en Gestión de Proyectos, Universidad de las Ciencias Informáticas, La Habana, Cuba
9. Komar R, Patil A (2023) Emerging trends in cloud computing: a comprehensive analysis of deployment models and service models for scalability, flexibility, and security enhancements. J Intell Syst Appl Data Sci 1
10. Zrenner J, Möller FO, Jung C, Eitel A, Otto B (2019) Usage control architecture options for data sovereignty in business ecosystems. J Enterp Inf Manag 32(3):477–495. https://doi.org/10.1108/JEIM-03-2018-0058
11. Mohamed N, Rao LS, Sharma M, Shukla SK (2023) In-depth review of integration of AI in cloud computing. In: 2023 3rd international conference on advance computing and innovative technologies in engineering (ICACITE), pp 1431–1434
12. Riofrío A, Caiza G (2021) Cloud computing architecture with dockers for an industrial process. In: 2021 International conference on engineering and emerging technologies (ICEET), IEEE, ISSN: 2409-2983. https://doi.org/10.1109/ICEET53442.2021.9659760. ISBN:978-1-6654-2714-2
13. VS DP, Sethuraman SC, Khan MK (2023) Container security: precaution levels, mitigation strategies, and research perspectives. Comput Secur 103490. https://doi.org/10.1016/j.cose.2023.103490
14. Oliveira F et al (2021) Software aging in container-based virtualization: an experimental analysis on docker platform. In: 16th Iberian conference on information systems and technologies (CISTI). IEEE. ISSN: 2166-0727. ISBN:978-989-54659-1-0. https://doi.org/10.23919/CISTI52073.2021.9476625
15. VasanthaKumari N, Arulmurugan R (202) Reorganizing virtual machines as docker containers for efficient data centres. In: 3rd EAI international conference on big data innovation for sustainable cognitive computing. Springer, Cham. ISBN 978-3-030-78749-3. https://doi.org/10.1007/978-3-030-78750-9_14

16. Kithulwatta WM et al (2022) Adoption of docker containers as an infrastructure for deploying software applications: a review. In: Advances on Smart and Soft Computing. ISBN:978-981-16-5558-6. https://doi.org/10.1007/978-981-16-5559-3_21, pp 247–259
17. Wen J, Chen Z, Jin X, Liu X (2023) Rise of the planet of serverless computing: a systematic review. ACM Trans Softw Eng Methodol 32:131:1–131:61. https://doi.org/10.1145/3579643
18. Roa PA, Morales C, Gutiérrez P (2015) Norma ISO/IEC 25000. Tecnología Investigación y Academia 3(2), 27–33. ISSN: 2344-8288
19. ISO (2018) ISO 16739-1:2018. Industry Foundation Classes (IFC) for data sharing in the construction and facility management industries—Part 1: data schema. ICS>25>25.040>25.040.40. https://www.iso.org/standard/70303.html
20. Babalola A, Musa S, Akinlolu MT, Haupt TC (2021) A bibliometric review of advances in building information modeling (BIM) research. J Eng Des Technol 21:690–710. https://doi.org/10.1108/JEDT-01-2021-0013
21. Royano García V, Gibert Armengol V, Serrat Piè C, Rapinski J (2023) Analysis of classification systems for the built environment: historical perspective, comprehensive review and discussion. J Build Eng 67. https://doi.org/10.1016/j.jobe.2023.105911

# Ecosystem for Construction of Hybrid Conversational Systems (BRasa)

**Yuniesky Orlando Vasconcelo Mir, Iliana Pérez Pupo, Pedro Yobanis Piñero Pérez, Luis Alvarado Acuña, and Rafael Bello Pérez**

**Abstract** The increasing maturity of artificial intelligence technologies related to conversational systems, such as machine learning algorithms, natural language processing, linguistic data summarization, and natural language generation, is changing the way in which users interact with information systems. In this sense, two trends stand out: the improvement of rule-based smart chatbots and Large Language Models. These two models have different characteristics, with advantages and disadvantages for different scenarios. In this work, the BRasa ecosystem is proposed, which allows the construction of conversational systems that exploit the advantages of the two models. The ecosystem combines different computational intelligence techniques with natural language processing techniques, among which the following stand out: linguistic data summary, fuzzy logic, and neutrosophic theory. As part of the validation of the proposal, using the ecosystem, two conversational systems are built for different environments. The experimental design consists of different tests that demonstrate the applicability and effectiveness of the proposal.

**Keywords** Conversational system · Chatbots · BRasa · Computational intelligence · Project management

---

Y. O. V. Mir · I. P. Pupo (✉) · P. Y. Piñero Pérez
Artificial Intelligence for a Sustainable Development Group, IADES, Havana, Cuba
e-mail: iliperezpupo@gmail.com; iades@businessredmine.com

Y. O. V. Mir
e-mail: iades@businessredmine.com

P. Y. Piñero Pérez
e-mail: pppyob@gmail.com; iades@businessredmine.com

L. A. Acuña
Departamento de Ingeniería de la Construcción, Universidad Católica del Norte, Antofagasta, Chile
e-mail: lualvar@ucn.cl

R. B. Pérez
Centro de Investigación en Informática, Universidad Central Marta Abreu de las Villas, Snata clara, Cuba
e-mail: rbellop@uclv.edu.cu



P. Y. Piñero Pérez et al. (eds.), *Computational Intelligence in Engineering and Project Management*, Studies in Computational Intelligence 1134,
https://doi.org/10.1007/978-3-031-50495-2_8

# 1 Introduction

Conversational systems are constantly evolving; it is a trend that they are used in various sectors [1]. Statistics show that many businesses and institutions adopt conversational systems as one of their main forms of interaction with their customers.

Authors such as [2] state that conversational systems constitute computer solutions that operate as an interface between human users and a software application, using spoken or written natural language as the primary means of communication. There are several names for these systems, among which stand out: conversational interfaces, conversational systems, smart chatbots, dialogue systems, chatbots, conversational artificial agents, intelligent virtual agents, virtual assistants, among others [3, 4]. In order to facilitate understanding, the authors of this paper will use the terms "chatbot" and "conversational systems" interchangeably to refer to the same concept.

The design of conversational solutions has evolved and has impacted the way of interaction between end users and computer systems [5]. For example, the study identifies that approximately 19% of the entities consulted had already introduced conversational systems in their operations, while 62% have inclusion plans and only 19% have no plans to introduce these technologies [6].

In addition, in a study carried out by [7] 38% of users preferred to wait for the human service representative, while 62% of consumers preferred to speak with a chatbot rather than a human agent [8].

These percentages also change according to the technological approaches of the different generations. About 20% de of Gen Z shoppers prefer to start their customer service experience with chatbots rather than speaking with human agents, compared to just 4% of boomers. In the study [9], it shows that 62% of the respondents from generation Z prefer the use of conversational systems to order food, 71% for product search, and 56% report that supplier companies should use chatbots [10, 11].

There are, different chatbot experiences and works in education [12, 13], government [14, 15], business, e-commerce [16], healthcare [17], entertainment [18], and many more [19, 20]. In the business field, chatbots have become very common because they reduce service costs and can manage many customers simultaneously.

There are two general approaches to Smart chatbots considering the artificial intelligence techniques adoption: rule-based and machine learning-based approaches.

Rule-based chatbots match user input against a pattern of rules, identify the user's intent, and select the response from a predefined set during a training process. These systems generally do not have the ability to generate new responses, they just scroll through the predefined set of responses. In this type of system, many rules are generally required to cover good behavior in a specific domain. An example of this system is the Rasa platform [21, 22].

To improve their responsiveness, rule-based chatbots generally incorporate the processing of conversations obtained from interaction with users. This adjustment in learning occurs gradually and often takes several months to achieve a smooth exchange in the conversation with users. The biggest difficulty with the rule-based approach is that the responses are repetitive and fail to fully simulate the originality and spontaneity of the human response. However, they have a faster response rate than other architectures because they do not rely on deep semantic or syntactic analysis for response construction.

Large Language Models (LLM) have significantly advanced the field of Natural Language Processing (NLP). These systems are useful tools for a wide range of applications. However, the direct application of LLM to solve sophisticated problems in specific domains faces many obstacles, caused by the heterogeneity of the domain data and the uniqueness of the domain goals [23, 24]. Elements such as social and cultural norms, religious beliefs, and ethical standards in each specific domain have a high impact on the applicability and acceptance of chatbots in specific scenarios [25].

In general, some of the limitations of conversational systems in their current state are [26–28]:

- Conversational styles are comfortable and have expanded rapidly in some sectors, such as commerce, while in other sectors they have been little introduced, such as engineering [29].
- Many of the chatbots' interactive training interfaces are unfriendly, making active learning difficult.
- LLMs previously trained from large corpora of web data contain different types of implicit knowledge and do not require human supervision for their exploitation [30]. However, their very nature, based on deep neural networks, makes them difficult to interpret and update.
- Both the learning models based on Deep Neural Networks (DNN) and those based on NLP are computationally time-consuming.
- In most smart chatbots, the construction of conversational systems for a specific domain is a costly process in terms of time and human effort. In addition, it has not been possible to systematize the learning capacity of most chatbots from interactions with humans [31].
- In the process of creating domain-closed and machine learning-based chatbots, large data sets of the specific domain are required [32]. In general, the generation of chatbots from domain models is a field open to research [33, 34].
- More recent research has focused on training a robot from labeled data training sets, but rarely on how the robot can improve through online interaction with humans [31].
- A major weakness of most chatbots is that they cannot learn new knowledge during the conversation process; their knowledge is fixed in advance and cannot be extended or updated during the conversation [35].

From this analysis, it is identified that it is necessary to create new models and platforms that facilitate the automatic or semi-automatic construction of conversational systems, taking advantage of the best experiences. The main objective of the proposal is to build an ecosystem for the construction of conversational systems that we will call BRasa. That allows you to create and manage multiple conversational systems and learn new knowledge continuously and interactively during conversations.

The rest of the work is structured into five sections. The third section presents the architectural model of the BRasa platform as services. Section 4 presents an analysis of the results of applying the platform to project decision-making. Finally, conclusions and future work are presented.

## 2 New BRasa Architectural Model Supported by Different Soft Computing Techniques

This section presents a new architectural model for building conversational systems for different solution domains. Based on this proposal, a platform as a service called BRasa is built, and its applicability in different sectors is demonstrated. In this proposal, the conversational system has a dialogue agent with the ability to extract new training examples from the conversations in which it participates. When each chatbot generated by the platform starts a conversation, the user's gratitude with their responses is estimated. When the conversation seems to be going well, the user's responses become new training examples. When the agent believes he has made a mistake, he asks for feedback that is, to learn to predict. Feedback further improves the dialogue capabilities of chatbots.

The principles that govern the construction of the proposal are listed below:

- Combination of different soft computing techniques focused on three fundamental processes: platform knowledge management, learning or training of the model for specific domains, tuning, and exploitation in real environments.
- Model that enhances interactive learning strategies and continuous learning.
- Introduction of different strategies for evaluating conversations and a strong connection with knowledge management. Application of elements of neutrosophic theory in evaluation.
- Multilingual approach of the proposal to promote transculturation.
- Integration with elements of explanatory artificial intelligence (XAI), to facilitate the analysis of causal relationships from knowledge management.
- Model that stores not only correct expressions but also incorrect expressions, integrated with Large Language Model models. In this way, when faced with unknown situations, the conversational system will be able to respond as rationally as possible.

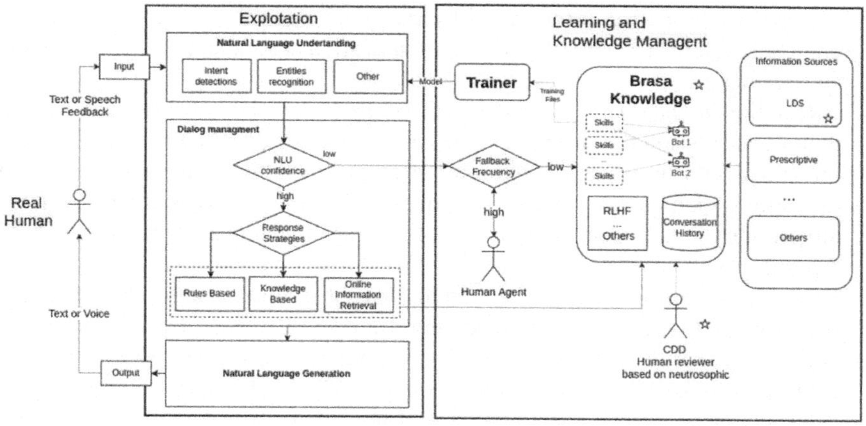


**Fig. 1** General diagram of the BRasa ecosystem

Figure 1 represents the general diagram of the proposed model.

The components that make up the proposed solution are listed below. The following subsections describe each of these components:

- The augmented LDS generation subsystem (BRasa_LDS): is one of the subsystems that contributes knowledge to the BRasa Knowledge subsystem by applying Linguistic Data Summarization techniques.
- BRasa Prescriptive Subsystem: it is one of the sources of information that facilitates the generation of prescriptive knowledge and its incorporation into the BRasa Knowledge. Subsystem for knowledge management (BRasa Knowledge): it is the environment that draws on multiple sources to orchestrate a large knowledge base. From this subsystem, it will be possible to manage knowledge bases from different domains to build different conversational systems.
- Conversational Drive Development (CDD) subsystem: it is oriented toward learning from interactions with users. It applies innovative information exchange techniques to facilitate engineering and knowledge management. It constitutes one of the sources of information in the BRasa Knowledge subsystem.
- Trainer subsystem: this component receives as input data the knowledge files generated by Brasa Knowledge, representing a bot. It generates a trained connectionist model, ready to be coupled to a RASA-type conversational system [36].
- DNN subsystem: it is for managing customer responses, it is the main component in the customer interaction process. It receives the utterances and issues or manages a response, a question, or a request, where it processes them and issues a response. In the architectural model proposed, this subsystem has a connection with three other subsystems: the BRIntelligent Services and Data Analysis

subsystem, the BRasa Information Retrieval subsystem, and the Brasa Knowledge Subsystem. The details of the communication are explained in the following subsections.

- BRIntelligent Data Analysis and Services Subsystem: this subsystem allows you to integrate the solution with a set of information retrieval and processing services integrated into the BusinessRedmine platform, facilitating the management of operations such as reservations or requesting options, among others. A set of algorithms for data processing to support management are also incorporated into this subsystem.
- BRasa Information Retrieval subsystem: this subsystem is particularly complex. It is activated when the pre-trained neural model does not have an answer to a customer question. Your responsibility is to manage a response to the customer from various sources, or pass the question to a human agent in case there is definitely no answer. It constitutes one of the information sources of the BRasa Knowledge subsystem.
- Conversations and Stories Management Subsystem: this subsystem aims to process the conversations carried out. Its task is to enable a human agent to evaluate the level of response of the conversational system using techniques from neutrosophic theory. It allows the construction of stories and scripts to increase the system's prediction capabilities in a conversation. It constitutes one of the sources of information in the BRasa Knowledge subsystem.
- UI Subsystem for Information Exchange: it is made up of a set of JavaScript solutions that facilitate the exchange of information with end users. In the current version of the ecosystem, it is made up of two solutions: the webchat.js environment, where the main form of communication is text, and another Brasa.js environment that is capable of managing text and audio for human-machine interactions.

The functionalities of each of the sub-systems are described in detail below.

## *2.1 BRasa Knowledge Subsystem*

It constitutes the main subsystem for knowledge management in the conversational system. It is made up of a structured database that allows the storage of information and a set of services for knowledge management.

The following elements characterize the information managed by BRasa_Knowledge:

- Management of information associated with the elements described in BRasa_prescriptive: intentions, entities, utterances, responses, rules, synonyms, lookups, and stories.
- The term competence is used to group the capabilities that the chatbot has on a specific topic. In this sense, each knowledge that is managed in BRasa_Knowledge

belongs to or is framed under a competence. An example of a skill could be "greeting", but also knowing about project management or tourism.

- This subsystem allows you to build specialized bots in different domains. Each bot would be made up of a set of skills. Therefore, the system would facilitate the construction of multiple bots with different combinations of skills.
- The knowledge base has the capacity to incorporate multiple sources of information and knowledge, some of which are described in detail in this work. For example, the following subsystems constitute sources and subsystems that feed knowledge management: BRasa_prescriptive, Brasa_LDS, BRasa_CDD, BRasa_IRL (information retrieval), and the Conversations and Stories Management subsystem.
- BRasa_Knowledge, in addition to managing knowledge, stores information on certainty, falsity, and neutrality associated with expressions and conversations. It integrates elements of fuzzy logic and neutrosophic theory. In this way, it achieves a better treatment of the uncertainty of the information.

## 2.2 Augmenting LDS Generation (BRasa_LDS) Subsystem

*BRasa_LDS* is one of the defined ways for learning and knowledge generation in the proposed architecture. This approach is supported by linguistic data summarization techniques (LDS) [37–39]. These techniques allow us to extract relevant information from a set of data and present it in a natural, easy-to-understand language. The linguistic summaries generated can express relationships between data, constructing clear and concise responses. This section proposes the construction of a new LDS algorithm specialized in the construction of summaries for conversational systems inspired by the algorithms proposed in [37, 40].

Below are the prerequisites for using this algorithm:

- There must be data sets organized in a structured way with a clear definition of attributes that will be involved in learning and will be part of the R filters and S summarizers of a linguistic summary.
- There must be a set of controlled natural languages in which linguistic expressions are required to be constructed.

### 2.2.1 Protoforms of Summaries that Are Generated

The linguistic summaries generated from these algorithms are structured considering the following protoforms "*Qy's are S*" and "*QRy's are S*" [37]. Each linguistic summary will express a causal relationship between the attributes located in the set of filters *R* and in the set of summarizers *S*. These causal relationships are not rules themselves but are presented in the form of sentences in natural language that facilitate the establishment of conversational styles.

An example of a simple summary of the form "*QRy's are S*" could be: *"Approximately 50% of the coffee shops that introduced rum-aged coffee increased their sales by 10% in the period analyzed"*. In this case, "*y*" is "*coffee shops*", *Q* is a quantifier "*Approximately 50%*", the filter *R* is "*they introduced coffee aged with rum*" and the summarizer is *"they increased their sales by 10% in the period analyzed"*.

But there are other variants of protoforms [41], where the authors combine protoforms with other terms such as; "*but*", "*while*" and "*especially*". These approaches aim to make summaries more expressive, reflecting sentences of contrast or emphasis. This approach constitutes an open line of research, especially in conversational systems. Below are different examples obtained from simulated data only, with the objective of exemplifying the structure of the summaries [39]:

Examples of contrast protoform:

*"60% of customers from Nordic regions prefer acidic coffees,"* **while** *"the majority of customers from Southern Europe prefer less acidic and more aromatic coffees"*.

Example of emphasis protoform:

*"The highest coffee sales are recorded in winter"*, **especially** *"Many of the sales in the second half of January"*.

The Fig. 2 shows an example of a possible conversation with customer participation where linguistic summaries associated with online sales through the conversational platform are involved.

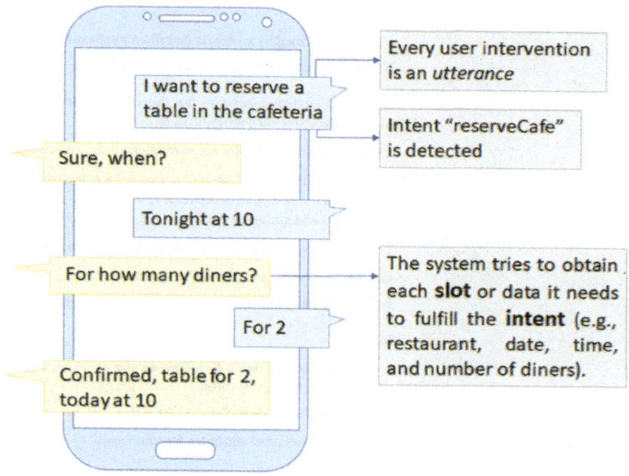


**Fig. 2** Example of interaction with a conversational system with online reservations for a cafeteria

### 2.2.2 Algorithm for Generating Summaries

The algorithm generates a set of candidate summaries. Each candidate summary constitutes an object *P(y, Q, R, S, Tm Summary)*, formed by the entity "*y*" the quantifier *Q*, the filter or qualifier *R*, the summarizers *S*, the set of quality indicators *T* and a linguistic summary that integrates the rest of the components.

Initially, the variables of the input data set are transformed into linguistic variables by applying the principle of maximum membership. Two sets of attributes are constituted that will represent the filters and summarizers of the linguistic summaries to be constructed. These sets are initialized with all the variables in the data set as independent subsets and are transformed during the search process. Subsequently, the dependence in degree k between groups of attributes that represent filters and summarizers is calculated; those that exceed the established threshold will form candidate summaries. Then, the candidate linguistic summaries are filtered and humanized using a controlled natural language, allowing the generation of the final linguistic summaries (see Algorithm 1). This algorithm constitutes an extension of the RST_LDS algorithm proposed in Iliana Pérez's doctoral thesis [37].

**Algorithm 1.**

*Input:*

*U:* data set formed by *n* objects described by the set of *A* attributes

*H:* set of attributes that describe the objects in *U* such that $p = |H|$.

$A_{LV}$: Set of linguistic variables that describe the attributes H, such that there is one linguistic variable for each attribute in *H*.

$LV_Q$: linguistic variable for the quantifiers of the summaries.

*LNC*: controlled natural language used, which includes the $LNC_{Grammar}$ grammar and the $LNC_{Dictionary}$ dictionary, with simple phrases that describe the variables and attributes of the problem in question.

*T:* traditional quality assessment indicators of linguistic summaries

$k_1$, alpha cut, which establishes the upper limit for the dependence in degree *k*.

$k_2$, threshold that establishes a lower limit for dependence in degree *k*.

*Begin*

1. *Transform U into a linguistic dataset considering fuzzy sets and maximum membership principles.*

2. *Create attribute sets R_Set = S_Set = A*

3. $S \subseteq A$ set of selected attributes for summaries.

4. *CandidatesSum = {},*

5. *TempStack = Stack.new*

6. *While TempStack is not empty*

6.1. *R_Set = TempStack.pop*

6.2. *for each* $B \subseteq R\_Set$

6.3. *for each* $X \subseteq S\_Set$: $B \cap X = \varnothing$

6.4. $k(X,B) = \frac{|POS_B(D)|}{|U|}$ $POS_B(D) = \bigcup_{X \in U/D} B_*(X)$, $B \Rightarrow_k D$, $k(B, X) \geq \alpha_{k2}$

6.5. *for each* $O_t(B,X) \in POS_B(D)$

6.6. *If* $(k(B, X) \geq \alpha_{k1})$ *then*

6.7. $CandidatesSum = CandidatesSum \cup O_t(B,X)$

6.8. *Else*

6.9. *If* $(k(B, X) \geq \alpha_{k2})$ *then*

6.10. *TempStack.push*($\{B \cup X\}$)

*End of the conditional started in step 6.9*

*End of the conditional started in step 6.6*

*End of the cycle started in step 6.5*

*End of the cycle started in step 6.3*

*End of the cycle started in step 6.2*

*End of the cycle started in step 6*

*Step.7.* *Summaries = generate_summaries_from_candidates(U, CandidatesSum,* $A_{LV}$, $LV_Q$, *T)*

*Step.8.* *calculate_quantifiers(Summaries,* $LV_Q$*)*

*Step.9.* *calculate_T(Summaries, α)*

*Step.10.* *LinguisticSummaries = personificate_summaries(Summaries,* $A_{LV}$, $LV_Q$ *, LNC)*

*Step.11.* *For each P(y, Q, R, S, T) in Summaries do*

Step.11.1. *Update BRasa_Knowledge Database*

*Step.12.* *return LinguisticSummaries*

In step 1, the information system *S* is transformed into a linguistic data set considering fuzzy sets and the principles of maximum membership.

In step 2, each element $a \in A$ is transformed into a set $\{a\}$ and inserted into a set R_*Set*, Example: $\{a_1, a_2, a_3\} \geq \{\{a_1\}, \{a_2\}, \{a_3\}\}$. Then in steps 3 and 4 the *S_Set* variable and the *Stackset* stack are initialized.

Step 6 constitutes the heart of this algorithm, in steps 6.2 and 6.3 pairs of sets $B$ and $X$ are created that represent possible filters and summarizers respectively. Then in step 6.4 the positive region $POS_B(D)$ is calculated. The condition $k(B, X) \geq \alpha_{k2}$, helps to find summaries between attributes of partial dependence, but considering the sets of attributes that have a minimum level of relationship. Then in step 6.6, $O_t \in U$, $O_t(B, X)$ represents the values of the attributes $a \in B \cup X$ in the object $O_t$; if there is dependency greater than the threshold $\alpha_{k1}$ in the context $O_t$, then the values of the attributes $O_t(B, X)$ will be used as candidate summaries. If the dependency found meets that $k(B, X) \in [\alpha_{k2}, \alpha_{k1}]$, the set is generated $Z = B \cup X$ and this new set is added to the stack to continue the search process, where this new set will become part of filters in new summaries.

In step 7, the pairs of sets that are in the *CandidatesSum* set of candidate summaries are traversed, each set representing a possible linguistic summary. In step 8, for each of the candidate summaries, the quantifier is determined. In step 9, the $T$ indicators are calculated, which allow the quality of the linguistic summaries obtained to be evaluated. Then, in step 10, Algorithm 2 is applied to generate the linguistic summaries, wanting them to have expressiveness.

Then in step 10 the final summaries are generated in *LinguisticSummaries* in the shape $P_{linguistic}(y, Q, R, S, T, summary)$. In this new format the values $y$, $Q$, $R$, $S$ constitute linguistic expressions. The "*summary*" component is the linguistic summary obtained from the integration of the $P_{linguistic}$ elements with the form "*Qy's are S*" and "*QRy's are S*" [37].

Step 11 each structured summary of the $P_{linguistic}(y, Q, R, S, T, summary)$ is used to update *BRasa_Knowledge Database,* with different combinations in close relationship with the traditional protoforms of linguistic summaries as shown in the Table 1.

**Table 1** Different intensions and knowledges learned

| Case | Linguistic summary obtained | Utterance or question | Intension | Answer |
|---|---|---|---|---|
| 0 | *QRy's* are *S* | Utterance deals with the certainty of linguistic summaries | *Know certainty* | *T certainty indicators* |
| 1 | *Qy's* are *S* | *Utterance asking about summary's quantifier* | *Know summary quantifier's behavior* | *Q behavior as answer* |
| 2 | *QRy's* are *S* | *Utterance asking about combining summary S value and filters R* | *Know summary quantifier's behavior* | *Q behavior as answer* |
| 3 | *QRy's* are *S* | *Open asking about "y" entity* | *About entity "y"* | *The complete linguistic summary QRy's* are *S* |

**Algorithm 2** personificate_summaries (CandidateSummaries, $A_{LV}$, $LV_Q$, LNC)

*Inputs*

*CandidateSummaries: list of candidate summaries*

*$A_{LV}$, $LV_Q$, LNC: parameters described in section 2.1 on notations.*

*Begin:*

*Step 1.* *summaries = []*

*Step 2.* *For each P(Q, R, S, T) in CandidateSummaries*

*Step 3.* *translations = []*

*Step 4.* *For each element de P(Q, R, S, T)*

*Step 5.* *element_translation=search_dictionary(element, $A_{LV}$, $LV_Q$, $LNC_{Dictionary}$)*

*Step 6.* *translations ← element_translation*

i. *End of step 4*

*Step 7.* *sentence = generate_sentence(translations, $LNC_{Grammar}$)*

*Step 8.* *summaries ← sentence*

i. *End of step 2*

*Step 9.* *Return summaries*

In this algorithm, the set of candidate summaries is traversed, where each summary has the structure $P(y, Q, R, S, T)$. The different elements that make up the summary are translated using the $LNC_{Dictionary}$ defined for the LNC in the specific application domain.

## 2.3 BRasa_Prescriptive

In the prescriptive environment, knowledge is introduced in the form of json files. These files are uploaded to the system and contain the different elements of the Smart Chatbot:

- Intentions: indicate user actions associated with the exchange of information.
- Utterance: include questions and other expressions associated with frequently asked questions that allow the user's intention to be identified.
- Responses: sentences or linguistic expressions that constitute responses or exchanges of information with the system.

- Entities: constitute objects, generally common nouns, associated with the application domain.
- Synonyms: are different terms with the same semantic meaning, they can even be phrases that indicate that we are talking about a particular entity.
- Lookups: they constitute prototypes of expressions that facilitate the construction of user questions with the same formats.
- Rules: that relate users' intentions to possible responses or actions to be executed.
- Stories: a script that simulates a sequence of dialogues between users and the chatbot based on previous experiences. These stories are used in learning the neural network to improve its prediction and response capacity.

This approach to feeding the knowledge base is the costliest human effort. But it can be combined with other technologies to achieve automatic or semi-automatic construction of feed files.

## *2.4 Conversational Drive Development (CDD) Subsystem*

This subsystem contributes to the learning process by feeding the BRasa_Knowledge knowledge base. The CDD system itself is itself a conversational system aimed at providing information to the knowledge base. Its behavior is governed by the following principles:

- The human agents that interact with the CDD assume behaviors aimed at transmitting information and knowledge rather than waiting for or receiving information.
- The exchange of audio messages is the fundamental style of information provision, facilitating agility in the transmission of information.
- The messages transmitted from the human agent to the conversational system can be presented in an imperative way to represent concepts and definitions.
- Information exchanges in this way do not introduce or consult services, they are fundamentally oriented towards the construction of ontologies in a certain application domain.
- Various forms are introduced in the exchange of information that facilitate the confirmation of the human agent's gratitude with the CDD, with a higher level than in the case of the conversational system oriented to end users.

Figure 3 is an example of the behavior of the CDD in the exchange of information with the human agent.

Fig. 3 Example of a conversation in BRasa in a reservation environment, responsible for BRIntelligent services

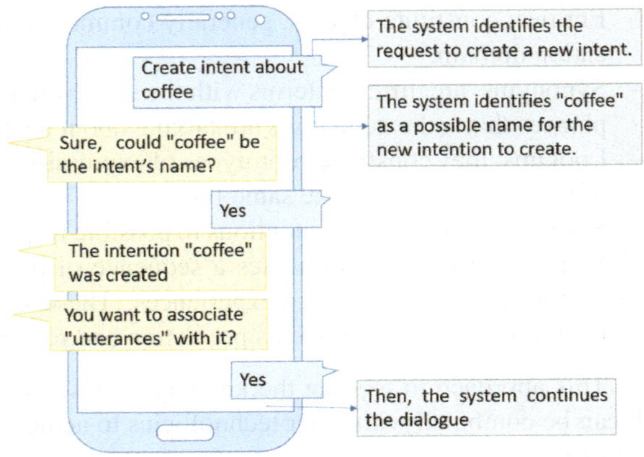


## 2.5 *Trainer Subsystem*

The trainer subsystem receives as input a set of files in yml format. These files contain the knowledge associated with the competencies collected in BRasa_Knowledge. The input files are made up of the aggregation of skills required to simulate a bot specialized in a certain domain.

Once the files have been received, a deep neural network training process occurs. As a result of this process, a trained neural network model is obtained, ready for interaction with human clients.

The trained model is then coupled to the customer response subsystem. This subsystem is made up of a deep neural network that quickly answers questions.

## 2.6 *User Response Model Subsystem*

This subsystem manages responses to the client. This system is made up of three fundamental components:

- The natural language processing and understanding component. This module allows the detection of intentions and the identification of the entities involved in the conversation.
- The dialogue management component is responsible for constructing responses following the three strategies explained in the following subsections.
- The natural language generation component.

In this subsystem, dialogue management is essential, which is based on three fundamental strategies:

1 The response model receives a question from the client that it is able to answer based on the training received. The response capacity of this strategy will depend on the technological infrastructure that supports the module.
2 The model receives a question from the customer that involves requesting a service, such as reservations, online purchases, etc. In this scenario, the request is sent to the BRIntelligent Services and Data Analysis Subsystem, which manages service interactions with the user. The basis of this module is its integration with the services of the BRasa_Knowledge subsystem.
3 The model receives a question that does not fit in the two previous scenarios and for which the pretrained system has no answer. In this case, the question asked by the user is transferred to the BRasa_IRL subsystem, which is capable of searching other sources of information.

The responses constructed from any of the three aforementioned strategies are processed by the natural language generation module. This module has the ability to humanize conversations or construct responses.

## 2.7 BRIntelligent Data Analysis and Services Subsystem

This subsystem allows the solution to be integrated into information retrieval and processing services integrated into the BusinessRedmine ecosystem. This ecosystem constitutes a highly configurable environment with applications in the following contexts:

- Management of portfolios, programs, and projects, facilitating schedule management, project file management, and human and non-human resources, among others. It provides numerous services that incorporate soft computing techniques for decision-making and automatic project evaluation [42].
- Management of virtual stores, including services associated with sales and advanced product search.
- Online reservation management, including services associated with reservations in hotels, restaurants, specialized cafeterias, or any other service.

## 2.8 BRasa Information Retrieval Subsystem

This subsystem has the function of answering questions that the rest of the subsystems have not been able to resolve. This subsystem is designed to retrieve information from different data sources, among which are:

- Consult LLM systems for the construction of answers [43].
- Search facilities on the websites of the entities and the URLs of specific domains.
- Furthermore, as part of the recovery mechanisms, this subsystem allows learning from text corpora.

## 2.9 Conversations and Stories Management Subsystem

The main function of this subsystem is to collect and store all conversations between the chatbot and the user in BRasa_Knowledge. Through this subsystem, the platform has the ability to extract new training examples from the conversations in which it participates. As our agent engages in a conversation, it also estimates user gratitude in their responses. When the conversation seems to be going well, the user's responses become new training examples to emulate. When the agent believes he has made a mistake, he asks for feedback; Learning to predict feedback further improves the Chatbot's dialogue capabilities.

In addition, it includes facilities for evaluating the system's responses to user requests. In particular, it combines elements of fuzzy logic and neutrosophic theory [44–48] and emits for each conversation three possible values: degree of certainty, degree of falsity, and degree of neutrality, see Fig. 4.

In this research, it is important to know some fundamental elements of the neutrosophic theory that are presented in [44, 45]. Definition 1 is presented below, where the neutrosophic set used in this research is formalized.

**Definition 1** *Neutrosophic set*: let *X*, be a universe of discourse, a neutrosophic set *A* over *X* is an object of the form: $A = \{\langle \mu_A(x), \tau_A(x), \sigma_A(x) \rangle : x \in X\}$, see Fig. 5, then:

- $\mu_A(x) \in [0,1]$ is a membership function that represents the degree of certainty of the membership of the value *x* to the set *A*, see Eq. 1.
- $\tau_A(x) \in [0,1]$ is a membership function that represents the degree of indeterminacy of the value *x* to the set *A*, see Eq. 2.

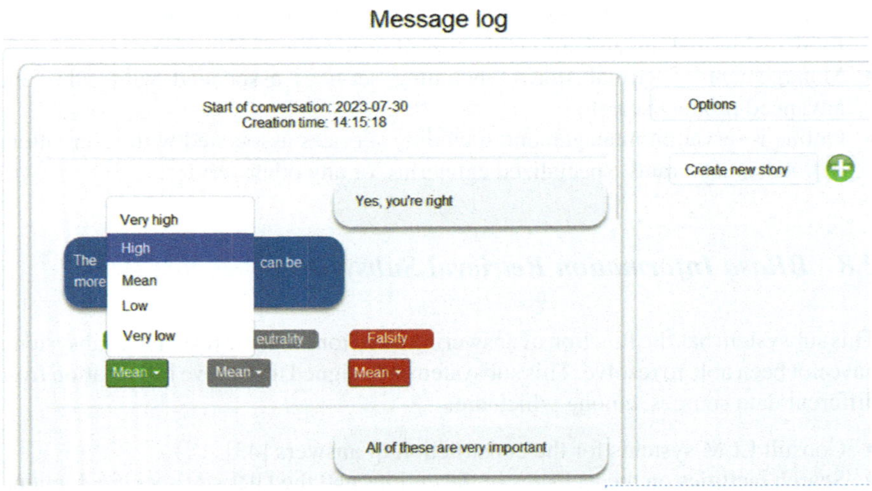


**Fig. 4** View of the evaluation of conversations between the system and users

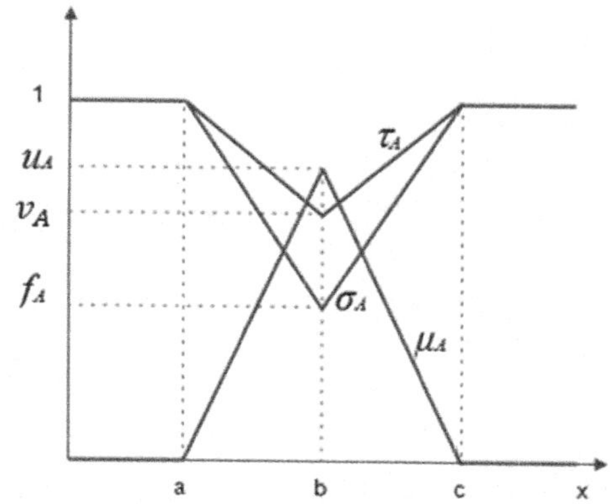


**Fig. 5** Neutrosophic set incorporating the values of truthfulness, neutrality, and falsehood

- $\sigma_A(x) \in [0,1]$ membership function that represents the degree of non-membership (or falsity) of the value $x$ to the set $A$, see Eq. 3.
- $0 \le \mu A(x) + \tau A(x) + \sigma A(x) \le 3 \forall x \in X$

$$\mu_A(x) = \begin{cases} (x-a)u_A/(b-a) & (a \le x < b) \\ u_A & (x = b) \\ (c-x)u_A/(c-b) & (b < x \le c) \\ 0 & \text{any other case} \end{cases} \tag{1}$$

$$\tau_A(x) = \begin{cases} (b-x+v_A(x-a))/(b-a) & (a \le x < b) \\ r_A & (x = b) \\ (x-b+v_A(c-x))/(c-b) & (b < x \le c) \\ 1 & \text{any other case} \end{cases} \tag{2}$$

$$\sigma_A(x) = \begin{cases} (b-x+f_A(x-a))/(b-a) & (a \le x < b) \\ f_A & (x = b) \\ (x-b+f_A(c-x))/(c-b) & (b < x \le c) \\ 1 & \text{any other case} \end{cases} \tag{3}$$

The use of neutrosophic theory allows the system's responses to be evaluated with the support of a human assistant. This conception makes it easier for the conversational system to simulate behavior that allows it to:

- Describe or predict the case in question, what was and what will be the case. Also, explain what is not the case and what could not be the case. The treatment of neutrality and falsity can help in this regard.
- The BRasa_Knowledge system would allow storing both correct and incorrect expressions. The treatment of falsity in evaluations makes it possible to identify as "impossible", the incorrect answers to certain questions. When answering a question, if we eliminate the "impossible", what remains as an answer will be what is rationally possible.

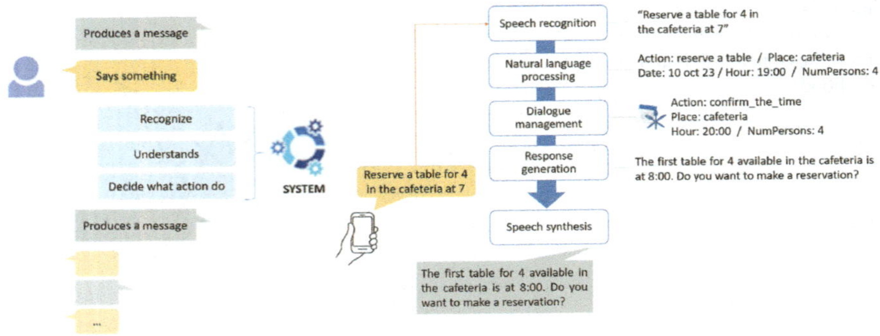


**Fig. 6** View of the behavior of the information exchange subsystems. Left: text-message-based environment. Right: Audio-message-based environment

Another important functionality of this subsystem is conversation management, which consists of saving the complete sequence of conversations between the conversational system and the user, as a script or story. These saved stories are incorporated into the BRasa_knowledge knowledge base, improving the ability to predict the responses to the user.

### 2.9.1 Information Exchange Subsystem

This subsystem is oriented to the exchange of information with the client. In this sense, two possible communication environments are established, see Fig. 6:

- Environment based on communication using text messages, see Fig. 6 at left
- Environment based on the exchange of audio messages, Fig. 6 at right, based on the principles proposed in [49].

The figure shows the flow of information and the sequence of information exchange in both cases.

## 3 Analysis of Results

The validation of the proposal is carried out based on empirical tests in two different scenarios: support projects for the ENZONA electronic commerce platform and support for integrated project management from the BusinessRedmine Ecosystem.

ENZONA is a software platform dedicated to electronic commerce, making it easier to carry out financial and commercial operations [50, 51]. It allows online payment for products and services in virtual and physical stores subscribed to the platform. For this platform, customer service through online support services that cover 24 h a day, seven days a week, is essential. Support for the platform was offered in person, by mail, and by telephone. ENZONA has grown significantly in terms of users and communication channels. In this context, traditional support systems are insufficient, affecting the perceived quality of users. As part of this work, a conversational system built from the basic structure of the architecture (BRasa) proposed in this work was introduced. The following sections show the application results.

## 3.1 *Definition of the Variables Used in Validation*

The quality of the system's behavior is measured by two variables: efficacy and efficiency, taking into account what is established in the ISO 9214 standard [52]. The evaluation period was 10 months, and efficacy was measured manually through a daily review of the dialogues between the chatbot and users by a human expert.

Table 2 describes the independent variables used in the experimentation, some of which are used in other works [53].

The dimensions and indicators of the dependent variable "efficacy" are described below (see Table 3).

To measure the dimension "Correct identification of intentions" indicators are used to evaluate classifier systems. These indicators are adapted to evaluate the quality of the responses and are based on the identification of the values: False Positive (FP), False Negative (FN), True Positive (TP) and True Negative (TN). From these primary indicators, the measures are calculated: accuracy (Eq. 4), precision (Eq. 5), recall (Eq. 6) and F1_score (Eq. 7).

- False Positive (FP): the conversational system identifies an intention and responds with a high degree of certainty incorrectly. It does not clearly identify the user's intention on a topic for which they were not previously trained.
- False Negative (FN): the conversational system does not recognize an intention for which it was previously trained. He acts incorrectly, indicating that he does not know the particular topic, although a correct identification was expected.

**Table 2** Definition of independent variables

| Variable | Dimensions | Indicators | Type |
|---|---|---|---|
| Efficiency | Size of database knowledge | Number of intensions | Numeric |
| | | Number of rules | Numeric |
| | | Responses | Numeric |
| | | Entities | Numeric |
| | | Custom actions | Numeric |
| | Messages per seconds | Basic technological infrastructure required | Numeric |
| | | Number of messages per seconds | Numeric |

**Table 3** Dimension of the dependent variable "efficacy"

| Variable | Dimensions | Indicators | Type |
|---|---|---|---|
| Efficacy | Correctly identified intentions | Accuracy | Numeric |
| | | Precision | Numeric |
| | | Recall | Numeric |
| | | F1_score | Numeric |
| | Level of interaction with users | Number of responses compared to other channels | % |
| | | Message length | Numeric |
| | Gratitude messages | Gratitude | % |
| | | Displeasure | % |

- True Positive (TP): when the conversational system receives an utterance or question and provides the expected response correctly.
- True Negative (TN): when the conversational system correctly recognizes a new topic and returns the answers associated with the unknown topic or intention.

Accuracy: a measure of the total quality, the result is better when accuracy is high, see Eq. 4.

$$\text{accuracy} = \frac{\text{TP} + \text{TN}}{\text{TP} + \text{TN} + \text{FP} + \text{FN}} \tag{4}$$

Precision: This measure is associated with the learning process. The best results are obtained when the conversational system makes no errors in solving known problems. A high value of this metric indicates good answers from the conversational system, see, see Eq. 5.

$$\text{precision} = \frac{\text{TP}}{\text{TP} + \text{NP}} \tag{5}$$

Recall: measures the number of correct answers when the system faces situations for which it was pre-trained. The result is better when the recall is greater, see Eq. 6.

$$\text{recall} = \frac{\text{TP}}{\text{TP} + \text{FN}} \tag{6}$$

F1_score: Used to combine the precision and recall measures into a single value. The smaller this indicator is, the better the response quality of the conversational system, see Eq. 7.

$$\text{F1\_score} = 2 * \frac{\text{recall} + \text{precision}}{\text{recall} * \text{precision}} \tag{7}$$

**Table 4** Evaluation of the independent variable "Structure of the Knowledge Base"

| Dimensions | Domain | Indicators | Valor |
|---|---|---|---|
| Size of database knowledge | E-Commerce domain | Number of intensions | 148 |
| | | Number of rules | 148 |
| | | Responses | 152 |
| | | Entities | 2 |
| | | Custom actions | 2 |
| | Project managent domain | Number of intensions | 292 |
| | | Number of rules | 298 |
| | | Responses | 294 |
| | | Entities | 284 |
| | | Custom actions | 36 |

## 3.2 *Results of the Evaluation of the Independent Variable "Efficiency"*

Regarding the independent variable Efficiency in the "Size of the knowledge base" dimension, Table 4 shows the behavior in both application domains.

The knowledge base used in the chatbot built for e-commerce was built using the prescriptive strategy. While the chatbot for project management was built by combining linguistic data summary techniques with semi-automatic text generation techniques, as Table 4 shows, the size of the knowledge base of the project management chatbot is significantly larger than that of e-commerce. Furthermore, the project management knowledge base was built in a much shorter time and with less participation from human resources. This demonstrates the viability of using linguistic data summary techniques in the construction of knowledge bases.

Regarding the required technological infrastructure, the chatbots in operation were mounted on a basic infrastructure of 16 GB of RAM and 8 vCPU cores. In this environment, it was demonstrated that the proposed architecture can process approximately 300 messages per second with a latency of less than 500 ms. With a performance similar to other chatbots reported in the consulted bibliography [54–56].

## 3.3 *Results of the Evaluation of the Dependent Variable "Efficacy"*

Regarding efficacy, a high level of system response is achieved in both application domains. A high level of "Correct identification of intentions" is achieved, as shown in Table 5.

A higher level of response is identified in the E-commerce domain than in the Project Management domain, although not significantly. This behavior is expected

**Table 5** Evaluation of the dependent variable "Quality of customer response"

| Dimensions | Domain | Accuracy | Precision | Recall | F1-score |
|---|---|---|---|---|---|
| Correct identification of intentions | E-commerce domain | 0.985 | 0.989 | 0.98 | 4.06 |
| | Project management domain | 0.91 | 0.955 | 0.937 | 4.23 |

considering that the size of the project management knowledge base is significantly larger than that of e-commerce in the application scenarios.

Regarding the dimension "Level of interaction with users" the following results are obtained.

The number of responses to users attended to as part of the support services is also compared. In this case, the proposed architecture is compared with other support channels established in the two domains discussed, see Table 6.

Table 6 shows that the responsiveness of chatbots on the built platform was significantly higher than that of other support channels. Thus, showing its viability. On the other hand, the need to raise the technological support conditions of the platform was identified to raise its level of availability and recovery from failures.

Regarding the "Length of conversations" indicator, the conversational systems built for each domain behaved as shown in Table 7.

This behavior is normal considering the complexity of e-commerce operations compared to project management operations. Furthermore, a directly proportional relationship is identified between the size of both domains and the length of the conversations.

The dimension "Level of gratitude expressed directly by users" was measured from the messages of gratitude or displeasure received by the system. No survey was carried out on users about their level of gratitude with the answers received to avoid

**Table 6** Percentage of responses answered to doubts and support requests in less than 24 h

| Dimension | Email | Phone calls | Chatbots with the proposed architecture |
|---|---|---|---|
| Number of responses compared to other channels (%) | 37,19 | 39,58 | 68.75 |

**Table 7** "Average number of messages per conversation" indicator

| Conversational UI Type | Number of messages | Percent (%) |
|---|---|---|
| Project management environment | 1–5 messages | 69 |
| | 6–10 messages | 22 |
| | more than 10 messages | 9 |
| Electronic commerce environment | 1–5 messages | 76 |
| | 6–10 messages | 16 |
| | More than 10 messages | 8 |

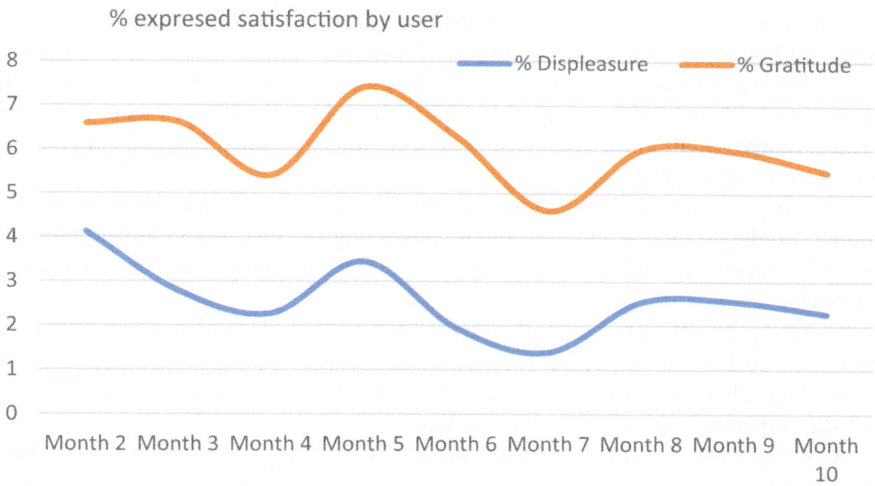


**Fig. 7** View of gratitude expressed directly by users

possible bias in the questions. The results are shown in Fig. 7. Gratitude levels were calculated in percentages, with 6.05 being the average value of gratitude, above the average value of expressed displeasure, 2.6. The standard deviation behaved similarly in both levels of gratitude and displeasure, with values of 0.81 and 0.82, respectively. In kurtosis, a significant difference was also identified, being 0.74 in displeasure and 0.39 in gratitude.

In general, is positive, that the level of gratitude is above the level of expressed displeasure. However, as future work, the authors identify the need to introduce greater diversity in the ways of measuring perceived quality. The need to achieve a higher level of humanization of the responses is also identified.

## 4 Conclusions

In the current development of conversational systems, two fundamental approaches are identified. Smart rule-based chatbots and Large Language Models. Both models incorporate advantages and disadvantages associated with the quality of the answers, the capacity of explain and continuous learning. The proposed architecture proposes the combination of both approaches, enhancing the advantages of each one.

The viability of the proposed architecture for the construction of multiple conversational systems, specialized in specific domains, was demonstrated. The platform is tested in the domains of e-commerce and project management.

In both application scenarios, a high level of response was achieved from the chatbots built using the proposed architecture. Furthermore, the level of gratitude expressed directly in the messages is higher than the level of displeasure.

Building conversational systems using prescriptive methods is costly in terms of time and human resources. In the proposal made, the need to include various ways of automatically or semi-automatically feeding the knowledge base was identified. In particular, the use of linguistic data summary techniques allowed us to build the project management knowledge base with a significantly greater scope than that of e-commerce and in less time.

The responsiveness of the chatbots built using the BRasa platform positively influenced the efficiency and efficacy of responding to users. A significantly higher response level was achieved than other customer support and communication channels.

To set up an efficient ecosystem associated with conversational systems, it is necessary to have a dedicated technological infrastructure with a high level of computing where several applications converge. The use of Docker technologies or other superior technologies is recommended.

As future work, the authors identify the need to introduce new ways of measuring the perceived quality of the end users of the system. The need to achieve a higher level of humanization of the responses and a greater linkage of reinforcement learning techniques is also identified.

## References


1. Brandtzaeg PB, Følstad A (2018) Chatbots: changing user needs and motivations. Interactions 25:38–43. https://doi.org/10.1145/3236669
2. Galitsky B (2019) Developing enterprise chatbots: learning linguistic structures. Springer International Publishing, Cham
3. O'Leary DE (2023) An analysis of three chatbots: BlenderBot, ChatGPT and LaMDA. Intell Syst Account Finance Manag 30:41–54. https://doi.org/10.1002/isaf.1531
4. Markets and Markets (2023) Conversational AI Market Size, Industry Share Forecast 2028. In: MarketsandMarkets. https://www.marketsandmarkets.com/Market-Reports/conversational-ai-market-49043506.html. Accessed 20 Aug 2023
5. Adamopoulou E, Moussiades L (2020) An overview of chatbot technology. In: Maglogiannis I, Iliadis L, Pimenidis E (eds) Artificial intelligence applications and innovations. Springer International Publishing, Cham, pp 373–383
6. Patel S (2019) Top 12 chatbots trends and statistics to follow in 2023. In: REVE Chat. https://www.revechat.com/blog/chatbots-trends-stats/. Accessed 4 Sep 2023
7. iAdvize (2023) 5 trends to watch in conversational AI. https://www.iadvize.com/fr. Accessed 4 Sep 2023
8. Følstad A, Nordheim CB, Bjørkli CA (2018) What makes users trust a chatbot for customer service? An exploratory interview study. In: Bodrunova SS (ed) Internet science. Springer International Publishing, Cham, pp 194–208
9. SoftWeb Solutions C (2022) Top chatbot trends to keep an eye on in 2023. In: Softwebsolutions. https://www.softwebsolutions.com/resources/chatbot-trends-2023.html. Accessed 20 Aug 2023
10. Deriu J, Rodrigo A, Otegi A, Echegoyen G, Rosset S, Agirre E, Cieliebak M (2021) Survey on evaluation methods for dialogue systems. Artif Intell Rev 54:755–810. https://doi.org/10.1007/s10462-020-09866-x
11. Ramesh A, Chawla V (2022) Chatbots in marketing: a literature review using morphological and co-occurrence analyses 57. https://doi.org/10.1177/10949968221095549

12. Kooli C (2023) Chatbots in education and research: a critical examination of ethical implications and solutions. Sustainability 15:5614. https://doi.org/10.3390/su15075614
13. Lin C-C, Huang AYQ, Yang SJH (2023) A review of AI-driven conversational chatbots implementation methodologies and challenges (1999–2022). Sustainability 15:4012. https://doi.org/10.3390/su15054012
14. Abdelghani R, Wang Y-H, Yuan X, Wang T, Lucas P, Sauzéon H, Oudeyer P-Y (2023) GPT-3-driven pedagogical agents for training children's curious question-asking skills. Int J Artif Intell Educ. https://doi.org/10.1007/s40593-023-00340-7
15. Ahuja AS, Polascik BW, Doddapaneni D, Byrnes ES, Sridhar J (2023) The digital metaverse: applications in artificial intelligence, medical education, and integrative health. Integr Med Res 12:100917. https://doi.org/10.1016/j.imr.2022.100917
16. Landim ARDB, Pereira AM, Vieira T, de B. Costa E, Moura JAB, Wanick V, Bazaki E (2022) Chatbot design approaches for fashion E-commerce: an interdisciplinary review. Int J Fash Des Technol Educ 15:200–210. https://doi.org/10.1080/17543266.2021.1990417
17. Wilson L, Marasoiu M (2022) The development and use of chatbots in public health: scoping review. JMIR Hum Factors 9:e35882. https://doi.org/10.2196/35882
18. Nicolescu L, Tudorache MT (2022) Human-computer interaction in customer service: the experience with AI chatbots—A systematic literature review. Electronics 11:1579. https://doi.org/10.3390/electronics11101579
19. Mohamad Suhaili S, Salim N, Jambli MN (2021) Service chatbots: a systematic review. Expert Syst Appl 184:115461. https://doi.org/10.1016/j.eswa.2021.115461
20. Tran DH, Sheng QZ, Zhang WE, Hamad SA, Khoa NLD, Tran NH (2022) Deep conversational recommender systems: challenges and opportunities. Computer 55:30–39. https://doi.org/10.1109/MC.2020.3045426
21. Sharma RK, National Informatic Center (2020) An analytical study and review of open source chatbot framework, Rasa. Int J Eng Res V9:IJERTV9IS060723. https://doi.org/10.17577/IJERTV9IS060723
22. Omar AA, Maginga TJ, Rutunda S, Nzanywayingoma F, Nsenga J (2023) Developing an IoT-based conversational AI recommender assistant for vital sign predicted anomalies. In: Proceedings of the 2023 8th international conference on machine learning technologies. Association for Computing Machinery, New York, NY, USA, pp 23–27
23. Chen S, Wu M, Zhu KQ, Lan K, Zhang Z, Cui L (2023) LLM-empowered chatbots for psychiatrist and patient simulation: application and evaluation
24. Cabrera J, Loyola MS, Magaña I, Rojas R (2023) Ethical dilemmas, mental health, artificial intelligence, and LLM-based chatbots. In: Rojas I, Valenzuela O, Rojas Ruiz F, Herrera LJ, Ortuño F (eds) Bioinformatics and biomedical engineering. Springer Nature Switzerland, Cham, pp 313–326
25. Fan L, Li L, Ma Z, Lee S, Yu H, Hemphill L (2023) A bibliometric review of large language models research from 2017 to 2023
26. Hancock B, Bordes A, Mazaré P-E, Weston J (2019) Learning from dialogue after deployment: feed yourself, chatbot!
27. Santos GA, de Andrade GG, Silva GRS, Duarte FCM, Costa JPJD, de Sousa RT (2022) A conversation-driven approach for chatbot management. IEEE Access 10:8474–8486. https://doi.org/10.1109/ACCESS.2022.3143323
28. Kapočiūtė-Dzikienė J (2020) A domain-specific generative chatbot trained from little data. Appl Sci 10:2221. https://doi.org/10.3390/app10072221
29. Saka AB, Oyedele LO, Akanbi LA, Ganiyu SA, Chan DWM, Bello SA (2023) Conversational artificial intelligence in the AEC industry: a review of present status, challenges and opportunities. Adv Eng Inform 55:101869. https://doi.org/10.1016/j.aei.2022.101869
30. AlKhamissi B, Li M, Celikyilmaz A, Diab M, Ghazvininejad M (2022) A review on language models as knowledge bases
31. Li J, Miller AH, Chopra S, Ranzato M, Weston J (2017) Dialogue learning with human-in-the-loop

32. Khatri S, Iqbal M, Ubakanma G, van der Vliet-Firth S (2022) SkillBot: towards data augmentation using transformer language model and linguistic evaluation. In: 2022 human-centered cognitive systems (HCCS), pp 1–9
33. Pérez-Soler S, Guerra E, de Lara J (2020) Model-driven chatbot development. In: Dobbie G, Frank U, Kappel G, Liddle SW, Mayr HC (eds) Conceptual modeling. Springer International Publishing, Cham, pp 207–222
34. Pérez-Soler S, Daniel G, Cabot J, Guerra E, de Lara J (2020) Towards automating the synthesis of chatbots for conversational model query. In: Nurcan S, Reinhartz-Berger I, Soffer P, Zdravkovic J (eds) Enterprise, business-process and information systems modeling. Springer International Publishing, Cham, pp 257–265
35. Mazumder S, Ma N, Liu B (2018) Towards a continuous knowledge learning engine for chatbots
36. Bocklisch T, Faulkner J, Pawlowski N, Nichol A (2017) Rasa: open source language understanding and dialogue management
37. Pérez Pupo I (2021) Algoritmos para la sumarización lingüística de datos para la ayuda a la toma de decisiones. Doctoral, Centro de Estudios de Gestión de Proyectos y Toma de Decisiones, Universidad de las Ciencias Informáticas
38. Pérez Pupo I, Piñero Pérez PY, Al-subhi SH, Mahdi GSS, Bello Pérez RE (2022) Linguistic data summarization with multilingual approach. In: Piñero Pérez PY, Bello Pérez RE, Kacprzyk J (eds) Artificial intelligence in project management and making decisions. Springer International Publishing, Cham, pp 39–64
39. Pérez Pupo I, Piñero Pérez PY, Bello Pérez RE, García Vacacela R, Villavicencio Bermúdez N (2022) Linguistic data summarization: a systematic review. In: Piñero Pérez PY, Bello Pérez RE, Kacprzyk J (eds) Artificial intelligence in project management and making decisions. Springer International Publishing, Cham, pp 3–21
40. Pérez Pupo I, Piñero Pérez PY, Bello R, Acuña LA, Vacacela RG (2020) Linguistic summaries generation with hybridization method based on rough and fuzzy sets. In: International joint conference on rough sets. Lecture notes in computer science, vol 12179, LNAI. Springer, pp 385–397. ISSN:0302-9743, ISBN:978-3-030-52704-4
41. A Ramos-Soto P Martin-Rodillab 2019 Ramos-Soto A, Martin-Rodillab P (2019) Enriching linguistic descriptions of data: a framework for composite protoforms. Fuzzy Sets Syst 26. https://doi.org/10.1016/j.fss.2019.11.013
42. Piñero Ramírez PE, Pérez Pupo I, Piñero Pérez PY, Marquez Ruiz Y, Fustiel Alvarez Y (2022) A software ecosystem for project management in BIM environments assisted by artificial intelligent techniques. In: Piñero Pérez PY, Bello Pérez RE, Kacprzyk J (eds) Artificial intelligence in project management and making decisions. Springer International Publishing, Cham, pp 191–212
43. Berti A, Qafari MS (2023) Leveraging large language models (LLMs) for process mining (Technical Report)
44. Pérez Pupo I, Piñero Pérez PY, García Vacacela R, Bello R, Santos Acosta O, Leyva Vázquez MY (2018) Extensions to linguistic summaries indicators based on neutrosophic theory: applications in project management decisions. Neutrosophic Sets Syst Univ N M 22:87–100
45. Pérez Pupo I, Piñero Pérez PY, Bello Pérez RE (2022) New indicators for the assessment of linguistic summaries considering a rough sets approach. In: Piñero Pérez PY, Bello Pérez RE, Kacprzyk J (eds) Artificial intelligence in project management and making decisions. Springer International Publishing, Cham, pp 99–120
46. Sahin M, Abdullah Kargin FS (2018) Generalized single valued triangular neutrosophic numbers and aggregation operators for application to multi-attribute group decision making. New Trends Neutrosophic Theory Appl II:51–84
47. Smarandache F (2010) Neutrosophic set-a generalization of the intuitionistic fuzzy set. J Def Resour Manag 1:107
48. Goguen JA, Zadeh LA (1973) Fuzzy sets. Inf Control 8(1965):338–353; Zadeh LA (1971) Similarity relations and fuzzy orderings. Inf Sci 3:177–200; J Symb Log 38:656–657. https://doi.org/10.2307/2272014

49. Janarthanam S (2017) Hands-on chatbots and conversational UI development: build chatbots and voice user interfaces with Chatfuel, Dialogflow, Microsoft Bot Framework, Twilio, and Alexa Skills. Packt, Birmingham Mumbai
50. Enzona (2023) Portal Web ENZONA. https://www.enzona.net/
51. XETID (2023) Portal Web de la XETID. In: Portal Web XETID. https://www.xetid.cu/es. Accessed 11 Sep 2023
52. Casas J, Tricot M-O, Abou Khaled O, Mugellini E, Cudré-Mauroux P (2020) Trends & methods in chatbot evaluation. Companion publication of the 2020 international conference on multimodal interaction. ACM, Virtual Event Netherlands, pp 280–286
53. Maroengsit W, Piyakulpinyo T, Phonyiam K, Pongnumkul S, Chaovalit P, Theeramunkong T (2019) A survey on evaluation methods for chatbots. In: Proceedings of the 2019 7th international conference on information and education technology. Association for Computing Machinery, New York, NY, USA, pp 111–119
54. Chaix B, Bibault J-E, Pienkowski A, Delamon G, Guillemassé A, Nectoux P, Brouard B (2019) When chatbots meet patients: one-year prospective study of conversations between patients with breast cancer and a chatbot. JMIR Cancer 5:e12856. https://doi.org/10.2196/12856
55. Segura C, Palau À, Luque J, Costa-Jussà MR, Banchs RE (2019) Chatbol, a chatbot for the Spanish "La Liga." In: D'Haro LF, Banchs RE, Li H (eds) 9th international workshop on spoken dialogue system technology. Springer, Singapore, pp 319–330
56. Fiore D, Baldauf M, Thiel C (2019) "Forgot your password again?": acceptance and user experience of a chatbot for in-company IT support. In: Proceedings of the 18th international conference on mobile and ubiquitous multimedia. Association for Computing Machinery, New York, NY, USA, pp 1–11

# Design of a Technological System with Artificial Intelligence to Manage Projects Through the Use of Knowledge Management and Lessons Learned

**Alejandro Andrés Galleguillos Rosales, Alfredo González León, and Ingrid Luisa Galleguillos Rosales**

**Abstract** Taking advantage of accumulated knowledge and lessons learned at the project level is essential to improving efficiency, making informed decisions and optimizing results. In this context, this study is presented through a system that seeks to collect and store valuable information generated during the project life cycle and will also use advanced analysis and machine learning techniques to extract significant knowledge. Its main objective is to provide professionals and project teams with an efficient and accessible tool to access lessons learned, best practices, and recommendations, thus facilitating informed decision-making and avoiding repeating mistakes. These goals are achieved with a robust and scalable architecture, integrating various data sources, such as project reports, technical documentation, historical data, and the experience of project consultants, to build a centralized repository of knowledge. In addition, the technical feasibility of natural language processing and machine learning algorithms to analyze, classify and label the stored information will be evaluated, allowing a precise search and an enriching user experience. The design of the AI system implies not only the implementation of advanced technological infrastructure but also the definition of a framework for the capture, validation, and continuous updating of the knowledge generated by the projects, promoting a culture of organizational learning where users will be encouraged to share their experiences and good practices, thus enriching the system and benefiting the project community as a whole.

---

A. A. Galleguillos Rosales (✉)
Blumell 0643, Población Playa Blanca, Antofagasta, Chile
e-mail: alejan-dro.galleguillos@pminortechile.cl

A. González León
Universidad Católica del Norte, Avenida. Angamos, 0610 Antofagasta, Chile
e-mail: agonzale@ucn.cl

I. L. Galleguillos Rosales
Las Vegas 689, Villa Caspana, Calama, Chile
e-mail: ingrid.luisa@pminortechile.cl

 
P. Y. Piñero Pérez et al. (eds.), *Computational Intelligence in Engineering and Project Management*, Studies in Computational Intelligence 1134,
https://doi.org/10.1007/978-3-031-50495-2_9

**Keywords** Artificial intelligence · Good practices · Knowledge management · Lessons learned · Project management

## 1 Introduction

Knowledge management in Chile has been gaining importance and has experienced progress in recent years. Although each organization may have different approaches and levels of implementation of knowledge management, in general, the following characteristics can be observed.

**Initiatives in the public sector**: the Chilean government has promoted implementing knowledge management strategies in the public sector. The Ministry of Education, in conjunction with the Ministry of Telecommunications of Chile [1, 2], is implementing the Digital Transformation Strategy, which forms an integral part of the Strengthening of Learning axis of the Reactivation Plan implemented by the Ministry of Education and seeks to ensure technological resources, connectivity, and the development of digital skills in educational communities.

**Investments in education and training**: Chile has invested in improving education and training on issues related to knowledge management [3]. Academic programs have been developed, and the training of professionals specializing in knowledge management and related areas has been promoted.

**Support for innovation and entrepreneurship**: Knowledge management is closely linked to innovation and entrepreneurship. Chile is among the first places to invest in innovation in Latin America, according to [4, 5]. They have established various programs and support entities to promote the creation and development of startups and the transfer of knowledge from academia and research centers to the business sector.

**Use of information and communication technologies**: Technology plays a fundamental role in knowledge management. According to UNESCO [6–8], in Chile, STEM (science, technology, engineering, and mathematics) careers have been promoted, which are associated with increasingly essential and valuable skills, such as solving complex problems and analytical and critical thinking, with increasing participation of women.

**Networks and collaboration**: Networks and communities of practice have been established in Chile to foster collaboration and knowledge sharing among organizations, institutions, and professionals. These networks allow the sharing of experiences, lessons learned, and best practices and promote mutual learning.

According to a study by Gregorio Pérez Arrau [9], the introduction of the knowledge management model in the organizational field has been relatively slow in Chile. Although Chile has indeed stood out in the region for its progress in the economic and social spheres, it is evident that there is an evident delay in organizations in terms of adopting advanced organizational practices.

## 2 Literature Review

### *2.1 Artificial Intelligence (AI)*

Below is a summary of those recognized as the first AI researchers, such as Alan Turing [10], considered one of the founders of computer science and artificial intelligence. In 1950, he proposed the famous "Turing Test" to assess a machine's brightness. Likewise, John McCarthy is another author considered the father of artificial intelligence. In 1956, he organized the Dartmouth Conference, where the term "artificial intelligence" was coined and laid the foundation for research in this field. Another AI pioneer was Minsky [11], who significantly contributed to machine learning and computer vision. Together with who significantly contributed to machine learning and computer vision. Together with John McCarthy, he co-founded the MIT Artificial Intelligence Laboratory. Along the same lines, other authors developed the first AI program capable of proving mathematical theorems [12, 13]. They also developed the first AI program capable of proving mathematical theorems [14]. They also created the Lisp programming language, which is widely used in artificial intelligence research. Similarly, Arthur Samuel was among the first to work on machine learning [15]. In the 1950s, he developed a program capable of playing chess autonomously and improving his performance as he gained experience.

On the other hand, it is also necessary to briefly review the prominent AI authors of recent times. Considered one of the fathers of deep learning, Sabour et al. [16] has been instrumental in advancing neural networks and machine learning. His research has laid the foundation for advances in speech recognition, image processing, and many other fields of AI. Additionally, Andrew Ng is a leading researcher in machine learning and artificial intelligence. He has made significant contributions in areas such as computer vision and robotics.

Additionally, he has been an advocate and promoter of the democratization of machine learning through online educational platforms. Along the same lines, Bengio et al. [17] is known for his pioneering work on convolutional neural networks (CNN) and their application in image processing. He has conducted significant research in the area of computer vision. He has contributed to developing algorithms and techniques that have driven object and pattern recognition advances, like Bengio [18], a prominent figure in deep learning. His research focuses on neural networks and machine learning algorithms, especially deep neural networks, and their applications in natural language processing and computer vision. Finally, Russell and Norvig [19] authorizes the influential book "Artificial Intelligence: A Modern Approach." He has contributed to the development of AI and reflected on its impact on society and ethics.

## 2.2 Knowledge Management

The theories that talk about knowledge management date back about fifty years. Among the essential knowledge management theorists, we can mention Peter Senge's book The Fifth Discipline [20], which focuses on group problem-solving using methodical thinking to transform companies into learning organizations. The disciplines are approaches (theories and methods) to developing three basic learning capacities: aspiration, reflective conversation, and understanding the complex.

But what is knowledge management? What is knowledge management looking for?

There is no single definition of this concept, and many authors have defined from their experience and their reality what they understand by knowledge management.

Davenport and Prusak [21] define "knowledge management as a logical, organized and systematic process to produce, transfer and apply in concrete situations a harmonious combination of knowledge, experiences, values, contextual information and expert appreciations that provide a framework for their evaluation and incorporation of new experiences and information." By establishing for Davenport and Prusak that knowledge management is a process, they imply that, like any process, it must have an input, then a transformation, to finish with the output or outputs.

On the other hand, for Nonaka and Takeuchi [22, 23], the primary or central thing for those Japanese companies that manage their knowledge is to take advantage of all the ideas, dreams, imagination, and even hunches of the workers and make them available to the company for testing and use.

However, In [24] the changes that are currently observed in society are inevitable, whether in the social, economic, religious, or other spheres, in the same way that the Industrial Revolution, which came to replace many workers with few machines, had a direct impact and effect on people, society and countries. Likewise, from a political and economic point of view, changes are observed that are often difficult to control.

Modern organizations depend on knowledge and its exploitation to maintain an advantage over time. This situation requires an accurate understanding of knowledge management (KM) processes, specifically how knowledge is created, shared/transferred, acquired, stored/retrieved, and applied throughout an organizational system. However, since the beginning of the new millennium, these KM processes, according to [18] IEEE (2020), have been profoundly affected and modified with the appearance of the fourth industrial revolution 4.0, involving machines' interconnection and ability to learn and share data autonomously.

Knowledge management seeks to identify, capture and document relevant knowledge held by individuals, teams, and the organization. This concept may include technical knowledge, work experience, best practices, lessons learned, and internal and external information. As explained by Donald Hislop in his book: "Knowledge Management in Organizations: A Critical Introduction" [25], which provides a critical introduction to knowledge management, addressing key concepts, theories, and practices in the field. It examines how organizations can capture, store, share, and use knowledge effectively to improve performance and competitiveness.

Knowledge management is also concerned with promoting a culture of continuous and collaborative learning within the organization and encouraging the active participation of employees in the creation, co-creation, capture and dissemination of knowledge. This process may include fostering collaboration, establishing communities of practice, holding knowledge-sharing meetings, training, and professional development.

From the above, it is crucial to visualize the knowledge management process. According to Pedraza-Rejas et al. [26], it comprises three stages [27]:

i. *Creating knowledge*: through the conceived interaction between individuals, the exploitation, combination, and discovery of knowledge are considered.
ii. *Knowledge Sharing*: In this stage of knowledge sharing, people transfer the acquired knowledge
iii. *Apply knowledge*: it is possible to convert knowledge into new products or services for the organization.

## 2.3 Learned Lessons

In the PMBOK® Guide—Seventh Edition [28], the Project Management Institute (PMI) defines lessons learned as the knowledge gained during a project that shows how project events were or should be addressed in the future to improve future performance.

Lessons learned may include information on technical, practical, management, and communication aspects. These lessons can emerge from situations where challenges have been faced, effective solutions have been found, or mistakes have been made that can be avoided in subsequent projects.

Capturing lessons learned involves documenting in a systematic and structured way the knowledge acquired throughout the project, including the description of the problem or challenge, the actions taken, the results obtained, and recommendations for the future. These lessons can be shared with other professionals and project teams, so they can learn from their experiences and avoid similar mistakes.

Managing lessons learned is a fundamental practice in project management, as it helps improve project efficiency, quality, and overall performance. By applying these actions, project management professionals can avoid unnecessary risks and take advantage of identified opportunities, best practices, and knowledge gained from previous projects.

In summary, according to the PMI, knowledge and experiences acquired during the execution of a project are documented in a structured way and used to improve performance and results in future projects.

# 3 Basis for the Design Development Proposal

According to Davenport [21], knowledge management is "a logical, organized and systematic process to produce, transfer and apply in concrete situations a harmonious combination of knowledge, experiences, values, contextual information and expert appreciations that provide a framework for their evaluation and incorporation of new experiences and information".

Davenport [21] establishes the importance of distinguishing between data, information, and knowledge.

- Data: Facts, information, statistics, is the minimum unit of information, known fact about something that happened.
- Information: Data whose meaning and usefulness are used for decision-making and reduce uncertainty about something.
- Knowledge: It is that information that adds experience, are truths or principles results of some study.

It is crucial to remember the above because this design proposal is based on said classification.

Specifically, this design seeks the development of technological functionalities that allow the following:

- Determine if a project in its initial configuration has the correct parameters.
- When the projects are in execution, they can alert based on the lessons learned if preventive actions should be taken.

Of the expected benefits we have:

- Have the functionalities that can be configured to any project, program, and portfolio based on the Project Management Institute's (PMI) best practices to reduce the time in which corrective actions must be taken.

This situational analysis of scenarios and alerts should focus on more than just time, cost, and scope data. It should also consider the aspects of procurement, stakeholders, risks, and human resources.

But considering having a limited development time for the development of a minimum viable product that allows obtaining data to evaluate the effectiveness of the proposed system, it has been decided that it is in this phase that the design and development focus on the following variables: Time and Cost.

## *3.1 Methodological Guidelines for the Design Proposal*

The purpose of the design proposal is to determine the technical feasibility of generating an "Expert Project Consultant," which is based on Artificial Intelligence and remarkably simulates the activities carried out by a consultant.

For the development of the system, the following guidelines will be taken into account:

- *Guideline 01—Consultant compartment simulation*: Given that it is unrealistic to try to simulate the cognitive behavior of a consultant with all its analysis variants, an abstraction of three types of activities has been carried out, which are expected to be technical and would make it possible to create a minimum viable product. These are:
    1. Improve time prediction: It refers to the ability of an expert consultant who, when analyzing a project, detects that the activities are underestimated in the effort.
    2. Addition or elimination of activities: Refers to the capacity of an expert consultant who, when analyzing a project, detects that essential actions are missing or that some activities do not contribute to the project.
    3. Propose a sequence of activities: This refers to the ability of an expert consultant who, when analyzing a project, detects that the activities could be sequenced in a non-optimal way.
- *Guideline 02—Developed guided by Low-Code*: Considering that you want to make a technological product that is not conditioned by programming teams and, as far as possible, the end customer himself can, in some functionalities, have the possibility of improving its configuration, the work will be based on the Integrated Project Management System (SISGIP), which is developed under the Low-Code philosophy (Low programming) and the best practices of PMI.
- *Guideline 03—Knowledge-based Artificial Intelligence*: For this design proposal, a very preliminary analysis indicates that the ideal technique to develop these functionalities is knowledge-based Expert Systems, where the crucial thing is to determine the most significant number of inference rules. Notwithstanding the preceding, it is feasible to have another approach if the feasibility analysis so indicates.

## 3.2 Integrated Project Management System (Sisgip)

This design proposal, as indicated in Guideline 02, does not start from the assumption that a platform will be built from scratch. Still, the development of new Artificial Intelligence functionalities will be incorporated into the SISGIP system, see Fig. 1. At the end of its implementation, it would have the following modules:

- **Project Management (Base Product)**: Allows you to manage Portfolios, Programs, and Projects, managing costs, scope, time, and risks. It allows control and monitoring of projects, and all project documentation can be centralized in a single place.
- **Improve Time/Cost estimation with Artificial Intelligence**: a module to improve cost and time estimates through Artificial Intelligence.

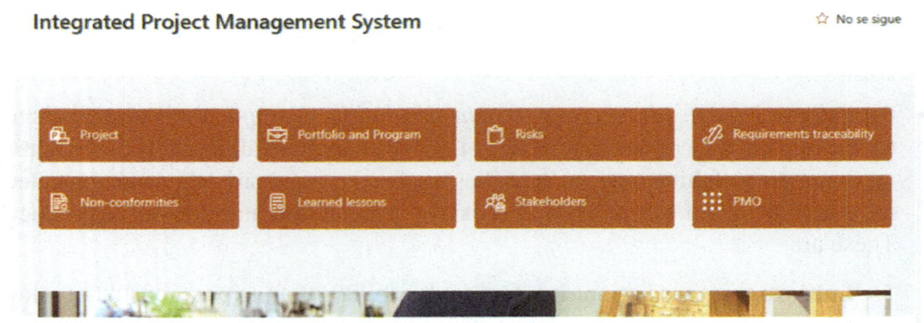


**Fig. 1** Integrated project management system

- **Knowledge Management with Artificial Intelligence and Mixed** Reality This functionality allows for improving knowledge management both at the level of capture and use, offering a modern user experience through Mixed Reality through an integrated Microsoft Hololens device.

The project's functionality allows you to manage the essential elements: time, cost, and scope, so that you can quickly monitor the status of projects through a color display, which indicates the status of the three variables mentioned above, as indicated in Fig. 2.

The project's functionality also allows managing project resources and their management to obtain them, whether internal or external, as shown in Fig. 3.

The functionality of Programs|Portfolios allows you to manage a group of projects centrally, based on the following definitions, see Fig. 4:

- Program: It is a group of related projects, subprograms, and program activities, whose management is carried out in a coordinated manner to obtain benefits that would not be accepted if managed individually.

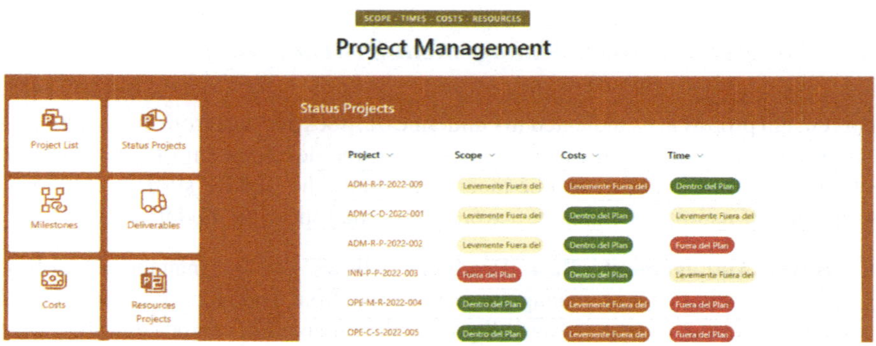


**Fig. 2** Quick tracking using project status colors

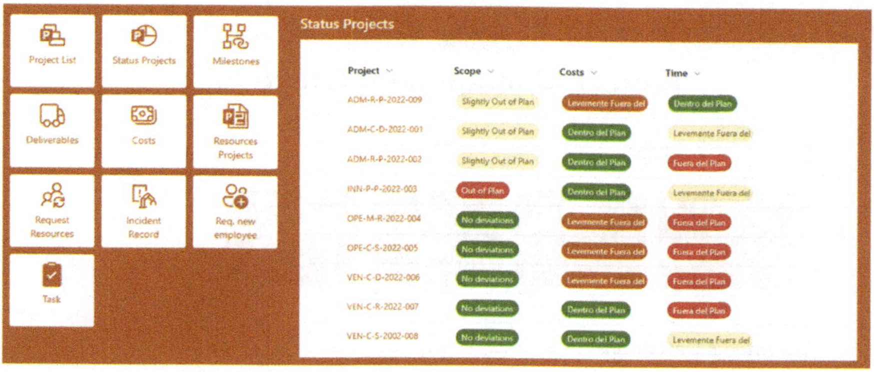


**Fig. 3** Quick tracking of resource management using project status colors

**Fig. 4** Strategies and scenarios for programs and portfolios

- Portfolio: A set of projects, programs, and other work grouped to facilitate effective management of that work to meet strategic business objectives.

The PMO functionality allows you to manage all the elements a Project Manager needs to execute his management work. In it, you can obtain, for example, the project codes, define the appropriate life cycle for each project, and identify the ideal project team according to the skills required. At the program and portfolio level in the PMO, those parameters that allow the implementation and guidance of meeting the company's strategic objectives can be defined, see Fig. 5.

## 4 Expected Design Results

This design proposal hopes to be able to deliver, upon completion of its technological implementation, a solution to the following problems:

**Fig. 5** Outline of the project management office (PMO)

**Loss of knowledge and experience**: With proper knowledge management, the knowledge and expertise accumulated by individuals and teams throughout projects can be recovered when people leave the organization or move to other areas. This situation can result in the repetition of past mistakes and the failure to take advantage of good practices, which negatively affects the efficiency and quality of future projects.

**Lack of organizational learning**: Knowledge management is fundamental to organizational learning, that is, the ability to learn from experience and apply that learning to improve. With a structured system to capture and share lessons learned, the organization may gain valuable opportunities to improve its processes, reduce risk, and increase efficiency in project execution.

**Decision-making difficulties**: Lack of access to prior knowledge and lessons learned can make it difficult to make informed decisions. Professionals and project teams may face the same challenges that others have faced and solved before, but they need access to that information. This situation can lead to suboptimal decisions, delays, and higher costs in project development.

**Lack of standardization and consistency**: Knowledge management provides a foundation for standardization and consistency in project management processes and practices. Without a system that facilitates the documentation and dissemination of best practices, each project can reinvent the wheel and follow different approaches, resulting in a lack of uniformity and difficulty in taking advantage of synergies and economies of scale.

**Difficulties in knowledge transfer**: When professionals retire or change roles within the organization, there may be difficulties in transferring their knowledge and experience to other team members. The lack of proper knowledge management can lead to the loss of critical ability and the need to relearn lessons already learned, which can cause delays and errors in the execution of projects.

## 5 Analysis Results

Regarding the analysis of the results, it is essential to define the segments of companies that would obtain benefits by implementing this proposal to design technological tools within their project management processes.

The main segments are:

(1) Small and medium-sized companies dedicated to generating and managing engineering projects under their own or international methodologies and standards.
(2) Small and medium-sized companies that are dedicated to the sale of services in compliance with service levels.
(3) Freelancers working as engineers/project managers.

However, this solution can be perfectly applied to large companies. Still, the entry barrier on the side of large companies occurs because they are captive to commercial agreements with large brands such as Microsoft and Oracle.

### *5.1 Economic Impact*

Implementing this solution in project management based on Artificial Intelligence can have various effects on the economies of these companies, both in terms of costs and benefits. Here are some key points:

**Automation of repetitive tasks**: Artificial Intelligence can take on repetitive and routine tasks that typically require human labor. This situation can lead to greater efficiency in project management, as resources and time are freed up for more strategic tasks. As a result, operating costs can be reduced, directly benefiting the company's economy.

**Improved decision-making**: Artificial Intelligence can quickly analyze large amounts of data and help make informed decisions. This improvement can allow small-medium companies to make better decisions in allocating resources, project planning, and identifying potential risks. As a result, the likelihood of costly errors and project delays is reduced, which can lead to significant savings.

**Resource Optimization**: Artificial Intelligence can analyze the availability of internal and external resources and the factors that affect project performance. With this information, companies can optimize resource allocation and avoid unnecessary costs, resulting in increased profitability for the project and the company.

**Prediction and prevention of problems**: Artificial Intelligence can anticipate potential issues in project management by analyzing historical patterns and real-time data. By identifying problems before crises arise, companies can take timely corrective action and avoid additional expenses associated with unforeseen problem resolution.

**Reduction of delivery time**: Implementing Artificial Intelligence can speed up the project management process by streamlining planning, scheduling, and monitoring. The timely delivery of projects can lead to greater customer satisfaction, translating into more business and a competitive advantage for the company.

## *5.2 Social Impact*

The adoption of artificial intelligence in project management in public companies can have a significant social impact, improving the efficiency of public services, increasing transparency and accountability, and addressing ethical and privacy challenges. However, it is essential to address ethical, social, and employment aspects to ensure that the implementation of AI benefits society at large and is inclusive of all citizens.

**Improvement of public services**: The implementation of Artificial Intelligence in project management in public companies can lead to a significant improvement in the provision of public services. By automating specific tasks and improving decision-making, services can become more efficient, accessible, and tailored to the needs of citizens. This action can have a positive social impact by increasing user satisfaction and improving their quality of life.

**Greater transparency and accountability**: Artificial Intelligence can facilitate the monitoring and evaluation of public projects, which improves transparency and accountability. Citizens can access up-to-date information on the progress of the projects and the results obtained. This greater transparency can increase citizens' trust in public institutions and strengthen democracy.

**Bridging the digital divide**: Adopting Artificial Intelligence can lead to greater digitization of public services. However, ensuring that this digital transformation is inclusive and includes those who have access to or the skills to use advanced technologies is essential. Digital inclusion and technology training are critical aspects of guaranteeing that all citizens can benefit from services based on Artificial Intelligence.

## 6 Design Challenges

The development of this design proposal will be conditioned by the following challenges: type, technological, legal, and data sources:

**Data quality**: Obtaining high-quality data is critical to building accurate AI models. The challenge lies in finding reliable sources and ensuring the data is clean, relevant, and labeled correctly. There may also be challenges in collecting unstructured data or needing more data for specific scenarios.

**Privacy and data security**: Protection of privacy and data security is a significant concern when collecting and using data sets for artificial intelligence. Complying with regulations and ensuring that personal or sensitive data is adequately protected is necessary.

**Data Labeling**: In many cases, AI models require labeled data for supervised training. Manually marking large amounts of data can be expensive and time-consuming. Additionally, there can be ambiguities in the labels, making it difficult to obtain a consistent and accurate data set.

**Data biases**: Data may contain inherent biases, such as sample representativeness or human subjectivity in collection and labeling. These biases can lead to biased or discriminatory AI models, raising ethical and equity challenges.

**Data Access**: In some cases, accessing the data needed to train an AI model can be challenging. This situation may be due to legal restrictions, privacy agreements, technological limitations, or a lack of cooperation from the data owners. Obtaining permits and partnership agreements can be a challenge in and of itself.

**Updating and maintaining data**: AI models often require up-to-date data to support their accuracy over time. Collecting and updating data can be expensive and require ongoing efforts to ensure that models adapt to changing patterns and trends.

## 7 Conclusions

Designing an artificial intelligence system for projects is a strategic and complex task requiring a comprehensive approach. Throughout this process, essential factors must be considered, such as the precise definition of the project objectives, the proper selection and preparation of the data, the choice of appropriate algorithms and models, the efficient implementation of the technological infrastructure, and the continuous evaluation of the system's performance.

A robust and efficient AI system can be achieved by addressing these aspects effectively. This action involves understanding the project's specific requirements and establishing clear evaluation metrics to measure the system's success.

In addition, it is essential to select and prepare the correct data, ensuring its quality, relevance, and availability. The choice of appropriate algorithms and models is based on a deep understanding of the data and the project's objectives and knowledge of best practices in artificial intelligence.

The efficient implementation of the technological infrastructure is another critical aspect. This implementation involves selecting the right tools and platforms to support the AI system, considering scalability, computational efficiency, and maintainability. In addition, proper integration with other existing systems and workflows within the project must be ensured.

Once the AI system is up and running continuous performance evaluation is essential. This task involves regularly monitoring and analyzing the results and adjusting the system accordingly. User feedback and performance metrics should be used to improve the system and optimize its performance continuously.

In conclusion, designing an artificial intelligence system for projects requires careful planning and consideration of several interrelated aspects. An efficient and successful system can be built that generates valuable results by adequately addressing the objectives, data, algorithms, infrastructure, and evaluation.

## References


1. Valdés KN, y Alpera SQ, Cerdá Suárez LM (2021) An institutional perspective for evaluating digital transformation in higher education: insights from the Chilean csase. Sustainability 13:9850. https://doi.org/10.3390/su13179850
2. Scorians CT, Marenzana G, Espinosa C, Herrero M (2022) Los destinos turísticos en el territorio digital. Competitividad y experiencias en el nuevo entorno. Caso: Ente oficial de turismo de Pucón (Chile)/Realidad. Tendencias y Desafíos en Turismo (CONDET) 20:1–20
3. Acevedo J, Diaz-Molina I (2022) Learning organizations in emerging economies: the effect of knowledge management on innovative culture in Chilean companies. Learn Organ 30:37–54. https://doi.org/10.1108/TLO-01-2021-0009
4. Elangovan N, Ismail WR, Zainol M (2022) A data-envelopment analysis-based systematic review of the literature on innovation performance. Heliyon (Elsevier) 8:1–15
5. Organization WIP (2022) Global innovation index 2022: what is the future of innovation-driven growth? WIPO
6. Rojas Zúñiga CJ (2023) Desempeño de mujeres en el desarrollo de software en el Departamento de Ciencias de la Computación de la Universidad de Chile
7. Galdames IS (2023) Ciencia Ciudadana: apropiación del conocimiento por la sociedad. Sci Comm Report 3:1–6. https://doi.org/10.32457/scr.v3i1.2152
8. Rincón Soto IB, Rengifo Lozano RA, Hernández Suárez CA, Prada Núñez R (2022) Educación, innovación, emprendimiento, crecimiento y desarrollo en América Latina. Revista de ciencias sociales 28:110–128
9. Arrau GP (2016) Gestión del Conocimiento en el sector público chileno. GESTÃO do conhecimento no sector 47
10. Pinar Saygin A, Cicekli I, Akman V (2000) Turing test: 50 years later. Mind Mach 10:463–518. https://doi.org/10.1023/A:1011288000451

11. Minsky ML (1967) Computation. Prentice-Hall Englewood Cliffs
12. Brcic M, Yampolskiy RV (2023) Impossibility results in AI: a survey. ACM Comput Surv 56:8–24. https://doi.org/10.1145/3603371
13. Ali M (2023) Brief overview on AI and robotics. BULLET: J Multidisip Ilmu 2:239–245
14. Davis E (2023) Mathematics, word problems, common sense, and artificial intelligence
15. Urinovich KA, Pakhritdinovna KD (2022) Development of artificial intelligence: characteristics of research directions. Texa J Mutl Stud 8:206–208
16. Sabour S, Frosst N, Hinton GE (2017) Dynamic routing between capsules. In: Advances in neural information processing systems. Curran Associates, Inc.
17. Bengio Y, Courville A, Vincent P (2013) Representation learning: a review and new perspectives. IEEE Trans Pattern Anal Mach Intell 35:1798–1828. https://doi.org/10.1109/TPAMI.2013.50
18. Bengio Y (2009) Learning deep architectures for AI. MAL 2:1–127. https://doi.org/10.1561/2200000006
19. Russell SJ, Norvig P (2010) Artificial intelligence a modern approach, 3rd edn. Pearson Education, Inc.
20. Senge PM (1997) The fifth discipline. Meas Bus Excell 1:46–51. https://doi.org/10.1108/eb025496
21. Davenport TH, Prusak L (1998) Working knowledge: how organizations manage what they know. Harvard Business Press
22. Hoe SL (2006) Tacit knowledge, Nonaka and Takeuchi SECI model and informal knowledge processes. Int J Organiz Theory Behav 9:490–502. https://doi.org/10.1108/IJOTB-09-04-2006-B002
23. Centobelli P, Cerchione R (2022) 23rd European conference on knowledge management, vol 2. Academic Conferences and Publishing Limited
24. Nieto ESD, Medina DIC, Rivas RT (2023) Leadership and organizational sustainability: the knowledge management approach. Taylor & Francis
25. Hislop D, Bosua R, Helms R (2018) Knowledge management in organizations: a critical introduction. Oxford University Press
26. Pedraja-Rejas L, Rodríguez-Ponce E, Rodríguez-Ponce J (2006) Knowledge society and strategic management: an integrative proposal. Interciencia 31:570–576
27. Takacs F, Brunner D, Frankenberger K (2022) Barriers to a circular economy in small- and medium-sized enterprises and their integration in a sustainable strategic management framework. J Clean Prod 362:132227. https://doi.org/10.1016/j.jclepro.2022.132227
28. PMI (2021) A guide to the project management body of knowledge: PMBOK

# Artificial Intelligence Contribution to the Development of Cuban Port Logistics Chains

**Claudia Bemelys Rodríguez Rodríguez, Deborah R. Galpert Cañizares, José A. Knudsen González, Andrés V. Silva Delgado, and Gilberto D. Hernández Pérez**

**Abstract** The technological development of logistics chains is one of the most essential aspects of modern business. This is based on the fact that they help improve resource management and minimize delivery times, errors and storage costs, among other aspects. All this justifies the use of artificial intelligence techniques in logistics and transport to achieve efficiency in these logistics processes. Currently, in the port logistics chains that operate through Cuban ports, efficiency is not achieved since there are delays in deliveries to final customers and high logistics costs, mainly due to the non-inclusion of the resources needed to stabilize the operation of these chains. This research is considered a contribution to the development of these port logistics chains, since it is possible to predict the necessary resources for their stable operation. For this, simulation is used as a technique that allows for the rational use of resources in the different logistics processes of these chains. The main result shows the different application scenarios of the designed simulation model that allow establishing the resource needs of the different members of the studied chain. One application is shown in the grain logistic chain, specifically the rice one at the Cienfuegos port.

**Keywords** Artificial intelligence · Port logistics chains · Simulation model

---

C. B. R. Rodríguez (✉) · D. R. G. Cañizares
Facultad de Matemática Física y Computación (MFC), Universidad Central "Marta Abreu" de las Villas (UCLV), Santa Clara, Villa Clara, Cuba
e-mail: clrodriguez@uclv.cu

D. R. G. Cañizares
e-mail: deborah@uclv.edu.cu

J. A. K. González · G. D. H. Pérez
Facultad de Ingeniería Mecánica e Industrial (FIMI), Universidad Central "Marta Abreu" de las Villas (UCLV), Santa Clara, Villa Clara, Cuba
e-mail: knudsen@uclv.edu.cu

A. V. S. Delgado
Empresa de Servicios Portuarios del Centro (ESPC), Cienfuegos, Cuba



P. Y. Piñero Pérez et al. (eds.), *Computational Intelligence in Engineering and Project Management*, Studies in Computational Intelligence 1134,
https://doi.org/10.1007/978-3-031-50495-2_10

## 1 Introduction

Supply chains operate in increasingly complex and uncertain environments. Adaptive planning and control in such environments are of utmost importance to ensure delivery to end customers [1]. The common goal of supply chain modeling is to optimize efficiency and reduce costs; but trade-offs of efficiency and leanness with flexibility and resilience may not be fully addressed [2].

Port logistics chains are those that operate through a seaport, including the different actors involved in the international trade process, such as importers and exporters, the port authority, terminal operators, customs, customs agents, transport companies (land and sea), and freight forwarders [3–6]. Port supply chains play a very important role in the process of economic globalization. The main factor restricting the development of port supply chains is a lack of trust in the mechanism. Blockchains have great potential to solve the trust problem among all participants in port supply chains [7].

The use of Artificial Intelligence (AI) in logistics chains enables competitive advantages to be obtained in the market, improving the efficiency of logistics processes. This is based on the possibility of becoming a key factor in strategic, tactical and operational decision-making in these chains, all of which serve as a basis for efficient resource planning. The contribution of AI to the development of port logistics chains can be seen through its integration into value chains, which makes it possible to invest in infrastructure and platforms, add value to supply chains, establish security standards for data protection, and improve the quality of life of end customers in these chains [8, 9].

Several studies have demonstrated the applicability of simulation in port systems [10, 11]. Simulation models make it possible to understand, analyze and refine many different scenarios at a high level of detail through hypothetical analysis. Their main advantage lies in their ability to handle complex problems without the intricate treatment of analytical formulas [12]. As an example, Swedish [13] developed a simulation model as a decision support tool for logistics management within a maritime distribution system to determine the fleet size and resource allocation needed to meet timely delivery requirements.

Triska et al. [14] developed a simulation-based method for port capacity assessment and expansion planning. Their method allows deciding, for a given set of resources in the simulation model, which expansions should be performed along the port throughput growth. Cimpeanu et al. [15] presented a discrete event simulation model to analyze the impact of equipment and infrastructure changes on berth productivity. Irannezhad et al. [16] presented a prototype of a decision support system (DSS) for horizontal and vertical cooperation between cargo agents involved in port logistics.

On the other hand, simulation is frequently used to manage congestion at container terminals [17]. In Colombia, a study was carried out using simulation models to analyze the unloading of goods at the port of Baranquilla using Arena software, and it was observed that queues are formed during weighing activities, thus allowing recommendations to be made for the management of congestion as: reconditioning equipment, preventive maintenance and revision, increasing the number of equipment, etc. [18].

In some countries, such as Spain, AI revenue forecasts have been made for global markets. The results of these forecasts point to a progressive increase in the coming years [19]. In Cuba, the research carried out by Acevedo Suárez et al. [20] can serve as a precedent for the present work, which proposes a simulation model to rationalize the use of resources throughout the chain. Also, Padilla-Aguiar et al. [21] try to achieve the integration of the assurance model that guarantees the service or product to the end customer.

The importance of the model designed is given by the possibility that the different actors in the chain will be able to know the resource needs that guarantee a continuous and uninterrupted flow of materials moving through the port logistics chains, seeking to add value to them.

The main results obtained show the resources needed by the main actors of the chain in the defined scenarios that guarantee the stable and coordinated operation of the port logistics chain. It is important to note that this research is part of the research project Management of territorial ports as integrated logistic platforms, which gave rise to the results presented in this publication and received funds from the Office of Funds Management and International Projects under code PN223VC014-007. In this paper, in Sect. 2: Simulation Procedure, we specify the procedure to build the model and apply it, and in Sect. 3: Simulation of the Rice Chain of the Cienfuegos Port, we present the application of this procedure in a study case.

## 2 Simulation Procedure

The purpose of the simulation model is to determine the resources required at each link of the port logistics chain. These resources must guarantee a stable and coordinated operation of the chain while complying with the unloading rate established by the vessel. It is important to clarify that the unloading rate is the rate (tons per hour) at which the vessel unloads to meet the time established for the vessel's stay in port. Failure to comply with the rate results in the payment of demurrage to the vessel involved.

The steps to be completed in the design and implementation of the model are based on the methodology proposed in [22] and specified in the next subsections.

## 2.1 STEP 1: Definition of the Scenarios for the Port Logistics Chain

The number of scenarios to be defined depends on the decisions to be made or analyzed.

SCENARIO 1: Current Situation of the Port Logistics Chain

This scenario is characterized by demonstrating the real possibilities of the chain under study with its current resources. Once this scenario has been analyzed, the following question must be answered: How long does it take to unload a vessel with the resources currently available in the port logistics chain?

SCENARIO 2: Ideal operation of the port logistics chain

The ideal functioning of the chain is characterized by valuing the inclusion of more resources depending on the needs. Once analyzed, the following question must be answered: How many resources are needed in the different processes of the port logistics chain in order to meet the discharge rate established by the vessel?

SCENARIO 3: Most likely operation of the port logistics chain

It is characterized by achieving an improved performance in relation to the current situation but without reaching the ideal performance. Once it has been assessed, the following question must be answered: How many resources are needed in the different processes of the port logistics chain to be able to come close to meeting the discharge rate established by the vessel?

In all scenarios, there will always be input and output variables, which may vary depending on the objectives defined in STEP 2.

The variables could be:

- Number of human resources required to unload the vessel.
- Quantity of handling equipment and estimated duration of the operation.
- Quantity of transport equipment to move the products to the intermediate points and to the final client.
- Duration of transportation cycles for all shipments.
- Maximum capacities of intermediate and final warehouses (if any).

## 2.2 STEP 2: Planning the Simulation Model

The aspects to be considered in this step are:

(a) Definition of objectives

The definition of the objectives of the simulation model will depend on the scenario selected in the previous step and the current conditions of the logistics processes of the chain under study. The objectives of the study are directed to the analysis of all the logistical processes, from the unloading of the vessel to the arrival of these products at the final customers.

(b) Definition of the scope of the model

The scope of the simulation model will also depend on the scenario defined and the variants or alternatives to be analyzed. It is important to define the scope of the simulation model, considering the mapping of the port logistics chain.

(c) Planning the experiment

The planning of a simulation model or the planning of experiments should consider a set of experimental tests, considering several replications of the model, so that the data generated can be analyzed statistically to obtain valid and objective conclusions about the established problem.

Reproducibility is very important in an experiment, i.e., being able to repeat the experiment. This provides an estimate of the experimental error and makes it possible to obtain a more precise estimate of the average effect of any factor.

## 2.3 *STEP 3: Data Collection and Processing*

This step basically includes three stages: collection and entry, processing and presentation. The analysis of the data will be aided by statistical packages and involves first entering the data into appropriate files. This involves defining "target" variables of interest, appropriately structured according to the software package to be used. The objective of the data analysis may be to characterize the variables in general (the frequency of their values, if discrete, or their central tendency and variability, if continuous). If the variables form an important part of the simulation, the characterization will be more demanding. It will be necessary to identify its distribution (its fit to a theoretical distribution or to a certain empirical distribution), compare the values of this variable between entities, etc.

## 2.4 *STEP 4: Model Construction*

(a) Definition of the list of activities

The model covers all logistics activities or processes occurring in the port logistics chain under study. This includes the operational variants used in these chains (direct variant and indirect variant), which predetermine the processes or activities that are carried out throughout the chain. This may include the availability of means for the execution of these activities, as well as other data that can later be used in the replications of the simulation model (equipment performance, average movement speeds, etc.).

(b) Planning of the interrelationship between modules

In this step, the characteristics of the simulation software to be used must be specified. This includes the possible limitations and adjustments to be made for the model runs. The modules to be created for the flowchart and the limitations or values of the variables used must also be established.

In relation to the interrelationship between modules, it is possible not to include it in this step since it can be detailed in later steps.

## 2.5 STEP 5: Model Verification and Validation

Verification and validation help make these results more reliable. Both provide a mechanism for developing a formal evaluation of the knowledge-based system. Any errors or inadequacies in the system must be discovered and then eliminated. It is important to distinguish between the terms "verification" and "validation" in an appropriate context.

Verification aims to ensure the existence of a relationship between the system specifications and what the system actually does. It consists of constructing the representation correctly. It also ensures that the system is free of errors. It is necessary and important since it implies the capacity of the model to obtain a result in accordance with the mathematical approach representative of the model.

The validation itself consists of constructing the correct representation. Its objective is to ensure that the results of the model are correct in relation to the behavior of the system under study. Pragmatic approximations can be made as long as they consider areas where the model is known to work successfully.

## 2.6 STEP 6: Experimentation

Experimentation is the method used to test the hypotheses regarding the object being studied. It is used to demonstrate how certain phenomena of interest occur. To do this, one must replicate these phenomena in the simulation software used by controlling all the variables of interest in order to demonstrate that a hypothesis is not a product of chance but of a universal law.

## 2.7 STEP 7: Presentation and Analysis of the Results

This step is based on the presentation of the parameters of interest. It constitutes the final and conclusive part of the simulation, where all the information that has appeared in the study is processed. The variables of interest are presented in an orderly and comprehensive manner and an attempt is made to reach the conclusions that these data give rise to.

The analysis of the behavior of each parameter of interest with respect to the elaborated plan has two moments: the first is to compare how far the indicator is from its ideal value, and the second is to detect the internal reserves that exist in the chain. Depending on the behavior, possible improvement measures are adopted.

To correct deviations, a plan of measures is drawn up, taking into account the dates for compliance and the responsible persons.

Through the dissemination and execution of the action plan, each member of the chain is informed of the measures to be adopted at all times to improve the performance of the port logistics chain and achieve total integration in its operation. Subsequently, the adopted action plan must be monitored.

As a result of this stage, several situations may arise, among which the following stand out:

- Changes in the characteristics of the port logistics chain.
- Changes in data and information gathered from studies, research or investments.
- Changes in strategic or tactical decisions.

# 3 Simulation of the Rice Chain of the Cienfuegos Port

For the application of the procedure, the port logistics chain for food grains was selected, in particular the bagged rice chain. This selection is based on the priorities currently established by the Internal Economy Port Transport Operation (OPTEI), among which are the basic food basket products destined for the population. It is important to clarify that in this application, the bags of rice are transported to wholesale food companies in Cienfuegos, Villa Clara and Sancti Spíritus. The main results of the practical application of this model are as follows.

**STEP 1: Definition of the scenarios for the port logistics chain**

<u>**Scenario 1**</u>: Current situation of the port logistics chain under study.

This scenario has six trucks and five forklifts (three for loading the trucks and two for the transfer to the warehouse). In this scenario, the 8477 t are unloaded in 325 h, which is approximately 14 days. The delay time at the door where the weighing and document formalization is performed follows a distribution of 1–1.5 h. A total of 3150 tons are transported to the Cienfuegos Wholesale Company, 2610 tons to Villa Clara, 1680 tons to Sancti Spíritus and 1050 tons to the port warehouse. In total there are 8490 t, which implies 283 trips (at a rate of 30 t/trip for each truck).

Scenario 2: Analysis of the desired (ideal) operation in the port logistics chain (compliance with everything agreed with the shipping company).

This scenario requires ten trucks and eight forklifts (four for loading the trucks and four for the warehouses). The latter are able to complete this operation in 24 h and can then be incorporated into the loading of bags onto the truck. The delay time at the door where the weighing and document formalization take place is reduced to a maximum of 0.5 h. In this scenario, a time of 152 h is achieved, which represents 6.33 days, while the norm given by the vessel is 122.55 h, which is 5.10 days; as can be seen, it is very similar to reality, with a difference of 29.45 h.

Scenario 3: Analysis of the most probable operation of the port logistics chain.

For the analysis of this scenario, the measures of the action plan that will be discussed below must be addressed.

**STEP 2: Planning the simulation model**

(a) Definition of objectives

Scenario 1: Evaluate, with available resources, the total vessel discharge time.

Scenario 2: To show the real resource requirements in order to meet the discharge rate agreed with the shipping company.

(b) Definition of the scope of the model

The study is limited to the operations of the unloading process of rice in bags from the vessel and its transfer, depending on whether it is direct (to the wholesale companies) or indirect (to the port warehouse).

(c) Experiment planning

An experiment with 100 replications is planned, where the input variables are: number of trucks, number of forklifts, process time, number of hours of the simulation, percentage of decision of entities to their destination; and as output, variables to be taken into account are: number of departures, queue waiting time per entity and per process, total time and number of entities distributed at the destinations. Non-parametric tests are performed to determine whether there are significant differences between the real data samples and the real scenario in the model.

**STEP 3: Data collection and processing**

The data required for a preliminary version of the simulation project are the activities that take place at the port, such as (i) the unloading of the rice bags by the crane, which places them on the move and is picked up by the forklifts until they are taken to the trucks (in the case of the direct route) or to the port warehouse (in the case of the indirect route), (ii) the filling of the trucks, (iii) the process that takes place at the gate, and (iv) until the rice is transferred to its different destinations. The collection of data also took into account arrival times, truck loading and unloading times, delays in processes such as transporting the bags of rice by forklifts and the documentation and weighing of the truck at the gate, the distances to the destinations, and the speed of the trucks. In addition to the capacity of the crane, the trucks and the quantities of tons to be received per day were considered (data that will be shown in more detail in the next step). It is important to highlight that in order to obtain the data, information from research carried out by Silva Delgado [5, 23], opinions of experts in the working group, and field studies were also taken into account. The data were processed using the statistical software R.

**STEP 4: Building the model**

## 3.1 *Definition of the Activity List*

The activities described below are given from the time the vessel arrives at the port to its destinations, with their direct and indirect variants:

- The vessel arrives with 8477 tons of rice. Each bag of rice contains 50 kg
- A single crane unloads the bags of rice onto the quay. Each unloading contains 20 bags of rice (equivalent to 1 ton) and this process takes 1 min (20 bags of rice arrive every minute).

Then the process is divided into direct and indirect operational variants:

Direct variant

This variant comprises the transfer of rice from the dock to its three as shown in Table 1. This process contains as resources three forklifts of 10 t capacity moving at a speed of 10 km/h and six trucks of 30 t capacity moving at a speed of 60 km/h.

- One forklift loads 20 bags on the dock and carries them to a truck. The loading and unloading take one minute each. The distance from the dock to the truck is

**Table 1** Direct variant destinations

| Destination | Distance (km) | Demand (t/day) | One-way movement time (h: min) |
|---|---|---|---|
| Wholesale food company of Cienfuegos | 8.0 | 140.0 | |
| Villa Clara food wholesale company | 78.9 | 140.0 | 1: 24 |
| Wholesale food company of Sancti Spíritus | 147.7 | 60.0 | 2: 4 |
| Wholesale food company of Cienfuegos | 8.0 | 140.0 | |
| Villa Clara food wholesale company | 78.9 | 140.0 | 1: 24 |

100 m and a forklift is moving at a speed of 10 km/h. The whole trip takes 0.03 h for a total of 0.09 h round trip. This process is carried out by the 3 forklifts in parallel. To fill a truck they need to make the described trip 10 times. The time it takes to fill a truck with 30 tons is 0.5 h, expressed in a uniform distribution of 0.5 - 0.65 h.

- The loaded truck takes three minutes to arrive at the gate and the formalization of documents and weighing at the gate takes between 1 and 1 h and 15 min.
- The truck then moves to its destination to deliver the sacks of rice. The same truck returns to be loaded again (the truck is working 24 h a day). Unloading the truck takes between 1 h and 45 min and 2 h.

Indirect variant

The transfer of 1000 t of rice in sacks from the dock to the warehouse at the port of Cienfuegos is carried out. This process uses two forklifts that move at a speed of 10 km/h. These two forklifts are simultaneously carrying the bags from the dock to the port warehouse, which is 200 m away. The warehouse has a capacity of 8,000 tons. The loading and unloading take one minute each. Each forklift takes 0.09 h each way. The two forklifts work simultaneously and take 1.35 h to move the 30 t that were taken as the equivalent quantity.

## 3.2 Approach to the Interrelationship Between the Modules

As Arena is not free software, the student version used has the impediment that the input of entities in the system cannot exceed 150. Therefore, we work with the assumption that for the program, 1 entity is equivalent to 30 t in the real system (which represents the total capacity of a truck), and thus the pertinent runs will be made until covering the number of real entities that are needed. The different modules of the flowchart used in the simulation are described below (see Fig. 1):

- Module create: Starts the process flow and represents the arrival of the entities to the system, in a time of 0.5 when representing the arrival of 30 tons.
- Station module: defines from where the truck transport starts.
- Module decide: It is an alternative or option to take within the flow process, it defines the variants that the entity is going to follow. The priority is given in percentages where the 34 entities that go to the warehouse represent 22.22% and the rest go to the trucks.

In the indirect variant, the modules process, record, dispose.

- Process module: the process in which the 2 forklifts take 30 tons, represented in 1 entity, to the warehouse is established. A single resource called 2 forklifts was established, which has the time of the two in reality, this process follows a uniform distribution of 1.35–1.40 h.

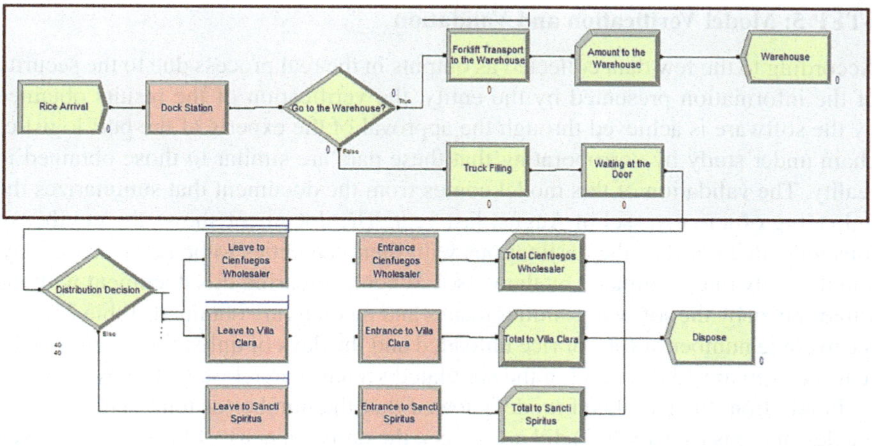


**Fig. 1** Simulation model in Arena

- Record module: Collects statistics, such as entity count or cycle time.
- Dispose module: It is the end point of the entities. It is the module that ends the process flow.

In the direct variant, the modules process, decide, enter, leave, record and dispose.

- Process module (truck filling): represents the process of the 3 forklifts filling the truck. One resource called 3 forklifts was taken, which contains the time it takes to fill a truck, which follows a uniform distribution of 0.7–1.2 h.
- Process module (waiting at the door): A door resource was defined to describe the process of weighing and formalizing documents that takes place there, which follows a uniform distribution of 1–1.5 h.
- Module decide: The percentages of the destinations to be followed by the truck with the rice were established, with the daily quantities to be received in Villa Clara and Cienfuegos equal and greater than in Sancti Spíritus, 40% was distributed in Villa Clara, 40% to Cienfuegos and the rest, which constitutes 20% to Sancti Spíritus.
- Module enter: describes the entry of the truck of a wholesaler, with a delay of 2 h in which the unloading is delayed.
- Module leave: Describes the departure of the truck from the wholesaler with a time of 0.20 h.
- Record module: It is used to count the departure of trucks.
- Dispose module: It is the end point of the entities. It is the module that ends the flow of the process.

**STEP 5: Model Verification and Validation**

According to the few data collected as outputs in the real process due to the security of the information presented by the entity, the verification of the results obtained by the software is achieved through the approval of the experts of the port logistics chain under study by corroborating that these data are similar to those obtained in reality. The validation of this model comes from the document that summarizes the unloading of a rice vessel in August last year. The document shows the number of tons unloaded per day, the destinations with their capacities to be received per day, and the daily rates estimated by the vessel. When comparing this document with the values given by the software, similar means and medians are obtained. Table 2 shows the average number of tons of rice unloaded and the days of unloading, which in the real scenario are 13 days and in the simulated scenario are close to 14 days.

In addition, the simulated model showed that the number of tons that arrived at the destinations (see Table 3) did not exceed the daily capacity of tons to be received per day (see Table 4).

**STEP 6: Experimentation**

Figure 1 shows the simulation model implemented in Arena considering the modules discussed above.

For the hypothesis testing, non-parametric tests of comparison of two independent samples were performed since the variables analyzed are not normally distributed according to the result of goodness-of-fit tests to this distribution. The selected test is the Mann-Whitney test and the Wilcoxon rank sum test. Given the random variables total time in the rice system, time in queue and number of tons of rice bags unloaded and distributed, where X and Y are the values of these variables for the actual and ideal variants, respectively.

Ho: The ranges of the values of the two variables do not differ significantly.
H1: The ranges of the values of the two variables differ significantly.

**Table 2** Comparison between means and medians of selected samples

| Quantity of simulated rice per day | Quantity of real rice unloaded per day | Quantity of real rice unloaded per day |
|---|---|---|
| Number of samples | 13.8 | 13 |
| Mean (t/day) | 637 | 647 |
| Median (t/day) | 643 | 625 |

**Table 3** Maximum number of tons per day to be received in reality

| Destination | Maximum quantity to be transported t/day |
|---|---|
| Cienfuegos | 140 |
| Villa Clara | 140 |
| Sancti Spíritus | 60 |

**Table 4** Number of tons per day distributed in the simulation

| Destination | Quantity transported t/day |
|---|---|
| Cienfuegos | 138 |
| Villa Clara | 140 |
| Sancti Spíritus | 59 |

If the significance value of the test is less than 0.05, Ho is rejected.

The test results showed in the three tests performed a significance of 0.00, so the null hypothesis is rejected and thus demonstrates that the real and ideal variants differ significantly, with the ideal variant being more convenient in terms of time.

**STEP 7: Presentation and analysis of the results**

After performing the 100 preliminary runs on both the real and ideal models, the significant results are shown in Table 5.

In summary, the main resources to be used in the ideal scenario to fulfill the plan are:

Empresa Servicios Portuarios Centro

For indirect variant

- Workforce: 27 men in the technological line [24].
- Two forklifts for the transfer of rice to the port warehouse.

For direct variant:

- Work force: 28 men on the technological line [24].
- Three forklifts for loading means of transport at the dock.

Motor transport companies (ORDER Cienfuegos)

6 trucks are needed daily at a rate of two for each destination:

**Table 5** Presentation and analysis of results

| Quantity of simulated rice per day | Quantity of real rice unloaded per day | Quantity of ideal rice unloaded per day |
|---|---|---|
| Average waiting hours of entities in the system | 28.65 | 0.81 |
| Average total hours of an entity in the system | 32.93 | 4.24 |
| Average waiting hours per entity at the door | 34.24 | 0.52 |
| Number of downloading days | 13 | 8 |
| % Utilization of resources:<br>Forklifts (trucks)<br>Forklifts (warehouse) | 0.45<br>0.30 | 0.63<br>0.49 |

- To make five trips to Cienfuegos.
- Make five trips to Santa Clara.
- Make five trips to Sancti Spíritus.

In all cases, the trucks are of 30 ton capacity, and the destinations are food wholesalers.

Once the application of the model has been completed, a plan of measures can be established [25], as shown in Table 6.

In order to have an accurate assessment of the social and economic impacts derived from this research, a cost–benefit analysis must be performed once the proposed measures have been implemented. However, it is possible to comment on some of the benefits obtained from the results of the Ideal Simulated Model. These include the following:

Socially:

- Increased sources of employment for dockworkers, drivers and other workers involved in the different links of the port logistics chains.
- Possibilities of increasing their salary income through compliance and over-compliance with established standards.
- Elimination of delays in the delivery of products to customers in the domestic economy.

Economically:

- Decrease in freight charges and delays in unloading ships.
- Possibility of achieving savings in the above expenses due to prompt clearance.
- Increased utilization of handling and transportation facilities.

**Table 6** Measures adopted to improve the operational planning of the packed rice port logistics chain

<table>
<tr><th>Deficiency</th><th>Measure</th><th>Responsible</th><th>Date of compliance</th></tr>
<tr><td>Prolonged stays of vessels in port</td><td rowspan="2">(a) Consider the quantities of labor force, means of handling and means of transportation according to the scenarios proposed</td><td rowspan="2">Member companies of the port logistics community</td><td>2023</td></tr>
<tr><td>Lack of knowledge of resource requirements to meet unloading rates</td><td>2023</td></tr>
<tr><td>Delays in deliveries to customers in the domestic economy</td><td>(b) Reduction of the delay time of the process at the gate</td><td>Central port service companies</td><td>2023</td></tr>
</table>

## 4 Discussion

Most of the research referred to in studies of port logistics chains [25, 26] focuses its attention on the application of simulation as a technique to optimize the ship unloading process, leaving gaps in the other logistics processes that occur in this type of logistics chain.

The main limitation reflected in the results of this research is related to the impediment of entity input in the Arena software. This implies working with equivalent units, which makes the capture and processing of information more complex.

Business Intelligence (BI) consists of the interpretation of data for its subsequent transformation into decision-making that generates the highest possible value for companies, which is generally oriented to the final consumer and always actively seeks the generation of value. This is precisely one of the fundamental objectives of port logistics chains for food imports.

Considering the above, the results obtained in this research can be improved by combining application software and programming software, which constitute Business Intelligence tools [9].

## 5 Conclusions

Once this research has been completed, the following conclusions can be drawn:

1. The simulation model applied for the planning of the bagged rice port logistics chain meets the expectations for which it was created since it allows planning the necessary resources by integrating the processes of the different actors in the chain, considering all their requirements.
2. The use of Arena software in the implementation of the simulation model in the two scenarios analyzed allowed for the obtaining of results that support decision-making. In the current scenario, the impossibility of complying with the established operating plan was demonstrated, and in the ideal scenario, the aforementioned resources should be adopted. These are complemented by the action plans developed for each of the members of the port logistics chain under study.
3. Delays in deliveries to customers are caused by long delays at the port gate. This implies the need to carry out studies in this regard in order to establish time limits for these activities.
4. The benefits discussed in this research demonstrate that the use of Business Intelligence tools in port logistics chains can contribute to improving the country's current economic situation.

## References


1. Rolf B, Jackson I, Müller M, Lang S, Reggelin T, Ivanov D (2022) A review on reinforcement learning algorithms and applications in supply chain management. Int J Prod Res. https://doi.org/10.1080/00207543.2022.2140221
2. Golan MS, Jernegan LH, Linkov I (2020) Trends and applications of resilience analytics in supply chain modeling: systematic literature review in the context of the COVID-19 pandemic
3. Yan Y, Chow AH, Ho CP, Kuo YH, Wu Q, Ying C (2022) Reinforcement learning for logistics and supply chain management: methodologies, state of the art, and future opportunities. Transp Res Part E: Logist Transp Rev 162:102712
4. Merchan Baque WE, Palacios Macías JE (2021) Gestión operativa en el área de descarga de pesca del puerto nacional de la ciudad de Manta. ESPAM MFL, Calceta. https://repositorio.espam.edu.ec/bitstream/42000/1400/1/TTAE19D.pdf
5. Silva Delgado A (2009) El diagnóstico operacional. Las brechas cuantitativas de desarrollo por línea de negocios, Empresa Servicios Portuarios Centro, Cienfuegos
6. Shao W (2023) Performance evaluation of port logistics informatization construction: a case study from China. Tehnički vjesnik 30(3):705–714. https://doi.org/10.17559/TV-20221217041443
7. Liu Y, Zhou Z, Yang Y, Ma Y (2022) Verifying the smart contracts of the port supply chain system based on probabilistic model checking. Systems 10(1):19. https://doi.org/10.3390/systems10010019
8. Karaś A (2020) Smart port as a key to the future development of modern ports. TransNav: Int J Marine Navig Safety Sea Transp 14(1). https://doi.org/10.12716/1001.14.01.01
9. Cárdenes Doctor J, G.F R (2022) La aplicación de Big Data e Inteligencia Artificial en logística y transporte para la optimización de procesos en empresas. Facultad de ciencias económicas y empresariales, in Universidad Pontificia, Madrid, España. https://repositorio.comillas.edu/xmlui/handle/11531/56434
10. Longo F, Huerta A, Nicoletti L (2013) Performance analysis of a Southern Mediterranean Seaport via Discrete-Event Simulation. Strojniški vestnik—J Mech Eng 59:517–525
11. Geng X, Lv Y, Zhao L, Wang Y (2023) Measurement and simulation of risk coupling in port hazardous chemical logistics. Int J Environ Res Public Health 20:4008. https://doi.org/10.3390/ijerph20054008
12. Figueira G, Almada-Lobo B (2014) Hybrid simulation-optimization methods: a taxonomy and discussion. Simul Model Pract Theory 46:118–134. https://doi.org/10.1016/j.simpat.2014.03.007
13. Swedish J (1998) Simulation of an inland waterway barge fleet distribution network. In: Winter simulation conference. Proceedings (Cat. No.98CH36274), vol 2, pp 1219–1221. https://doi.org/10.1109/wsc.1998.745982
14. Triska Y, Frazzon EM, Silva VMD (2020) Proposition of a simulation-based method for port capacity assessment and expansion planning. Simul Model Pract Theory 103:102098
15. Cimpeanu, R., DEVINE, M.T., and O"BRIEN, C., A simulation model for the management and expansion of extended port terminal operations. Transp Res Part E: Logist Transp Rev 98:105–131. https://doi.org/10.1016/j.tre.2016.12.005
16. Irannezhad E, Prato CP, Hickman M (2020) An intelligent decision support system prototype for hinterland port logistics. Decis Support Syst 130:113227
17. Ma N, Zhou C, Stephen A (2021) Simulation model and performance evaluation of battery-powered AGV systems in automated container terminals. Simul Model Pract Theory 106:102146
18. Pacheco Bolaño D, Troncoso-Palacio A (2021) A discrete event simulation model for analyzing the unloading of goods at a port. In: Prosiding Hasil-Basil Penelitian Tahun 2019. Dosen-Dosen Universitas Islam Kalikamantan. https://doi.org/10.20944/preprints202107.0169.v1
19. Gobierno de España (2020) Estrategia nacional de inteligencia artificial. https://www.lamoncloa.gob.es/presidente/actividades/Documents/2020/ENIA2B.pdf

20. Acevedo Suárez JA et al (2010) Modelo de Referencia de Redes de Valor para un desarrollo sostenible. Revista de Investigación Agraria y Ambiental 1(2):29–49. https://hemeroteca.unad.edu.co/index.php/riaa/article/view/903.
21. Padilla-Aguiar D, Acevedo-Suárez JA, Urquiaga-Rodríguez AJ, Acevedo-Urquiaga AJ, Sablón-Cossío N (2022) Aplicación del modelo de aseguramiento del proceso a la logística de servicios. Ingeniería Industrial XLIII(2):4320. ISSN:1815-5936. http://www.rii.cujae.edu.cu
22. Kelton WD, Sadowski RP, Zupick NB (2015) Simulation with Arena, 6th edn. McGraw-Hill Education, 2 Penn Plaza, New York, NY, p 10121
23. Silva Delgado A (2019) El diagnóstico operacional. Las brechas cuantitativas de desarrollo por línea de negocios, Empresa Servicios Portuarios Centro, Cienfuegos
24. Empresa de Servicios Portuarios Centro ASPORT, C., ESPC 2019. Manual de procesos tecnológicos (2019)
25. Párraga MM, Gonzalez-Cancelas N, Soler-Flores F (2016) Estrategia de planificación del puerto de aguas profundas de la ciudad de Manta: inclusión de metodología DaDED (dafo-delphi-estadística descriptiva). Revista Transporte y Territorio 14:39–60
26. Lee PT-W, y Lam JSL (2015) Container port competition and competitiveness analysis: Asian Major Ports. In: Handbook of container transport logistics. making global supply chain effective edition international series: operations research & management sciences, pp 97–136. https://www.researchgate.net/publications/2702869832015

# Decision-Making in Project-Oriented Organizations Supported by Computational Intelligence's Techniques

# Digital Transformation in Project Oriented Organizations, Supported by Intelligence Ecosystems

**Pedro Yobanis Piñero Pérez, Luis Alvarado, Iliana Pérez Pupo, Yosvani Márquez Ruiz, and Pedro E. Piñero Ramírez**

**Abstract** Digital transformation is a necessary step in the evolution of organizations. Boosting the competitiveness of organizations by improving effectiveness and efficiency in their management. Particularly in project-oriented organizations, digital transformation processes must put the generation of value and internal and external customer satisfaction at the center. These types of organizations have historically worked using traditional methods. However, the development of BIM technologies and the adoption of agile methodologies have changed the way things are done. This work presents a model for the digital transformation and improvement of project-oriented organizations that combine computational intelligence techniques and business architectures. The validation of the proposal is carried out based on a case study in an organization oriented to the development of software projects. The validity of the proposal and the gradual improvement in the organization's value generation chain are demonstrated.

**Keywords** Computational intelligence · Digital transformation · Enterprise architectures · Project management

---

P. Y. Piñero Pérez (✉) · I. P. Pupo · Y. M. Ruiz · P. E. P. Ramírez
Artificial Intelligence for a Sustainable Development Group, IADES, Havana, Cuba
e-mail: pppyob@gmail.com; iades@businessredmine.com

I. P. Pupo
e-mail: iades@businessredmine.com

Y. M. Ruiz
e-mail: ymarquezr@uci.cu; iades@businessredmine.com

P. E. P. Ramírez
e-mail: iades@businessredmine.com

L. Alvarado
Universidad Católica del Norte, Av. Angamos, 0610 Antofagasta, Chile
e-mail: lualvar@ucn.cl

Y. M. Ruiz
Dirección Técnica, Universidad de las Ciencias Informáticas, Habana, Cuba

 
P. Y. Piñero Pérez et al. (eds.), *Computational Intelligence in Engineering and Project Management*, Studies in Computational Intelligence 1134,
https://doi.org/10.1007/978-3-031-50495-2_11

# 1 Introduction

Since ancient times, man has had the need to organize human and non-human tasks and resources in the construction of cities. Over the years, practices have been perfected, giving rise to the development of project engineering. The constant effort to improve the organizations or entities that developed the projects has allowed the development of new techniques and a constant improvement in project management.

The process of improving project management gave rise to its current stage, which we call "Modern" and which began between the 50s and 60ss. The beginning of this stage had its origins in the development of military projects. Bernard Schriever is considered the father of modern project management. Bernard was the lead project manager for the development of Polaris ballistic missiles. Mr. Schriever developed the concept of concurrency, which gave birth to the Program Evaluation Procedure (PEP) and modern project management. One of the main changes introduced by Schriever was an integrated and centralized approach to planning and budgeting concepts. In addition, the parallel work of different project teams was introduced, beginning integrated project management.

But the greatest advance in the development of the discipline of project management has been the creation of schools or institutions dedicated to formalizing and developing new methods of organization and work. Among these institutions, we find the Project Management Institute (PMI) [1, 2], el International Project Management Association (IPMA) [3], both founded in 1965, with the Polytechnic University of Madrid represented by Rafael Heredia, standards ISO and the Software Engineering Institute (SEI) [4, 5]. Since their founding, these entities, among others, have been developing and publishing professional standards for project management. The most relevant are: PMBOK (Project Management Body of Knowledge) [6], ISO 21500 [7], ISO-10006 del 2003 and CMMI [8]. In general, project management has evolved to be an interdisciplinary area of knowledge where management techniques, technical knowledge of the specific area of application, psychology, human resource management, logistics, mathematical sciences and computer sciences converge [9].

Regarding the phases of a project, some authors and schools, such as the PMI they recognize that there are four phases, among which stand out initiation, planning, execution and closure. This approach analyzes the life cycle of projects isolated from business management.

Other authors, such as Rafael Heredia analyze projects and their phases as inseparable entities of business strategic planning. Surveys of project-oriented organizations show a gradual decrease in the number of failed projects. However, the percentage of satisfactory projects has not grown. Only the percentage of renegotiated projects has grown. In general, the dynamism in the current development of projects has motivated a group of changes and opportunities for improvement. Some of the improvement opportunities experienced by project-oriented organizations are listed below:

- Trends associated with the adoption of agile methods in value generation processes. Greater linkage of the strategic management processes of organizations with the value chain. New indicators and visualization of these in the balanced scorecard based on the value chain processes.

- There is a need to incorporate artificial intelligence and simulation techniques, streamlining decision-making processes. Increased use of information; previously, around 95% of the data was underused.
- Increased trends in project management from cloud platforms. Increase in software solutions in the forms of Infrastructure as Services (IaaS), Software as Services (SaaS), Platforms as Services (PaaS) and increase in mobility and facilities for accessing information. Increase in sustainable resource management strategies aimed at reducing resource underutilization rates, different references [10, 11] indicate that in construction projects, they account for around 34% of material resources.
- BIM (Building Information Modeling) technologies are being introduced quickly in the construction sector. Based on virtual construction, they facilitate the simulation and management of changes in construction processes.

In this work, the authors propose a model for the digital transformation and improvement of project-oriented organizations. The proposed model combines computational intelligence techniques and business architectures. In addition, the lessons learned from applying the proposed model and a set of good practices in different project management contexts are shown. The work is organized in the following sections. In the second section, the proposed model is presented. In the third section, the application results in different project-oriented environments are presented. Finally, the conclusions and future work are presented.

## 2 Model for Digital Transformation and Improvement of Project-Oriented Organizations

The proposed model is characterized by the following elements:

- Focused on generating value for customer satisfaction by improving processes and systems simultaneously. Close link between strategic management.
- Agile management of lessons learned and evidence from portfolios, programs and projects.
- Agile planning, control and monitoring based on objective indicators. The calculation of the indicators combines traditional methods and indicators with computational intelligence techniques for managing uncertainty. See Definition 1. These indicators are built automatically and allow the evaluation of professional skills, portfolios, programs and projects.
- Decision making supported by computational intelligence techniques that improve the treatment of uncertainty, vagueness and inconsistencies.
- Development ecosystem that works as a Platform as Services (PaaS), the use of Docker and Kubernates technologies.

In the context of this research, the terms "artificial intelligence" and "computational intelligence," among other terms associated with the treatment of uncertainty, are frequently used. Each of these terms is described below to facilitate their understanding and scope throughout the document.

**Artificial Intelligence**: a set of techniques dedicated to the creation of hardware and software that simulate human thinking and intelligence in natural systems to model solutions to problems for which there are no algorithms or for which those that exist have non-polynomial complexity. It is not intended to replace men but rather to support him in complex, unstructured decision-making processes [12–14].

**Computacional Intelligence**: many authors use the terms computational intelligence and soft computing interchangeably to refer to the same set of techniques as a branch of artificial intelligence. In this sense, Computational Intelligence is presented as the branch of AI that encompasses various techniques aimed at simulating human tolerance in decision-making processes in environments with uncertainty and imprecision. Among the precursor techniques of computational intelligence are: evolutionary algorithms, artificial neural networks, fuzzy set theory and fuzzy systems. They are also part of the computational intelligence techniques: rough set, linguistic data summary, conversational systems, fuzzy cognitive maps, collective intelligence techniques, and neutrosophic theory, among other extensions.

**Uncertainty**: refers to the veracity of the knowledge used by the system and is produced by the lack of certainty of the information.

**Inconsistency**: existing data in the application domain may be contradictory. For example, two practically identical examples belong to different classes.

**Indeterminacy**: is understood as the degree that characterizes the elements that do not have absolute membership in a set (certainty) and that it cannot be assured that they do not belong to the set itself absolutely (falsehood). That is, indeterminacy includes the elements that are between the opposite sets <A> and <antiA>, and is denoted by <neutA> [15, 16].

**Incomplete information:** Consider the possibility that not all the data required for an argument is available.

The following subsections present the elements of the proposed model in detail.

## 2.1 *Focused on Clients in Close Connection with Strategic Projection*

It is important to understand that the different parts of the model describe processes or tasks that may give the impression of a heavy or formal model. But the entire presentation of the model is based on the divide and conquer principle, which facilitates both the understanding and application of the proposal. For this reason, the tasks are presented separately, and the agile development of activities is insisted upon. In this way, the model helps to focus the work in an organized way, "agile models are not in contradiction with the organization."

The strategic planning of agile methods is frequently analyzed separately by several authors. However, regardless of the methodology or way of executing projects, organizations must be clear about their strategic projection.

We understand strategic planning as the establishment of policies and plans to achieve the organization's objectives within the framework of its mission. The plans are organized in the form of portfolios, programs and projects. These plans are developed under a conscious and active process, organized by management, and aimed at modifying and generating growth for the organization.

The basis of the entire process are projects designed to develop a unique product or service, and under this concept, the conception and development of projects are applicable to both project-oriented companies and entities that require improvements in their processes or facilities. In all cases, they are organized with the objective of contributing to the achievement of the goals proposed in the strategic planning of the organizations.

The objective of strategic planning is, first and foremost, the continuous improvement and growth of the organization. This planning must be objective and orderly, with a clear focus on prioritizing tasks and projects based on value generation.

In the proposed model, strategic planning is organized into the following processes (see Fig. 1).

Process I. Diagnosis of the organization, identification of the current state (As-is Model) based on the analysis of different dimensions, among which: business, financial, technological infrastructure, human resources, suppliers, data and internal control of the organization. Identification of Strategic Business Units.

Process II. Identification of the ideal state (To-be Model) that represents the state of the organization with the desired goals in each of the dimensions. From this

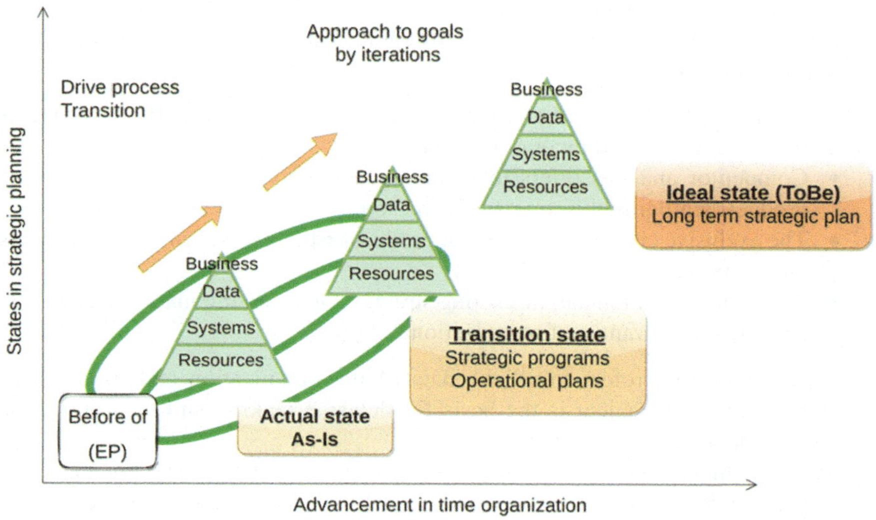


**Fig. 1** Iterative model of the evolution of organizations according to strategic planning

state, the general strategic objectives to be achieved are identified. The effectiveness indicators that represent the achievement of the objectives must also be identified.

Process III. Definition of transition states that indicate intermediate states of the organization on the way to the ideal state. From these transition states, specific or partial objectives can be identified. The planned state for the effectiveness indicators must be identified in correspondence with the specific objectives.

Process IV. Design of the solution path to lead the organization through the transition states from the current state to the ideal. These solution paths are made up of three types of plans: long-term strategic plans, medium-term programs, and short-term operational plans.

Process V. Conceptualization of projects to comply with the planned plans. Organization of projects by programs.

The organization's diagnosis process would be determined by the following tasks:

1. Refinement of the mission and vision of the organization. Identification of interested parties.
2. Identification of the dimensions of the analysis and the Strategic Business Units.
3. Application of techniques for collecting information and analyzing strengths and weaknesses.
4. Analysis and summary of the diagnosis.

The execution of these tasks must be carried out in the most agile way possible, depending on the size of the organization that carries out its strategic projection. It is important to identify that, to simplify the complexity of the analysis, it is convenient to organize the diagnosis by dimensions, as shown in Fig. 1.

The four fundamental dimensions are described below:

1. Business Dimension: defines the scope of the organization's services and products that are to be achieved from the perspective of meeting customer needs. It is oriented towards the generation of the value chain. In this dimension, it is also important to analyze:

    - The products or services on the market that can be substitutes for the products or services provided by our organization.
    - Competitor analysis refers to other organizations that provide services or products similar to the company's mission.
    - The exclusive characteristics of our product distinguish it from other products on the market.
    - The communication and marketing strategy for accessing market sectors that constitute a guarantee of organizational growth.

2. Data Dimension: refers to all the data of the organization and its management. Data management is the basis for decision-making and the continuous improvement of the organization.
3. Systems Dimension: In this dimension, the legacy systems of the organizations and the need for new systems are analyzed. BusinessRedmine software ecosystems are proposed for portfolio, program and project management. The introduction of this software ecosystem is the basis for agility in project management.

The incorporation into this ecosystem of novel artificial intelligence techniques is highlighted in the analysis and calculation of indicators and conversational styles to facilitate the updating of information.

4. Resources Dimension: defines the management of human and non-human resources, among which the following stand out: financial resources, human resources, material resources and technological infrastructure for development. Technological infrastructure in particular refers to all elements related to hardware, networks and all the equipment required for the execution of organizational processes.

Note that to get from the current state (To-Be Model) to the ideal state, there are transition states. How does the organization go from one state to another? Well, for the organization to evolve through the states, the form proposed by the model is projects. These projects can be of two types:

- Internal development projects of the organization, generally aimed at the development of its own technologies to improve the organization's services and products.
- Customer-oriented projects that, through development projects, enhance the growth of the organization. Following trends in project development.

The identification of the projects that form the solution path must be carried out using the strategy expressed in Algorithm 1.

**Algorithm 1: Definition of solution path projects, divide and conquer strategy**

1. Understanding the problem and identifying the project environment.
2. If (problem is simple and solvable).
3. Then Propose the conceptualization as a project and finish.
4. else Division of the problem into as many subproblems as necessary.
5. While (not all subproblems have been analyzed).
6. Analyze the subproblem using this same algorithm based on the divide and conquer strategy.
7. Integrate the solution of all subproblems into a single solution.

The creation of solution paths causes different projects to be executed simultaneously. Many of these projects share organizational resources. In this context, we talk about shared resource management, and two concepts for grouping projects are established, programs and portfolios.

Project programs constitute logical structures for grouping projects that share human and non-human resources. The programs are created to facilitate the management of the projects they contain. Program and project portfolios constitute logical structures for grouping projects that can be associated with different classes or concepts. For example, a grouping by portfolio could be associated with the concept of clients to whom the project is directed. Another example of a portfolio could be the organization of projects by theme.

## 2.2 *Agile Management of Lessons Learned from Portfolios, Programs and Projects*

In general, we can say that projects do not really start in the initiation phase of their development. The projects have their origin in the strategic projection of the organization and the approach to the problem that gives rise to them, as well as a response to the fulfillment of strategic objectives by the entity that generates them. The life cycle of projects and the results they produce are governed by the following moments:

Moment 1. Identification of a real problem: this is obtained from the strategic objectives of the main entity interested in the development of the project. An idea for a solution is developed.

Moment 2. Design of a preliminary project: A project is designed to solve the identified problem. This preliminary project is aimed at obtaining the products or services provided for in the project objectives. In the preliminary project, the resources necessary for the execution of the project and an execution plan are planned as accurately as possible.

Moment 3. Evaluation of the feasibility of the preliminary project: a feasibility analysis of the project is carried out from different dimensions, among which the following stand out: economic feasibility, social feasibility, technical feasibility, environmental feasibility and commercial feasibility. As part of the feasibility analysis, a decision-making process is carried out to analyze the different alternative solutions and select the best one. This process in the system is supported by word computing techniques that facilitate evaluation by experts and the aggregation of their preferences.

Moment 4. Search for financing and contracting: at this moment, we proceed to identify and manage the sources of financing for the development of the project and its contracting.

Moment 5. Project development: this is the moment where the actions planned in the project are executed and the planned products or services are obtained. The execution of the project contract formally begins until its closure. Here, the intensive use of agile methods is essential for both planning and execution, control and monitoring of projects.

Moment 6. Introduction of products and services: at this moment, the results obtained as a result of the project are introduced into social practice, and the analysis of the impact of the project is carried out.

Moment 7. Periodic analysis of the feasibility of the products and services introduced and identification of the need to deactivate them.

From the analysis of these moments, it is identified that the life cycle of the project is different from the life cycle of the products or services that are obtained from it (Fig. 2). The life cycle of the project ends when the project closure phase ends. While the products or services obtained as a result of the project cease to exist, long after, for some reason, they become obsolete, and it is decided to deactivate them.

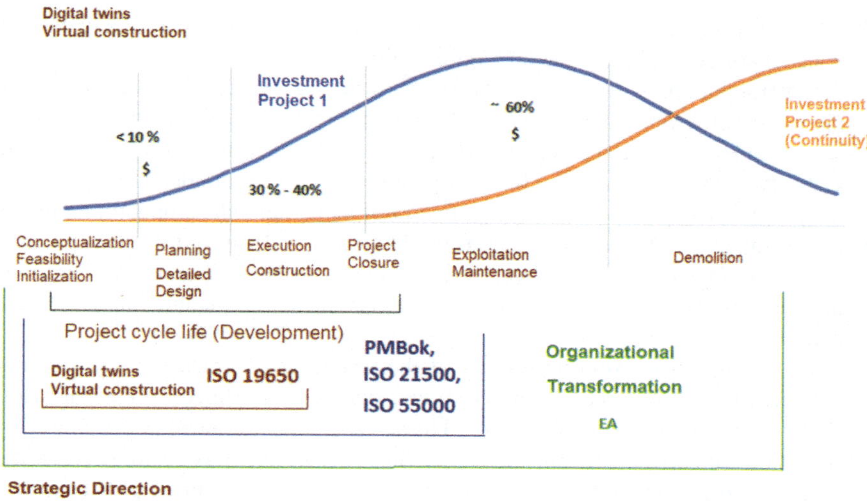


**Fig. 2** Project life cycle aligned with business strategic planning

In general, the life cycle of projects is often visualized as a transition through the following phases: "Initialization and feasibility", "Planning", "Execution" and "Closure". In formal development models, these phases are inextricably linked to changes in the project's status. But in the context of agile methods, the phases only respond to logical organizations that reflect the relative state of progress of the projects.

The transition of a project from one phase to the next is not a discrete process but a continuous one without well-defined limits. Placing projects in one phase or another is, above all, a simple way to understand the status of project execution. For example, a project in the initiation phase is clear that it is in an initial state and that no significant results have been obtained at this moment, but when it is stated that the project is in the closing phase, it is understood that it has obtained the main results planned in the project and that there is little time left for its final conclusion.

Finally, in the proposed model, the transit of projects by phases is identified with the fulfillment of the project milestones. In any case, understanding the phases is positive. The representation of the phases of the proposed model is described below:

- Conceptualization and contracting phase: this stage is characterized by the achievement of three fundamental milestones: the construction of the preliminary project, its feasibility analysis and the contracting of the project. Inception phase: the most representative events of this phase are the signing of the project initiation document and the formation of the project management team. At this time, a refinement of the objectives to be achieved in the project in question occurs.
- Planning phase: the project objectives and the desired scope are refined, and the project planning is updated based on the preliminary project prepared previously. We also proceeded to acquire most of the personnel required for the execution

of the project. It is important to highlight that there is no divorce between the concepts of general planning and agile methods.

- In general planning, you have an overview of the systems and high-level requirements. The first version of the project backlog is prepared. It is important to always have a general plan and an estimated budget, with a clear focus on the economic viability of the proposal. The basis of planning focuses on understanding that as the project progresses, planning is refined following agile development methods.
- Execution phase: Most of the actions to obtain the products and services planned as results of the project are executed. This execution also focuses on the iterations that, as explained before, last from 2 to 6 weeks in an inextricable union with the planning.
- The control and monitoring checks of the project are carried out with greater intensity, partial reports are prepared; and planning is refined by making decisions based on situations that arise. During all phases, control and monitoring activities are carried out, but these activities reach their maximum effort in the execution phase (see Fig. 4). It is recommended to follow principles based on agile methodologies such as SCRUM for control and monitoring.
- Closure phase: the results obtained are delivered, the final report is prepared, and the introduction process is monitored in its first phase. An important element in this phase is the management of the lessons learned in the project through a workshop or closing activity and the storage of the project file.

In the case of software projects, most methodologies refer to project phases such as the requirements phase, analysis phase, design phase, implementation phase and testing phase. Actually, the actions of these phases are included in the initiation, planning, execution and closure phases explained above (see Fig. 3). It is important to note that software development projects follow the same life cycle as all the others, the essential differences lie in the techniques they use to obtain products or services. It is essential to point out that software development projects require, like any other project, the existence of a development plan and a control and monitoring process based on indicators that depend on whether the activities planned therein are fulfilled.

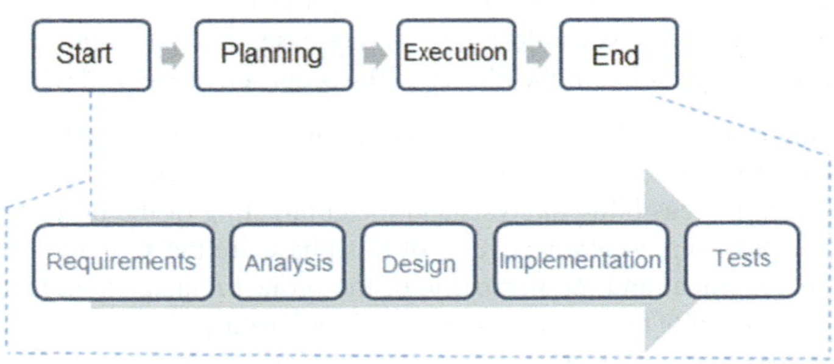


**Fig. 3** Homologation of phases in software projects and project life cycle

This charter presents the general theory of projects, and many of the examples presented are inspired by software projects, thus demonstrating that software projects are governed by the same general laws of project management.

### 2.2.1 What Information is Stored Associated with Lessons Learned?

The management of lessons learned and information in project files is important and does not contradict agile methods either. In this sense, the proposed model establishes the following elements:

- The management of information in portfolios, programs and projects is important because it guarantees future project planning processes and cost and risk prediction.
- The management of lessons learned constitutes an element that facilitates the credibility of the organization and certification. This has a notable impact on improving market opportunities.
- One of the ways to measure the level of maturity of the organization is the management of files and lessons learned.

Figure 4 graphically exemplifies the relationship that exists between the maturity of organizations and the management of project files.

As explained previously, the five phases identified in a project constitute logical transitions that represent the evolution of the project. But these transitions occur during the execution of groups of processes and tasks. In line with what has been proposed, the groups of fundamental processes are:

- Group of conceptualization and feasibility processes
- Initiation process group
- Group of planning processes
- Group of execution processes

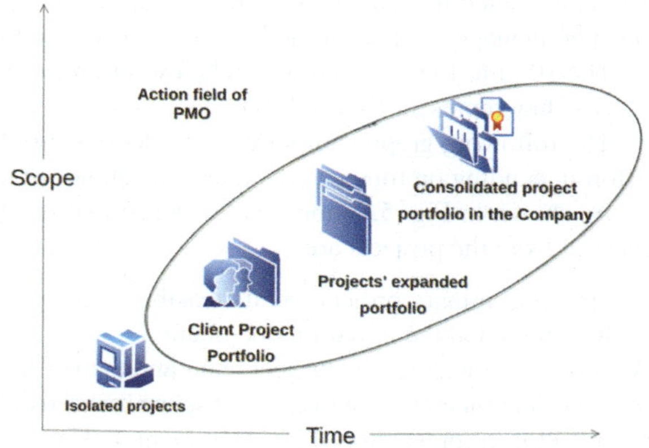


**Fig. 4** Maturity map of organizations, view from project files and lessons learned

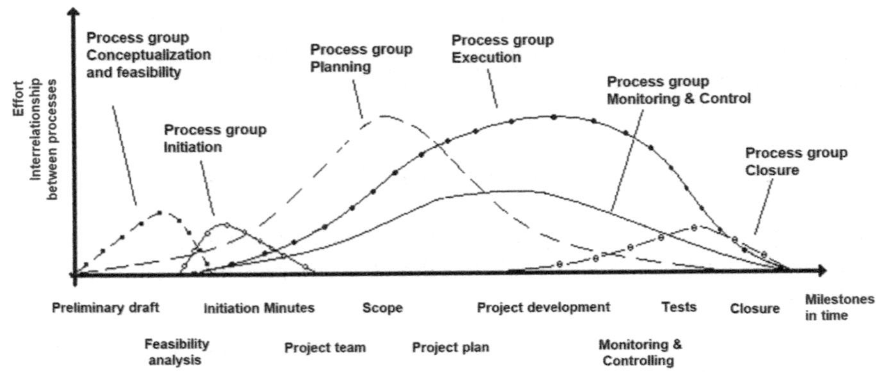


**Fig. 5** Chart of process groups and their effort as a function of time and the main milestones of the project

- Group of closing processes.

There is a close relationship between the phases and the groups of processes previously explained, with the particularity that the group of control and monitoring processes is executed practically throughout the entire life of the project, affecting all phases (see Fig. 5). It is common to think that the identification of processes, traditionally used in formal methodologies, is in conflict with agile methods.

But it's not like that. The processes simply collect the best practices for the execution of the activities to be carried out during the project life cycle. The processes and management of files will be as agile as the organization is capable of executing the projects with full customer satisfaction.

But the essence of process groups is to develop the milestones and tasks of the projects. Under an agile approach, the authors of this work present below the ten fundamental milestones of a project. These ten milestones are the essence of project development that managers should not forget. The deliverables that are generated from these ten milestones constitute the essence of the minimum documentation that must be managed as part of the lessons learned (see Fig. 5).

The 10 milestones are presented below, arranged by groups of processes and the phases they represent (See Table 1).

The following graph shows the interrelationship of the process groups and their effort depending on time and the main milestones of the project, see Fig. 5.

As shown in Fig. 5, among the evidence that must be built and used as lessons learned from the projects are:

1. The preliminary project and its feasibility analysis.
2. Contract and the initiation document.
3. Information about the project team and their competencies.
4. The documentation of the project scope focused on the construction of the project backlog (representing the requirements). The construction of user stories that

**Table 1** Main milestones by phases in the life cycle of a project

| Process groups in which they occur | Project milestone |
|---|---|
| Conceptualization and contracting | Milestone 1. Idea and conceptualization of the preliminary project as part of the solution to a specific problem of social practice |
| | Milestone 2. Feasibility analysis of the project and selection of the best alternative solution for contracting |
| Initial | Milestone 3. Signing of the project start document with the responsibilities of the parties, the preliminary scope and the prerogatives of the project manager |
| Planning | Milestone 4. Definition of the project management team and the development team |
| | Milestone 5. Scope Definition. Project Backlog, User Stories |
| | Milestone 6. Development of the project plan (schedule) |
| Execution | Milestone 7. Development of the project, obtaining results |
| | Milestone 8. Control and monitoring of the project |
| | Milestone 9. Quality acceptance tests of the products or services obtained |
| Closing | Milestone 10. Approval and closure of the project, lessons learned |

describe the requirements and the logical grouping of the requirements represent the different components of the technologies developed by the organization.
5. The project plan, which is essentially the project schedule.
6. The lessons associated with the development of the project, and relevant documents such as the architecture of the solutions.
7. The system of control and monitoring indicators.
8. Evidence of tests and project acceptance and closure documents.

An important element is that these elements do not need to be documents that are filed away without use. The evidence and lessons learned must be managed directly through the information systems. In the particular case of our model, the BusinessRedmine Ecosystem integrated into the BRasa platform is proposed as the information system. These two systems will be explained in the following sections.

The essence of the project is not to delight in writing many documents, but to obtain the specific result that is expected. All project tasks must be directed toward achieving the project objectives and scope.

## 2.3 Agile Planning, Control and Monitoring Based on Objective Indicators

In the development of a project, it is identified that there are regularities and that it is absolutely necessary to decompose a project into smaller components for both planning and better control and monitoring of its execution. These small parts of development are called iterations (see Definition 1).

**Definition 1 "Iterations"**: in this work, we define iterations as a temporary effort between two and six weeks in which a set of project activities are planned and executed with the objective of obtaining partial or final results and aligned with the general objectives of the project. At the end of each iteration, a project check must be carried out (project cut, see Definition 2) and the quality of the partial results obtained must be controlled. The time of the iterations may depend on the size of the results to be obtained, but they cannot be longer than the proposed time because otherwise they become very long periods with little capacity to adapt to changes, affecting communication with the project's stakeholders. In the model proposed, planning, control and monitoring are closely related. During the control and monitoring of a project, project cuts are established as a relevant point of control, see Definition 2.

**Definition 2 "A project cut"**: constitutes a project control point where its status is evaluated. It is generally done in the form of periodic meetings in which the main stakeholders of the project participate and in correspondence with the stage at which the project is located.

Two other fundamental concepts in the project planning process are free slack and waiting slack.

**Definition 3 "Waiting slack"**: is the period of time between the completion of a task and the start of its successor. This slack is introduced because the start of a task depends on the completion of all its predecessors, and, therefore, a waiting time is created between the task and its predecessors that have already finished (see Fig. 5).

**Definition 4 "Free slack"**: is the period of time between the completion of a task and the scheduled date to make a project cut. This slack is planned as part of the reserve time for risk management processes. During the execution of the project, this slack allows necessary adjustments to be made in the schedules in the event of unforeseen delays (see Fig. 6).

### 2.3.1 Planning, a Single Project Plan, a Single Schedule

It is recommended that there be a direct relationship between the iterations and the versions of the products resulting from the project. That is, there is a direct relationship between the planning of the iterations and the planned scope of the project. It is recommended that even this relationship be one-to-one, indicating that a version of

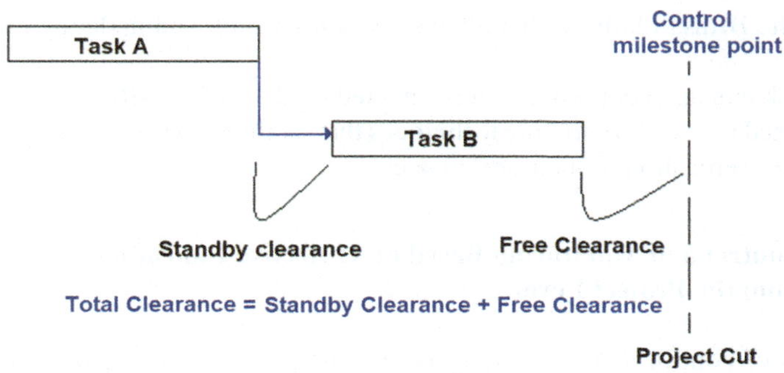


**Fig. 6** Cut view of projects, waiting clearance, free clearance and total clearance

the product will be obtained for each iteration of the project. The application of this principle will facilitate an alignment of the project with the principles proposed by agile development methodologies. Another essential element that marks the agility of the models is the recognition of the following statement: "All plans must be merged into a single plan, the project schedule, stored and managed from the information system." This statement stated by the authors of this work has the following implications:

- Project managers and quality certification entities frequently suggest that a project should have a quality management plan, a risk management plan, plans for human resources management, and plans for each of the areas of knowledge. However, the proposed model suggests that all these plans must be constituted as tasks or activities integrated into a single plan, "the project schedule." When you want to access any of the specific plans mentioned above, this can be achieved by filtering the project tasks in the particular plan.
- The risks of a project are not part of the schedule, but the tasks to avoid or mitigate the risks are part of the plan.
- Defects or change requests should not be part of the schedule, but the tasks that resolve the defects should be. In addition, assurance tasks are also part of the schedule.
- All schedule information must be managed from the information system, saving all changes to the baselines that occur. The recording of changes in schedules constitutes lessons learned for future planning.

The planning process is supported by tools based on artificial intelligence, among which the following stand out:

- The BusinessRedmine ecosystem, which includes facilities for constructing the schedule and its visualization in different ways.
- The BRIntelligent library and service package allow the optimization of project schedules based on metaheuristics. Includes a conversational system developed

from the BRasa Platform that allows the construction and updating of schedule tasks.

- The risk management process implemented on the BusinessRedmine platform is supported by word computing techniques that facilitate expert assessment of risks and the aggregation of their preferences.

### 2.3.2 Control and Monitoring Based on Objective Indicators from the Project Level

The basis of control and monitoring is the calculation of objective indicators for each project. One of the benefits of the model proposed in this work is the system of indicators shown below.

The model's system of indicators establishes that these are calculated from the work in the organization's project management information systems. They only require basic information on the fulfillment of tasks and associated resources.

On the other hand, the indicators are integrated into the scorecard for the organization's decision-making. But they are viewed gradually, from the most basic and general to the most specific. The basic ones are shown from the main interfaces with the greatest access by interested parties, accompanied by links to interfaces that show the specific indicators in greater detail of the status of a project. We call this process of gradual access to information the control and monitoring cascade. Some of the indicators are based on the PMBoK and agile methods. Furthermore, the indicators are focused on the generation of the value chain and are subdivided into three groups as proposed by the ISO 56000 standard for innovation management systems:

- Operational or execution indicators: allow control at project levels.
- Strategic indicators: which focus on the evaluation of programs and project portfolios. Support or support indicators: include indicators that complement and support the control of portfolios, programs and projects.

*Operational or execution indicators (project level)*

These focus on the evaluation of programs and projects and are subdivided into three groups: Scope and time control indicators and indicators associated with costs.

Time and Scope indicators:

- PLAN: indicates the percentage of tasks that must be completed at the time of the project cut-off (time of calculation of the indicator), see Eq. 1.

$$\text{PLAN} = 100 \cdot \frac{\text{Amount of tasks with due date before cut}}{\text{Total of Task's Project or Program}} \tag{1}$$

- REAL: indicates the percentage of tasks that are actually completed at the time of the project cut-off (time of calculation of the indicator), see Eq. 2.

$$\text{REAL} = 100 \cdot \frac{\text{Amount of closed tasks}}{\text{Total of Task's Project or Program}} \tag{2}$$

- EXECUTION IN PROCESS (EP): indicates the percentage of progress of the tasks declared by the project team that are at the time of the project cut-off (moment of calculation of the indicator), see Eq. 3.

$$\text{EP} = \frac{\sum_{i=1}^{n} \% Progress_{Task_i}}{100 \cdot \text{Total of Task's Project or Program}} \tag{3}$$

- QPI: quality of the tasks is evaluated based on the evaluation criteria of the system tasks, the tasks can be evaluated regarding their quality as satisfactory, regular, or poor. The evaluations of the tasks add up to those projects with the greatest number of satisfactory tasks obtaining better results from this indicator that takes values between 1 and 100, see Eq. 4.

$$QPI = 100. \frac{\sum_{i=1}^{p} \text{Quality Index}_{Task_i}}{\text{p} \cdot \text{Ordinal Best Evaluation Expected}} \tag{4}$$

  where
  p is the total number of project tasks that closed before the cut-off date.
  $\text{QualityIndex}_{\text{Task}_i}$ ordinal of quality evaluation of i-ésima tasks.
  Ordinal of Best Evaluation Expected.
- DQPI: Planning quality is calculated from the number of tasks with all the data necessary for planning. This indicator includes the number of unmanaged tasks. An unmanaged task is one that does not have an end date; therefore, its real progress status cannot be known with certainty.

**Indicators associated with project costs**

These indicators are based on the earned value technique, highlighting the BAC, PV, EV. AC, SPI, CPI, SV, and CV as defined by the PMBoK guide. All these indicators are calculated from basic information about the tasks and their associated resources. These indicators, depending on the organization, may be calculated by representing costs as monetary values or as effort based on man-hours.

BAC: indicates the budget or total effort of the program or project; for its calculation, the ascending estimation technique is applied; see Eq. 5.

$$\text{BAC} = \begin{cases} \left(\sum_{i}^{n} \text{Cost}_{\text{task}_i}\right) & \text{if monetary cost interest} \\ \left(\sum_{i}^{n} \text{Hours}_{\text{task}_i}\right) & \text{if hours effort interest} \end{cases} \tag{5}$$

where $n$ is the total number of tasks in the project.

***PV***: indicates the effort or cost of the program or project until the cut-off date (the moment of calculation of the indicator), for its calculation, the ascending estimation technique is applied, see Eq. 6.

$$PV = \begin{cases} \left( \sum_{i=1}^{m} \text{Cost}_{\text{task}_i\text{\_due\_date\_before\_cut}} \right) \text{ if monetary cost need} \\ \left( \sum_{i=1}^{m} \text{Hours}_{\text{task}_i\text{\_due\_date\_before\_cut}} \right) \text{ if hours effort need} \end{cases} \quad (6)$$

where m is the total number of project tasks with a scheduled closing date before the cut-off date.

***AC***: indicates the effort or cost of the real program or project calculated from the tasks closed before the cut-off date (the moment of calculation of the indicator), for its calculation, the bottom-up estimation technique is applied, see Eq. 7.

$$AC = \begin{cases} \left( \sum_{i=1}^{p} \text{Cost}_{\text{task}_i\text{\_close\_date\_before\_cut}} \right) \text{ if monetary cost need} \\ \left( \sum_{i=1}^{p} \text{Hours}_{\text{task}_i\text{\_close\_date\_before\_cut}} \right) \text{ if hours effort need} \end{cases} \quad (7)$$

where $p$ is the total number of project tasks that closed before the cutoff date.

***EV***: multiplication of the percentage progress of the project by the BAC and indicates what amount of the project represents the work that has been completed on the project until the cut-off date, see Eq. 8.

$$\boldsymbol{EV = \%Progress \cdot BAC} \quad (8)$$

***SV***: planning variation a negative value indicates a delay in the expected planning, see Eq. 9.

$$\boldsymbol{SV = EV - PV} \quad (9)$$

***CV cost variation***: a negative value indicates a delay in actual execution and may mean an overdraft in project costs or an over-declaration of actual hours worked, see Eq. 10

- $\boldsymbol{CV = EV - AC}$ (10)

***SPI***: Scheduling Performance Indicator indicates the status of the project compared to what was planned, see Eq. 11. This indicator indicates that the project is on time when it reaches a value of 1. Values greater than 1 mean that the project is ahead, while values less than 1 indicate a level of delay in the execution of the project.

- $\boldsymbol{SPI = EV/PV}$ (11)

***CPI***: Costs Performance Indicator indicates the status of the project with respect to costs. This indicator may be affected by different situations. Therefore, the analysis must be done carefully and complemented with other indicators, see Eq. 12.

- This indicator indicates that the project is on time when it reaches a value of 1.
- Values greater than 1 mean that the project has spent less than expected, in accordance with what was planned. This situation can mean savings, in which case it would be positive. But it may mean that the necessary resources for the following stages have not been acquired, which will cause a delay in the future. Also, it may mean that lower-quality resources have been purchased where savings are made, but the quality of the product or service could be affected.
- Values greater than 1 mean that the project has spent less than expected, in accordance with what was planned. This situation can mean savings, in which case it would be positive. But it may mean that the necessary resources for the following stages have not been acquired, which will cause a delay in the future. Also, it may mean that lower-quality resources have been purchased where savings are made, but the quality of the product or service could be affected.

$$\boldsymbol{CPI} = \boldsymbol{EV/AC} \tag{12}$$

*Support or Support Indicators*

These focus on the evaluation of support elements for management.

***HPI***: takes values between 1 and 100 and is an indicator of the effectiveness of the work of human resources. It is calculated from the PLAN and the ACTUAL of the tasks associated with HR human resources at the decision-making level being analyzed.

$$HPI_{RH} = 100 \cdot \frac{\sum_{\mathrm{i}}^{\mathrm{q}} \mathrm{Real}_{\mathrm{RH\_task_i}}}{\sum_{\mathrm{i}}^{\mathrm{q}} \mathrm{Plan}_{\mathrm{RH\_task_i}}} \tag{13}$$

where the RH subindex represents the human resources of the level at which the indicator is to be calculated and q the number of tasks associated with that resource.

- If you want to show this indicator at the level of a person, this subindex will represent the tasks associated with the particular person.
- If you want to show this indicator at a project level, this subindex will represent the tasks associated with the human resources members of the project.
- If you want to show this indicator at the level of a program, this subindex will represent the tasks associated with the human resources members of the projects associated with the program.

***LOGPI***: takes values between 1 and 100 and constitutes an indicator of the effectiveness of logistics and the timely delivery of the resources requested for the project, see Eq. 14.

$$LOGPI_{Resources} = 100 \cdot \frac{\sum_{i}^{n} \mathrm{Real}_{\mathrm{Resources\_task_i}}}{\sum_{i}^{n} \mathrm{Plan}_{\mathrm{Resources\_task_i}}} \tag{14}$$

where n is the total number of tasks in the project.

RISKPI: measures the level of exposure of the organization in its programs and projects.

$$\mathrm{RISKPI} = 100 \cdot \frac{\sum_{i}^{r} \mathrm{Evaluation}_{\mathrm{Risk_i}}}{\mathrm{r}} \tag{15}$$

where *r* is the total number of project risks

$\mathrm{Evaluation}_{\mathrm{Risk_i}}$ indicates the risk exposition of i-risk.

CONTRACT_KPI: index of contracts with execution on time.

$$\mathrm{CONTRACT\_KPI} = 100 \cdot \frac{\sum_{i}^{z} \mathrm{Real}_{\mathrm{Contract_i}}}{\sum_{i}^{z} \mathrm{Plan}_{\mathrm{Contract_i}}} \tag{16}$$

where z represents the number of contracts in the organization.

*Strategic indicators*

These focus on the evaluation of programs and portfolios based on the following characteristics:

- As explained before, the projects are grouped for management by the organization in programs and portfolios.
- The same project can belong to more than one portfolio, that is, there can be different custom fields, and each of them can represent the belonging of a project to a different portfolio category.

Once the indicators have been calculated at the project level, an aggregation process occurs to calculate program or portfolio indicators. This process complements the decision-making process at the organizational level. Algorithm 2 summarizes the essence of this process.

**Algorithm 2** Agregación de indicadores para la toma de decisiones en la organización

*Step 1. Receive Indicators per project (PLAN, REAL, EJECUCIÓN EN PROCESO, SPI, CPI, BAC, PV, EV, AC, SPI, CPI SV, CV).*
*Step 2. Application of aggregation operators, considering priority projects.*
*Step 3. Application of OWA operators for aggregation.*
*Step 4. Evaluation of programs and portfolios based on aggregate indicators.*

In this work, OWA operators are used to perform information aggregation. These operators allow weights to be assigned to the data to achieve compensation depending on their values and the representation of fuzzy concepts. Another advantage of these operators is the great flexibility they provide in their decision schemes to model many aggregation rules. Its mathematical definition is given in the following way:

**Definition 5** An OWA operator is defined as an n-dimensional function of the type F: $R^n \rightarrow R$ to which a vector is associated $W = [w_1, w_2 \ldots w_n]^{T\,wi}$ with length $n$, where $w_i$. $w_i \in [0, 1]$ y $\sum_i^n w_i = 1$.

**Interpretability of indicators**

Fuzzy logic techniques are used to interpret the indicators. Each indicator has an associated linguistic variable (see Definition 6) adjusted to the domain of the indicator.

**Definition 6** **Linguistic Variable**: A linguistic variable groups together fuzzy sets

associated with the same variable. A linguistic variable is characterized by a fivefold (x, T(x), X, G, M) in which x is the name of the variable, T(X) is the set of linguistic terms, X is the universe of discourse, M is a semantic rule which associates Z with each linguistic value its meaning M(Z), where M(Z) denotes a fuzzy set in X and G is the set of syntactic rules for generating compound terms, from the atomic terms that configure the sentences that give rise to each linguistic value. An example of a variable is presented below, see Fig. 7.

The name of the variable (x), associated with the indicator that you want to measure REAL, and T(REAL) the set of linguistic terms that describe the variable. T(ACTUAL) = {Very late, Late, On time}. The ideal number of terms to achieve fine granules is 5 or 7. In the case of the proposed model, a set of three terms was decided to simplify the interpretation and link these three terms with the colors of the traffic light that facilitate their interpretability.

Universe of discourse (X) will comprise a range of the appropriate indicator to define the fuzzy values significant for the problem in its entirety (between 0 and 1) Fig. 7. The syntactic rules assume three primary terms as primitive: Very late, Late, On time; In this way, terms like slightly Behind arise from the grammar that makes up the set of syntactic rules.

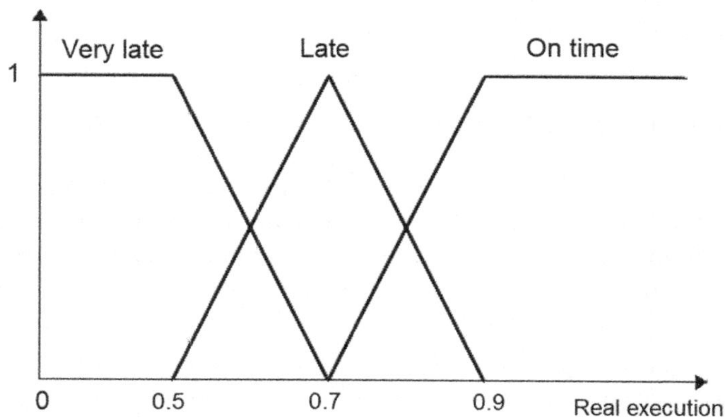


**Fig. 7** Representation of the linguistic variable to measure the REAL indicator

## 2.4 Decision-Making is Supported by Computational Intelligence Techniques

The indicators calculated in the control processes show a summary of the status of portfolios, programs and projects. These indicators will have a level of analysis and visualization depending on the level of decision-making where you are:

- If the user is at the level of a simple project without subprojects, the indicators that summarize the current behavior of the tasks and resources of the project in question will be displayed.
- If the user is at the level of a project with subprojects, the indicators that summarize the current behavior of all the subprojects and tasks in the project tree will be displayed.
- If a user is at the program or portfolio level, then indicators that summarize the current behavior of all projects and tasks in the program will be displayed.

Once a project check or cut is in place, the status of the indicators is evaluated as shown in Algorithm 3. And the decision-making process begins.

**Algorithm 3** Control and monitoring based on indicators

*Step 1. Calculation of indicators.*
*Step 2. Review of the quality of the work product.*
*Step 3. Analysis of indicators determining the areas of knowledge with difficulties.*
*Step 4. If there are no problems, finish the checkup and congratulate,*
*Step 5. Go down the waterfall, analyze the causes of the difficulties in each area.*
*Step 6. Once the causes have been identified, proceed to decision-making.*

As a final step of Algorithm 3, the decision-making processes are identified based on the calculated indicators. In the proposed model, as part of the BRIntelligence platform, access is provided to a fuzzy inference system that aids decision-making. In the following subsection, some of the rules that regulate the behavior of the proposed system are presented.

### 2.4.1 Decision-Making Supported by Knowledge-Based Systems

The decision-making process is supported in the proposed model by a fuzzy inference system (SIB). The use of the fuzzy inference system for the evaluation of program and project portfolios contributes to the treatment of information uncertainty in decision-making. In particular, the proposed system contributes to mitigating the adverse effects of the following phenomena:

- Mitigate vagueness in risk and human resource assessment.
- The inaccuracy that occurs from the declaration of the progress of the tasks and the time spent.

- Uncertainty that generally has its source in the lack of certainty in the declaration of the percentage progress of the tasks.

This system is not intended to replace decision-makers but rather to support them in decision-making, based on the following principles:

- There are no rules of thumb for all problems, but measures should focus on identifying and resolving the causes of delay.
- The systems suggest measures based on the analysis of the indicators, but the decision-makers have the last word.
- The measures taken by decision-makers must be aimed at:
  - Maintain focus on generating value with a focus on customer satisfaction and compliance with commitments.
  - Guarantee commitment of management and involvement of all members in the search for solutions.
  - Meet the quality requirements defined with the client.

The decision support fuzzy inference system has the following characteristics:

- It is an inference system, it is a zero-degree sugeno type system; see Fig. 8.
- As input, the system receives the indicators PLAN, REAL, EXECUTION IN PROCESS, SPI, CPI, BAC, PV, EV. AC, SPI, CPI SV, CV, RISKPI, LOGPI.
- Previously and prescriptively, fuzzy inference rules were established, which constitute the knowledge base of the system.
- SIB output classes are labels that determine possible project problems and their causes.
- A set of good practices is associated with each of these output labels that are presented as the output of the ecosystem to decision makers.
- This system, based on the analysis of indicators, is capable of evaluating the status of portfolios, programs, projects and proposing a set of measures.

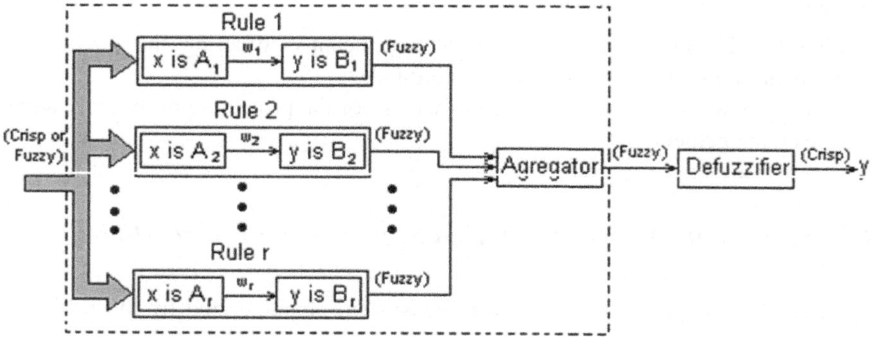


**Fig. 8** Diagram illustrating the operation of the fuzzy system

**Below are a set of possible output labels and the measures that the system recommends**

Output label associated with delay with respect to time

- Measure may be to plan overtime for the team.
- Generally, to resolve a backlog, bringing in a lot of new staff causes more backlog.
- Concentrate main human resources, freeing them from other tasks that affect their performance.
- Prioritize tasks, develop them based on priorities, renegotiating if necessary.
- When the delay exceeds the total slack and cannot be recovered, renegotiate with the client (do not lie).
- Review which tasks can be parallelized and proceed with other human resources on less prioritized tasks (outsource).
- Identify the competencies of the organization's human resources, reorganize the teams based on competencies.
- Apply better methods of staff motivation, rewards and penalties.

Output label associated with delay caused by resource issues:

- Move human and non-human resources from less prioritized projects to more prioritized ones. Stop lower priority projects if necessary.
- Review of contracts and adjustment of logistics management methods, prioritizing projects appropriately.

Output label associated with scope deviations:

- Identify better alternatives and apply alternative selection techniques.
- Prioritization of the subsystems of the project scope based on the client and renegotiation, if necessary, with all interested parties.
- Differentiate non-conformities from change requests. new features = new contracts.

Output label associated with cost deviations:

- Identify cheaper alternatives without significantly affecting the quality and application of alternative selection techniques.
- Identify new sources of financing or income for the project, appropriately adjust sales management.

## 2.5 Development Ecosystem that Supports the Entire Model

The model is supported by a software ecosystem characterized by the following elements:

- It is designed to increase the productivity and competitiveness of the developed solution.

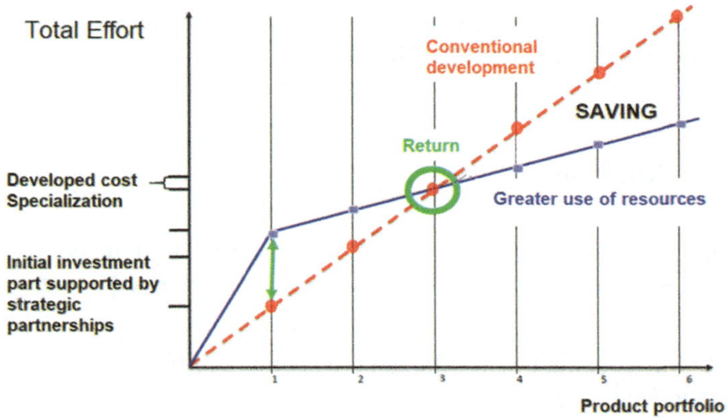


**Fig. 9** Reducing costs in software ecosystems with a production mode based on LPS and the cloud

- It is supported by a reference architecture and platforms that facilitate the integration of solutions and the exchange of information and assets in general.
- The application domain of the proposed solution is integrated project management and decision-making in project management based on the PMBOK v5 [2] y la ISO 21500 [7]. In addition, these standards are complemented by a group of indicators required for the control and monitoring of projects. It allows the development of product families in this domain and provides cloud services.
- The evolution of the ecosystem is supported by agile development methods and the conceptualization of a product line that develops solutions to provide cloud services. Existence of an asset repository that enhances reuse.

A significant element of software ecosystems is the considerable reduction in costs as the organization's product family increases, as shown in Fig. 9.

The development ecosystem that supports the model is made up of three large components. Each of them constitutes an ecosystem in itself due to the diversity of the technologies involved. The three components are briefly described below.

Component 1. BusinessRedmine ecosystem for business decision-making. It includes a balanced scorecard and specialized control boards to help make decisions: financial, accounting, sales management, people, logistics, project management, production management and task management. It is a platform with high added value that includes integration with BIM, conversational systems and artificial intelligence techniques. Registered Product 1212-05-2023.

Component 2. BRasa ecosystem for creating conversational systems and providing cloud services. Based on an architecture aimed at exploiting the advantages of rule-based smart chatbots and LLM models. It brings agility to the environment by providing the ability to update project data and consult statuses using conversational styles.

Component 3. BRIntelligence Platform as a Service (PaaS) that includes access to analytics services and data visualization, taking advantage of the availability of cloud resources. The services of this platform are presented by providing integrated services to the BusinessRedmine and BRasa platforms.

# 3 Results Analysis

This section explains the results of applying the proposed model in a project management scenario. The specific scenario is characterized by the following elements:

- Entity A with a central office with numerous offices or branches. Which needs to be computerized based on the following set of software solutions.
    1. Office 1 (INM) procedures management subsystem.
    2. Office 2 procedures management subsystem (MER).
    3. Office 3 (SA) control management subsystem.
    4. Web solution for data transfer between offices (SAWeb).
    5. Accounting and financial management subsystem (CON).
    6. Human resources (HR) management subsystem.
- Project-oriented entity B interested in developing a macro-project for the development of solutions guaranteeing the computerization of entity A.
- Difficulties arise in the relationships between entity A and entity B from multiple errors by Entity B.
- It is decided to start an improvement project for entity B to solve the initial problem.

## *3.1 Results of the to-be Model Diagnosis in Entity B*

A diagnosis of organization B was carried out and the following problems were detected.

1st Error: Poor initial selection of development tools, unilaterally by the project director. Architectural decisions were not consulted with the team.

This resulted in the need to change the data manager twice. First, he switched from SQL Express—Oracle Express and then from Oracle Express to the Oracle SE solution. In addition, it caused an increase in internal and external pressure, delays, and the need for replication treatment by the development team. Figure 10 shows that between November 4 and November 20, three project outages occurred where there was practically no delivery of results.

2nd Error associated with the lack of transparency on the part of the project director, who did not show either the technical project or the contract to the management team, architects, or analysts. The consequences of this situation were:

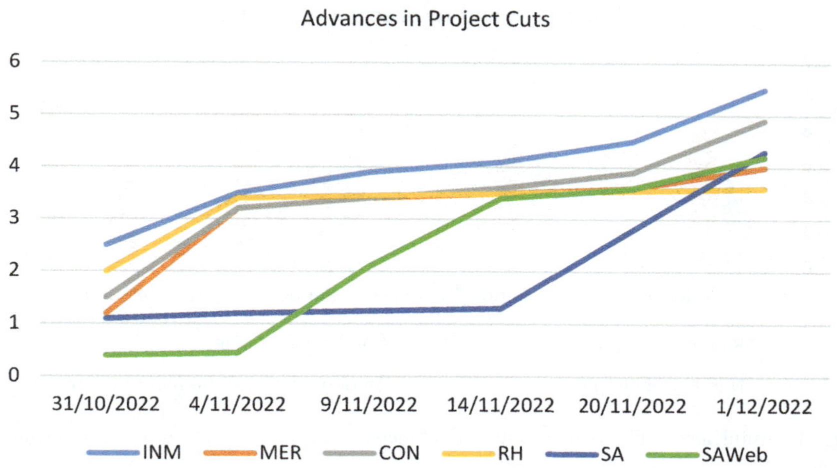


**Fig. 10** Visualization of the delays in the development of solutions between the months of October and the scope of entity A

- Functionalities that the analysis and architecture team had access to late, in the advanced stages of the project.
- Caused rescheduling, delays and customer dissatisfaction.

3rd Error associated with poor management of collective commitments, insufficient development deadlines, and contracts committed on time. The consequences of this error were:

- Increase in pressure, an attack on quality, and an increase in defects in solutions.
- Reduction of spaces for debate and integration of solutions. This caused rescheduling and little pushback on solutions between teams.

4th Error associated with work overload and poor management of individual commitments. The consequences of this error were an overload of human resources and increased pressure on the project.

5th Error establishing a formal work model with very few personnel for collect requirements process. Insufficient effectiveness in change management processes (Fig. 11).

The consequences of this error were the need to repeat the requirements gathering processes. In particular, entity B had to carry out four stages of requirement gathering.

6th Error by adopting formal methods and delaying meetings with the client, causing distancing with the client and a lack of trust. The consequences of this error were:

- Difficulty in deliveries and increased pressure.
- Reprogramming.

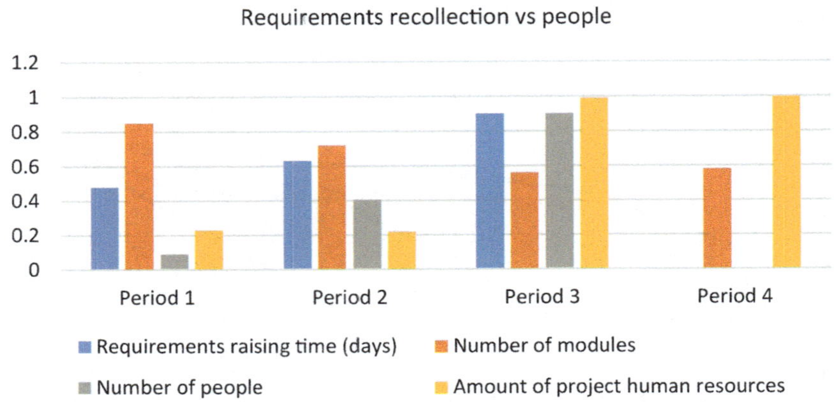


**Fig. 11** Insufficient staff combined with formal methods caused rework and difficulties with the client

7th Error associated with insufficient management and exploitation of software configuration elements. This situation had the consequences:

- Difficulties in managing non-conformities, the quality team reviewed parts of the product and configuration elements that were not updated in their latest version.
- Detection of numerous defects, which could be corrected during development,
- It was identified that company B also managed the work of 12 teams with a lot of diversity in version control technologies. Of them, 3 teams used the GITLAB platform, but each with independent servers. Two other teams used SVN while the rest did not use any technology.

Regarding the task and project management platforms, the teams of entity B behaved in the following way. Three teams managed their tasks using Redmine-based solutions, 1 team used the Wekan platform; and eight teams did not use any version control tools.

Given this situation, it was decided to suggest a transformation of entity B, improving processes and systems simultaneously.

## *3.2 Results of the Introduction of the Proposed Model in Entity B*

An ecosystem is established consisting of the following subsystems:

- Ecosystem for project management BusinessRedmine as a cloud server providing project management services for all work teams. In this way, project management was standardized and more than 200 projects were controlled and managed through the platform.

- The platform for centralized GITLAB version control is standardized for all teams. Allowing greater control of software as an intangible asset of the organization. At the time of writing, more than 80 COTS components were managed as independent projects in the centralized GITLAB environment. In addition, multiple forms of integration were created between the GITLAB platform and BusinessRedmine.
- A conversational system based on BRasa technology is created to improve agility in decision-making and task management processes. For the integration between BRasa and BusinessRedmine, the use of services was established.
- The BusinessRedmine platform was integrated with the RocketChat solution to facilitate communication between team members and even support issues.
- An image repository is established for the management of entity B's solutions and to complement the management of software configuration elements.
- The incorporation of continuous integration and Devops development techniques is gradually improving.
- There was a gradual improvement in process management in entity B with best practices for agile project management and the SCRUM methodology. The duration of the iterations in the development of Entity B's products was reduced to iterations of 4 weeks and with cuts every 10 days with deliveries to customers.
- The increase in deliveries to customers and meetings with customers were established every seven days, improving the level of communication with customers and their perceived quality. The organization of the version control process in the products of entity B allowed it to deliver different versions of products to entity A in 1 year of work. In particular, the following updates 4.2 were delivered to the BusinessRedmine ecosystem as the base integration platform. 6, 5.0.2, 5.0.3 and 5.0.4 which mean major changes that include security improvements. Additionally, 38 minor updates were delivered generally within 12 days of each other.

## 4 Conclusions

The proposed model significantly improved the management of the organizations where it was implemented. In particular, the control and monitoring of the projects and software assets developed by these institutions was improved.

The development of a work method inspired by agile development but that integrates the strategic projection of organizations with the projects and their tasks is achieved. The concept is proposed that all plans must converge in a single space, the project schedule, and that quality and risk plans are simply dimensions of schedule analysis.

The simultaneous improvement of processes and systems is established as the main element to achieve agility in management. This synergy is achieved in the model by considering that all the data that is managed in project-oriented organizations is recorded in the organization's information systems. Furthermore, these systems must be complemented with other technologies that facilitate access and updating of information, such as conversational systems. The capacity provided by the integration of project management solutions with version control systems is demonstrated.

Numerous computational intelligence techniques are incorporated into the environment but on the basis of complementing the work of project management. These novel techniques are incorporated into the solution not as the center of the problem but as powerful tools for the treatment of uncertainty that complement project management methods and facilitate the work of human decision-makers.

As future work, the need to continue perfecting the introduction of agile models is identified from the integration of management solutions with conversational models. Increase the methods of prediction and evaluation of competencies based on evidence.

## References

1. Jaramillo RLR, Yauri HYM (2022) Cambios en la Guía del PMBOK del Project Management Institute, su Certificación y aplicación en la Gestión de Proyectos: Una revisión sistemática de literatura. SCIÉNDO 25:437–443. https://doi.org/10.17268/sciendo.2022.055
2. PMI (2021) A guide to the project management body of knowledge: PMBOK, 7th edn. Project Management Institute, (®) Guide. In Project Management Institute. Pensilvania
3. Lukianov D, Mazhei K, Gogunskii V (2019) Transformation of the international project management association project managers individual competencies model. In: 2019 IEEE international conference on advanced trends in information theory (ATIT), pp 506–512
4. Pesante LH (2003) Software engineering institute (SEI). Encyclopedia of computer science. Wiley, GBR, pp 1611–1613
5. O'Regan G (2015) Software engineering institute (SEI). In: O'Regan G (ed) Pillars of computing: a compendium of select, pivotal technology firms. Springer International Publishing, Cham, pp 195–205
6. Institute PM (2021) A guide to the project management body of knowledge, 7th edn. Project Management Institute, Newtown Square, Pennsylvania
7. Takagi N, Varajão J (2021) ISO 21500 and success management: an integrated model for project management. Int J Quality Reliabil Manag 39:408–427. https://doi.org/10.1108/IJQRM-10-2020-0353
8. Ferdinansyah A, Purwandari B (2021) Challenges in combining agile development and CMMI: a systematic literature review. In: Proceedings of the 2021 10th international conference on software and computer applications. Association for Computing Machinery, New York, NY, USA, pp 63–69
9. Magano J, Silva C, Figueiredo C, Vitória A, Nogueira T, Pimenta Dinis MA (2020) Generation Z: fitting project management soft skills competencies—A mixed-method approach. Educ Sci 10:187. https://doi.org/10.3390/educsci10070187
10. Mohammed CM, Zeebaree SRM (2021) Sufficient comparison among cloud computing services: IaaS, PaaS, and SaaS: a review. Int J Sci Bus 5:17–30
11. Rashid A, Chaturvedi A (2019) Cloud computing characteristics and services: a brief review. Int J Comput Sci Eng 7:421–426. https://doi.org/10.26438/ijcse/v7i2.421426
12. Piñero Pérez PY, Pérez Pupo I, Piñero Ramírez PE, Marquez Ruiz Y, Fustiel Alvarez Y (2022) Project management repository for decision-making researches. In: Piñero Pérez PY, Bello Pérez RE, Kacprzyk J (eds) Artificial intelligence in project management and making decisions. Springer International Publishing, Cham, pp 303–317
13. Verdecia Vicet P, Piñero Pérez PY, Pérez Pupo I, García Vacacela R, Villavicencio Bermúdez N (2022) Combining artificial intelligence and project management techniques in ecosystem for training and innovation. In: Piñero Pérez PY, Bello Pérez RE, Kacprzyk J (eds) Artificial

intelligence in project management and making decisions. Springer International Publishing, Cham, pp 259–275

14. Pérez Pupo I, Piñero Pérez PY, Bello Pérez RE, García Vacacela R, Villavicencio Bermúdez N (2022) Linguistic Data Summarization: A Systematic Review. In: Piñero Pérez PY, Bello Pérez RE, Kacprzyk J (eds) Artificial intelligence in project management and making decisions. Springer International Publishing, Cham, pp 3–21
15. Smarandache F (2021) Indeterminacy in neutrosophic theories and their applications. Infinite Study
16. Pérez Pupo I, Piñero Pérez PY, Bello Pérez RE (2022) New indicators for the assessment of linguistic summaries considering a rough sets approach. In: Piñero Pérez PY, Bello Pérez RE, Kacprzyk J (eds) Artificial intelligence in project management and making decisions. Springer International Publishing, Cham, pp 99–120

# Sustainability Management Framework: Case Study of a Cuban IT Project Organization

**Juan Antonio Plasencia Soler, Fernando Marrero Delgado, Miriam Nicado García, and Yasmany Aguilera Sánchez**

**Abstract** In recent decades, sustainability has been gaining presence in the political, institutional, business and academic spheres; with a notable increase in regulations, institutional guidelines, indicators and indices developed by international organisations, which provide a guide for the incorporation of this issue in the management and administration system of companies and institutions. The aim of this research is to develop a framework for the management of sustainability in entities of the Information and Communication Technologies sector in Cuba from the perspective of strategic management. The framework is composed of three dimensions of articulation that provide the strategic, business process and normative ethical approach; and five dimensions of evaluation: institutional, economic, social, ecological and technological innovation, which allows the management and evaluation of sustainability in the organisation under study. The implementation of the framework makes it possible to diagnose and design the strategic direction for sustainability, prioritise and classify the sector's Critical Success Factors, identify relevant processes, design ethics and regulatory compliance programmes based on risks, in one of the sectors classified as a priority for the national economy, contributing to the improvement of the level of sustainability.

**Keywords** Sustainability · Business process management · Indicators · Integrated indices · Multi-criteria methods

J. A. P. Soler (✉) · Y. A. Sánchez
University of Informatics Sciences, Road to San Antonio de los Baños, Km. 2 1/2. Torrens, La Lisa, Havana, Cuba
e-mail: juanps@uci.cu

F. M. Delgado
Central University, "Marta Abreu" of Las VillasRoad to Camajuaní Km, 5 y 1/2, Santa Clara, Villa Clara, Cuba

M. N. García
University of Havana, San Lázaro and L Street, Vedado, Havana, Cuba



P. Y. Piñero Pérez et al. (eds.), *Computational Intelligence in Engineering and Project Management*, Studies in Computational Intelligence 1134,
https://doi.org/10.1007/978-3-031-50495-2_12

## 1 Introduction

The term "sustainable development" gained popularity in a Report of the World Commission on Environment and Development, called "Our Common Future", where it is defined as: "development that meets the needs of present generations without compromising the ability of future generations to meet their needs" [1].

In recent years, the international scientific community has become increasingly interested in the study of the sustainability of the systems that make up society, largely due to a growing awareness of the need for development that does not compromise the existence of future generations.

However, the authors [2, 3] agree that there is a lack of management systems that incorporate the three dimensions of sustainable development, and that social aspects are less well developed in the proposals. On the other hand, a review of more than 50 studies by [4] found a lack of integration between management systems and evaluation instruments for sustainability.

According to [5], the proposed frameworks are mostly designed for the strategic level and therefore need operational translation. On the other hand, according to [6], most of the consulted frameworks do not take into account the innovation dimension to assess their progress on sustainability.

The Cuban model of sustainability is consistent with international trends in sustainable development and its three pillars: social development, economic development, and environmental protection, but it has its peculiarities, as illustrated by the Plan for Economic and Social Development until 2030 (PENDES 2030), which defines the following strategic axes: effective and socialist government and social integration; productive transformation and international insertion; infrastructure; human potential, science, technology and innovation; natural resources and environment; and human development, justice and equity.

The first of these is that Cuba's interpretation of sustainable development is people-centred, which translates into qualitative improvements and expansion of human potential. The second peculiarity is given by the search for social justice and equity as values promoted by the state. Finally, the third peculiarity is given by the culture of saving the country's scarce natural resources [7].

Today, there is an industry that is set to influence sustainable development like no other: Information and Communication Technologies (ICTs). The ICT sector is directly responsible for 2% of global $CO_2$ emissions; despite this it is believed that ICTs have the capacity to influence the reduction of the remaining 98% of global carbon emissions [8].

The increase in emissions is related to the increased demand for ICT goods and services, which has led the industry to make significant efforts in technological innovation aimed at sustainability seen only from the economic and environmental dimensions through continuous improvement of business processes [9].

However, several authors [10, 11] agree that both the role of ICT organisations and the integration of sustainability in the management system of the technological enterprise is an area that has not been analysed in depth and requires research.

In Cuba, ICTs are considered to be one of the priority sectors of the economy and in recent years the government has allocated efforts, actions and policies for the informatisation of Cuban society and the reorganisation of ICT enterprises.

The organisation of the computer applications industry in line with informatisation priorities; the implementation of science and technology parks; the development of export capacities; the creation of strategic alliances; the promotion of socialist state enterprises; the participation of non-state forms of management in the sector; and the establishment of certification systems are some of the actions promoted by the government and included in the "Integral Policy for the Improvement of the Informatisation of Society in Cuba" [12].

However, a study carried out by the authors of this research, through interviews and questionnaires, involving 21 senior managers of the Grupo Empresarial de la Informática y las Comunicaciones (GEIC), with an average of 25 years of experience in the sector and more than 15 years in management positions, identified the following problems in the ICT sector entities:

- insufficient incorporation of strategic objectives and indicators aimed at social and ecological aspects in the strategic management process;
- insufficient elements of innovation management in business strategies and processes;
- insufficient monitoring by companies of the impacts of their management on the surrounding communities;
- business processes lack accreditation and international certification to support them;
- there are no environmental management systems in the companies, and there is a lack of awareness among workers of the impact of their operations on the ecology;
- insufficient budget for investment in innovation and development;
- insufficient use of automated systems for resource control and smart management of facilities and laboratories; and
- insufficient use of practices such as teleworking, videoconferencing, laptop use, reconfiguration of equipment, among others.

The objective of the research is to develop a framework for the management of the sustainability of the entities of the Information and Communication Technologies sector in Cuba from the strategic management process.

The research is structured as follows. In the first section, the advantages and limitations of the frameworks for sustainability management consulted by the authors of the research are discussed; in the second section, the framework for contributing to the sustainability of ICT entities in Cuba through strategic management is presented; then the results of the validation of the proposed framework are shown; and finally, the conclusions of the research are presented.

## 2 Brief Conceptual Framework

In [13] states that in order to deploy the principles of sustainability in organisations, it is necessary to address strategic, regulatory and operational aspects. Therefore, the authors of the research will argue these three aspects set out by Baumgartner, starting with strategic management.

In the last decade there has been an increase in scientific publications on the integration of sustainability into strategic management. However, a literature review, comprising 114 articles, conducted by [14] on the integration of sustainability into strategic management revealed that most of the literature combining both fields of research is associated with theoretical and conceptual frameworks, lacking empirical studies (quantitative and qualitative); it also corroborates the lack of a common or unifying basis for addressing the issues that influence the integration of sustainability into strategic management.

In what is considered a higher stage for the integration from strategic management of sustainability principles, the authors [15] call Sustainable Strategic Management as a consequence of the evolution towards a new stage of traditional strategic management, which integrates sustainability principles, contributing to making organisations more socially and ecologically responsible. Processes are economically competitive, socially responsible and in balance with the cycles of nature, according to the researchers.

Since strategy is an expression of organisational assumptions [16] the investigation of the normative assumptions of organisational culture related to sustainability may be necessary, and in the following the authors of this research will refer to the second aspect: the normative.

In more recent decades, organisations have formalised the management of their compliance responsibilities as the complexity of regulations has increased, as well as the complexity of business operations and transactions. Moreover, this management has become integrated into the programmes and actions that are implemented as part of the organisation's culture, resulting in the two terms, ethics and compliance, becoming integrated and complementary to each other.

Ethics and compliance management (ECM) has become an important and pervasive topic [17]. The implementation of an ECM is part of a moral reflection and has strategic relevance for organisations [18].

On the other hand, today's businesses operate in the presence of multiple stakeholders, divergent values and conflicting beliefs [19], making it necessary for elements of the ECM to combine regulatory compliance with an emphasis on ethical behaviour.

According to [20], some of the common elements of ECM are: compliance risk management; internal and external information and communication systems; internal control systems; and the development of a compliance culture.

The first of these elements, risk management, is defined as the set of coordinated activities to direct and control the organisation in relation to risk [21] by reducing or mitigating its negative effects [22]. The most common stages of risk management are context identification, risk assessment and risk treatment [23, 24].

Compliance risk management does not differ in its implementation from the management of other risks [25], it is necessary to take into account the organisation's strategic and operational objectives [26], its internal and external context, as well as the use of techniques for its assessment and treatment [27]. Risk is usually defined as the effect of uncertainty on objectives [21], while compliance risk is associated with the probability of occurrence and the consequences of non-compliance with requirements that an organisation is required to comply with or chooses to comply with voluntarily [28].

In terms of strategy operationalisation, the recent incorporation of sustainability into business processes has led to the emergence of Sustainable or Green Business Process Management. According to [29] SBPM is defined as the understanding, documentation, modelling, analysis, simulation, execution and continuous change of business process taking into account the environmental consequences of business processes.

This definition is largely based on the GMP life cycle and is focused on addressing and improving environmental aspects. The literature in the field of GBPM shows contributions on the impacts of business processes on the environment. The full incorporation of sustainability into processes also requires research that focuses on the social dimension [30].

On the other hand, it is important that sustainability frameworks incorporate assessment tools. Undoubtedly, the most internationally influential model for assessing sustainability [31] was popularised by John Elkington in 1994 [32, 33] and is known as the "Triple Bottom Line": TBL or 3BL.

This model argues that a company pursuing sustainable development is evaluated by its stakeholders not only in terms of its economic benefits, but also for its achievements in equity and social justice and environmental protection [34–36].

According to [32] the main benefit of this broader view is the potential impact of how the role of organisations is perceived and constructed by stakeholders. This model is easy to implement, allowing it to be adapted to different types of organisations.

The second most influential model was developed by the United Nations Commission for Sustainable Development (CSD) when it adopted the Work Programme on Indicators for Sustainable Development in 1995. These indicators were divided along what the UN called the Four Pillars of Sustainability: social, economic, environmental and institutional [37].

The CSD thus included, in addition to the aforementioned economic, social and ecological dimensions, the institutional dimension, which contained in its structure aspects related to strategic implementation, access to information, infrastructure for communication, science and technology, among others [38].

However, considering that innovation is considered as an internal driver of sustainability; they present the opportunity for companies to reposition their internal competences around more sustainable technologies, they provide strategic advantages for

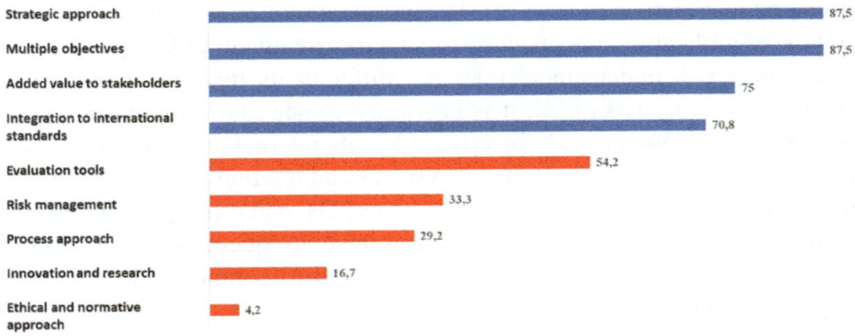


**Fig. 1** Presence of aspects in the research consulted by [44]

organisations [39], the authors of the research incorporate a technological innovation dimension to the institutional, social, economic and ecological dimensions.

On the other hand, according to [40] it is difficult to measure the achievements of business model innovation, especially in the area of sustainability, where many metrics are still under development, so only two researches [41, 42] address dimensions of innovation, technical or technological in the literature consulted for the study.

Then, according to [43] multi-criteria methods and aggregate indicators are the tools most commonly used by researchers to assess sustainability. In the case of multi-criteria methods, the most used were the Analytic Hierarchy Process (AHP) and Analytic Network Process (ANP), respectively.

Finally, a study by [44] that included more than 250 scientific articles related to sustainability management frameworks concluded that strategic aspects, the variety of objectives, added value to stakeholders and integration to standards were the most observed aspects in the consulted research, while evaluation tools, risk management, process approach, innovation and normative ethical aspects were scarcely present (see Fig. 1).

For the above reasons, the authors of this research propose a framework for sustainability management. The main components of this framework are presented in the following section.

## 3 Framework for Sustainability Management

The framework for the management of the sustainability of ICT entities in Cuba from the perspective of strategic management was structured on the basis of the following methodology:

1. Critical analysis of the state of knowledge and practice on sustainability and its incorporation into the leadership and management systems of ICT entities.

2. Study of other methodologies for the design of frameworks, proposed by [39, 45–48].
3. Analysis of the specific problems for achieving sustainability in the Information and Communication Technologies sector in Cuba.
4. Consultation with specialists in strategic management and organisational sustainability.

The framework is structured through eight dimensions; three management dimensions for their articulation; and five result or impact dimensions for their evaluation. The management dimensions allow the integration of all activities, resources and people in a systemic and coherent way, being the basis for the deployment of the model in an organisation to achieve sustainability, and are as follows:

- Strategic dimension: refers to the stages of the strategic management process, with its defined values, strategic objectives and measurement criteria, in full relation to the micro and macro environment, incorporating the attributes of sustainability.
- Sustainable business processes dimension: takes into account the operationalisation of the strategy, shared values and ethical and regulatory compliance, through the orderly, logical and structured sequence of business processes, while aligning them towards the achievement of sustainability.
- Ethical and regulatory compliance dimension: this is manifested in ethics and regulatory compliance programmes based on stakeholder requirements, the established ethical and legal framework, and risk management for each of the business processes defined in the organisation.

The five impact dimensions, on the other hand, are associated with the results of the organisation and its effects on the different stakeholders. These dimensions form the basis for the assessment of sustainability in the organisation and are defined by the authors as follows:

- Institutional dimension: refers to the inclusion of sustainability aspects in the strategies, objectives and indicators, and in the business value proposition, as well as the communication, information and risk management systems, and the relationships established with stakeholders.
- Economic dimension: takes into account the economic results of the organisation on stakeholders and its contribution to local, regional or national economic development.
- Social dimension: this is expressed through the impacts of the organisation on the social systems in which it operates, both internally (employees and management) and externally (customers, communities and regions).
- Ecological dimension: addresses the impacts of the organisation on living and non-living natural systems, including land, air, water and ecosystems.
- Technological innovation dimension: refers to the organisation's performance in the design, production, marketing and dissemination of new and improved products, processes and procedures in the field of science and technology.

These dimensions are in constant interaction with the environment, where the organisation develops and on which it exerts its influence in each of its factors: economic, political, technological, ecological, socio-cultural, demographic and legal. The following are input variables to the model:

- The strategic management process;
- the processes defined by the organisation;
- stakeholder requirements; and
- the behaviour of key organisational performance indicators.

The framework presents the following output variables:

- The level of sustainability of the organisation;
- the sustainable business processes designed; and
- ethics and compliance programmes.

The methodological framework of the proposal is implemented through four phases that correspond to the Alter Shewar Cycle or Deming Cycle: Plan, Do, Check and Act. The authors reflect the relationships between these dimensions, their input elements and expected results in Fig. 2.

In the following, the phases for its deployment are briefly described.

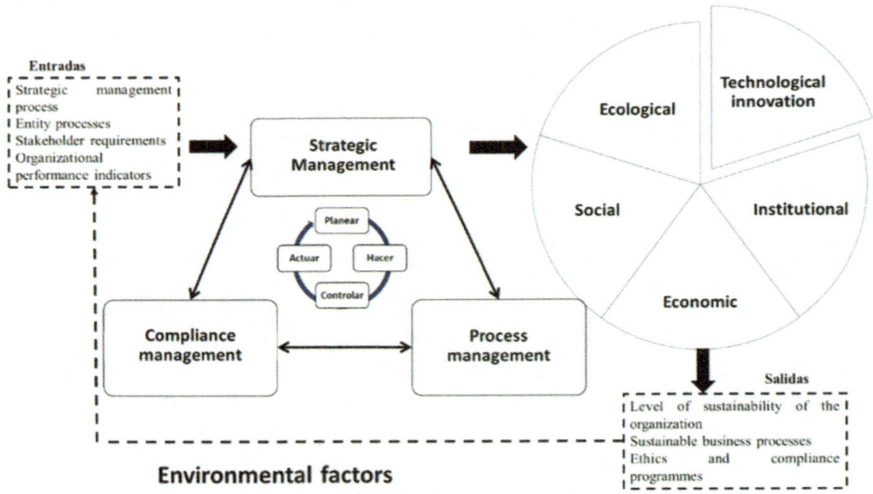


**Fig. 2** Graphical representation of the sustainability management framework

### 3.1 Phases for the Implementation of the Framework

The first phase is planning. In this phase, the conditions in the organisation are prepared for the implementation of the model, and the strategic direction and business processes are redesigned based on the analysis of the current situation of the steering and management system.

It is important to highlight in this phase the work with the Critical Success Factors (CSF). The CSFs are evaluated according to the impact on the objectives (Io), the probability of success (Pe) and the measurability (Pm) as shown in Eq. 1:

$$\text{FCE} = \text{f}(\text{Io}, \text{Pe}, \text{Pm}) \tag{1}$$

A panel of experts evaluates each variable on a scale of ordinal values, as proposed by [49].

Also in this phase, the relevant business processes of the organization are selected. For this, the processes are evaluated on the basis of three criteria: impacts on strategic objectives (SO), their impact on stakeholders (IG), and the dimensions of sustainability assessment (SA). A matrix is structured for each decision criterion, with the rows containing the processes ($\text{P} = \text{Pi}\ \forall\ \text{i} = 1, 2, \ldots, \text{m}$) and the columns containing each of the strategic objectives ($\text{SO} = \text{SOj}\ \forall\ \text{j} = 1, 2, \ldots, \text{n}$), the dimensions of sustainability assessment ($\text{SA} = \text{SAj}\ \forall\ \text{j} = 1, 2, \ldots, \text{n}$), and the identified stakeholders ($\text{IG} = \text{IGj}\ \forall\ \text{j} = 1, 2, \ldots, \text{n}$). The impacts are assessed by the consensus of the working team through an ordinal scale from 1 to 5, where a weighting of 5 is high impact of process (i) on criterion (j) and a rating of 1 means no impact.

The average values of the impacts of each process (Pij) on the three defined criteria and the weights (ωj) of the criteria ($\text{C} = \text{Cj}\ \forall\ \text{j} = 1, 2, 3$) are then calculated. The priority or relevance index of the process (Irpi) is then determined through Eq. 2:

$$\text{Irpi} = \sum (i = 1)^n Pij * \omega j \tag{2}$$

The processes are classified according to their Irpi taking into account the scale proposed by [6].

To the defined relevant processes, compliance risks are associated, calculating their criticality through an index that includes the impact dimensions of the framework as shown in Eq. 3:

$$\text{Ici} = \text{Oi} * \text{Cmaxi}(\text{Insi};\ \text{Ecoi};\ \text{Soci};\ \text{Ecoli};\ \text{Inni}) * \text{Di} \tag{3}$$

where:

- Oi — Probability of occurrence of risk (i)
- Cmaxi — Maximum value of the consequences after assessing risk (i) in the dimensions: Institutional (Insi), Economic (Ecoi), Social (Soci), Ecological (Ecoli) and Technological innovation (Inni)
- Di — Possibility of risk detection (i).

Once the criticality index has been calculated, the limit value Ici $\geq$ 12 is defined, taking into account the opinion of the work team, the studies carried out by [50].

In the second implementation phase, the planned actions and programmes are carried out, the defined strategic projection is implemented taking into account the FCEs, the defined relevant business processes and the risk-based compliance programmes.

In the control and continuous improvement phases, the most relevant aspect is the calculation of the organisation's level of sustainability through the impact dimensions. For this purpose, a hierarchy is created with the sustainability level (C_s) as the target, to which the dimensions (dimk), the Critical Success Factors (fcei) and the indicators (indj) defined in the initial phase are associated, as shown in Eq. 4:

$$\mathrm{Cs} = \mathrm{Cs}\ (\mathrm{dim}(\mathrm{fce}(\mathrm{fce}(\mathrm{ind})))) \tag{4}$$

After constructing the hierarchy, the author proposes to use the Analytic Hierarchy Process (AHP) developed by Tomas L Saaty in the seventies of the last century to determine the weights of each component. The AHP assigns weight to criteria through the construction of a pairwise comparison matrix, where experts, using a numerical scale, indicate how many times one element is more important than another [51]. It is recommended that the "Super Decisions" software be used at this stage for greater agility and reliability in the calculation of the weights.

The next step is to start from the value of the evaluation of the indicators and their ideal value defined in the initial phase of the framework. Due to the diverse nature of the values assumed by the indicators, they must be standardised and homogenised, which is why the author suggests using the distance-to-target standardisation method proposed by several researchers in recent years [52, 53]. For values that relate to indicators where the aim is to maximise the result, Eq. 5 is used:

$$\mathrm{Njik} = \mathrm{VRjik/VIjik} \tag{5}$$

For values that relate to indicators where the result is to be minimised, Eq. 6 is used:

$$\mathrm{Njik} = \mathrm{VIjik/VRjik} \tag{6}$$

where:

Njik Normalised vector of indicator (j) associated with FCE (i) of dimension (k).
VIjik Ideal value of indicator (j) associated with FCE (i) of dimension (k).
VRjik Actual value of indicator (j) associated with FCE (i) of dimension (k).

The impact of the dimensions (Dimk) is then calculated, using Eq. 7:

$$\mathrm{Dimk} = \sum (i = 1)^n \mathrm{FCEik} * \mathrm{Wik} \text{ for } i = 1 \ldots n \tag{7}$$

where:

FCEik Critical Success Factor (j) of dimension (k).
Wik weight of the CSF (i) associated with dimension (k).

Equation 8 is used to determine the FCEik:

$$\text{FCEik} = \sum (j = 1)^n \text{Njik} * \text{Wjik for } j = 1 \ \ldots \ n \tag{8}$$

where:

Wjik weight of indicator (j) associated with FCE (i) of dimension (k).

Step 4. Determination of the sustainability coefficient.

The sustainability coefficient (Cs) of the organisation is then determined, taking into account the impacts of the dimensions, as shown in Eq. 9:

$$\text{Cs} = \sum (i = 1)^n \text{Dimk} * \text{Wk for } k = 1 \ \ldots \ 5 \tag{9}$$

where:

Wk: weight associated with dimension (k).

The values of the variables composing the hierarchy and the weights are expressed, after normalisation, in values between 0 and 1, then the sustainability coefficient (Cs) is evaluated between 0 and 1 depending on the sustainable behaviour of the organisation. The closer the Cs value is to the value of one, the higher the level of sustainability of the assessed organisation.

The main results and their discussion in the application of the framework proposed in this research in an organisation in the ICT sector in Cuba are shown below.

## 4 Results and Discussion

In this section, the authors will present the main results of the application of the framework in an organisation that develops IT projects for export in Havana. The results achieved by the management dimensions; and the impact dimensions, following the phases of planning, implementation, control and updating, will then be discussed. In all cases, results will be shown for at least two periods, an initial period (1) and a final period after the framework has been applied (2).

In the planning phase, the strategic direction of the organisation is redesigned taking into account the FCEs validated by the consulted experts. The FCEs were evaluated with very high and high criticality according to the ordinal scale proposed by [49], as shown in Fig. 3.

For the sustainability diagnosis and taking into account the assessment dimensions of the framework, experts, interviews and a checklist with the variables defined for each impact dimension are used. Table 1 shows the results of the application of a questionnaire on green technology initiatives in the organisation in the two periods.

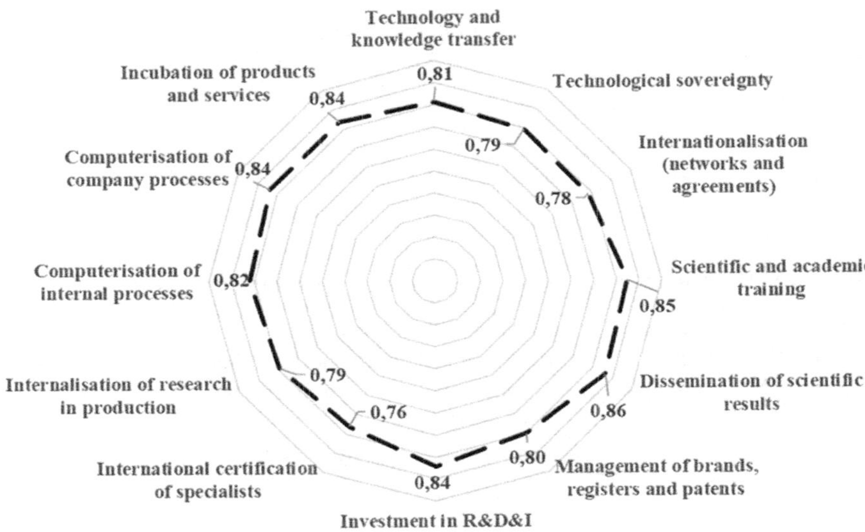


**Fig. 3** Evaluation of the FCE of the technological innovation dimension

The selected experts had to choose between the options strongly disagree (1) and strongly agree (5).

Table 1 shows that the performance of the variables evaluated by the experts has improved with respect to the diagnosis made in the first period. However, there are a group of variables that still perform below the desired level: automation and intelligent systems; the use of laptops and reconfiguration of equipment to save energy; and the extension of equipment life cycles.

**Table 1** Assessment of green technology initiatives

| Aspect | Period 1 | Period 2 |
|---|---|---|
| Awareness in reducing energy consumption | 3,60 | 4,00 |
| Ecological criteria in the purchase of technologies | 3,33 | 3,94 |
| Teleworking and videoconferencing | 3,20 | 4,06 |
| Virtualisation of processes | 3,07 | 3,88 |
| Extension of equipment life cycles | 3,60 | 3,76 |
| Control of energy expenditure | 3,27 | 3,82 |
| Use of laptop and reconfiguration of equipment for energy savings | 3,00 | 3,75 |
| Centralised management of technology supplies | 3,40 | 3,94 |
| Automation and intelligent systems for resource control | 3,00 | 3,94 |
| Automation and intelligent systems | 2,47 | 3,13 |

The case study organisation redefined and prioritised processes according to expert criteria through the 〖Irp〗 _i, as shown in Table 2. This ranking allows top management to focus their efforts on the processes that add most value to the sustainability of the organisation, and to prioritise their design and allocation of limited resources.

Another important aspect of the study is compliance risk management. Figure 4 shows the behaviour of the criticality index of the risks (see blue series) in the organisation's production and marketing process taking into account the periods evaluated. According to the limit value proposed in the previous section, the following are still risks to be addressed with actions: specialists without the necessary knowledge and skills (R0303), customer dissatisfaction (R0304), impact on the organisation's prestige (R0305) and energy inefficiency (R0315), among the most important (see red series).

Continuing with the analysis of the production and marketing process, the evaluation of the selected indicators of the process in three different periods, taking into account the defined FCE, show an improvement in the behaviour between the initial period and the final period evaluated. However, as can be seen in Fig. 5, there are indicators that need improvement actions, such as the % of specialist departures, the automation of laboratories, and the successful completion of IT projects.

Finally, it is possible to assess the sustainability of the organisation under study taking into account the impact dimensions defined in the framework (see Fig. 6). It can be seen that the worst evaluated dimensions were the technological innovation dimension and the institutional dimension.

In the case of the technological innovation dimension, the FCEs that most influence this behaviour are the dissemination of scientific results, the management of software and patent registrations, and the scientific and academic training of employees.

**Table 2** Classification of business processes according to the 〖Irp〗 _i

| Processes | $Irp_i$ | Ranking |
|---|---|---|
| Human resources management | 4,55 | Very high relevance |
| Science, technology and innovation | 4,54 | Very high relevance |
| Production and marketing | 4,40 | Very high relevance |
| Quality management | 4,37 | Very high relevance |
| Strategic management | 4,30 | Very high relevance |
| Undergraduate and postgraduate training | 4,06 | High relevance |
| Communication management | 4,03 | High relevance |
| University extension | 3,76 | High relevance |
| Economic and financial management | 3,73 | High relevance |
| International relations | 3,52 | High relevance |
| Internal control system | 3,51 | High relevance |
| Computerisation | 3,46 | High relevance |
| Management of community services | 3,36 | Medium relevance |

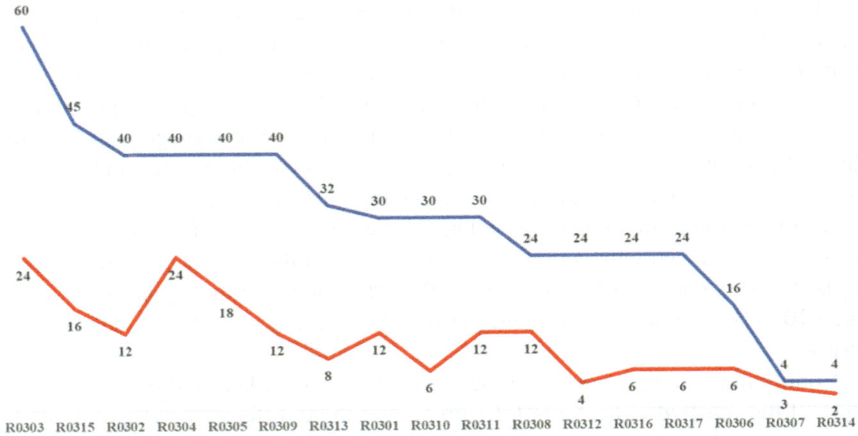


**Fig. 4** Behaviour of the criticality index of the risks in the production process

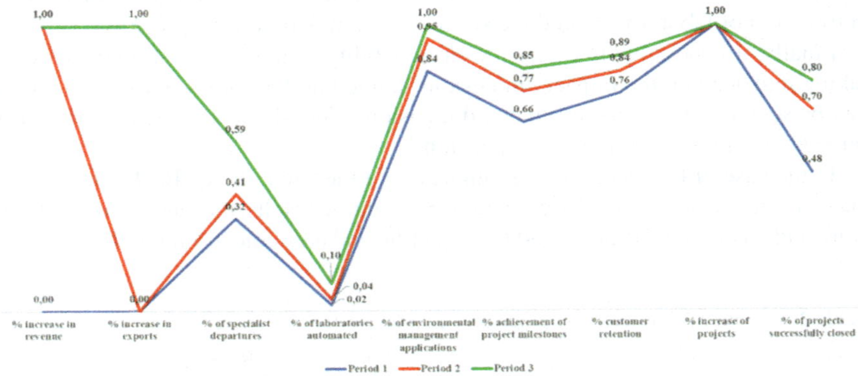


**Fig. 5** Behaviour of selected indicators of the production and commercialisation process

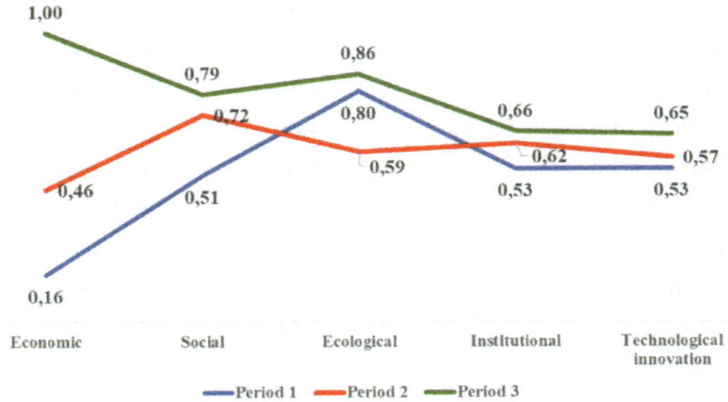


**Fig. 6** Behaviours of the evaluation dimensions of the proposed framework

On the other hand, the behaviour of the institutional dimension is influenced by the FCEs related to the certification and accreditation of processes.

## 5 Conclusions

The assessments made in the conceptual framework of the research revealed that, although there is a wide range of studies conducted in recent years that focus on embedding sustainability in organisations, they lack an integrated approach to sustainability management and assessment that takes into account the operationalisation of strategy through business processes, while also considering the management of ethics and compliance and technological innovation.

The proposed framework for sustainability management, through eight dimensions: three management dimensions that allow the model to be articulated through strategic management, with a focus on business processes, integrating ethical and compliance management through risk management; and five dimensions of economic, social, ecological, institutional and technological innovation performance that allow sustainability to be integrated into the governance and management system of Information and Communications Technology entities.

The results obtained from the application of the framework in an entity of the Information and Communication Technologies sector in the province of Havana, allowed the identification of the dimensions and variables that most affected its sustainable performance, being the basis for the development of objectives, lines of action, design of processes and ethics and regulatory compliance programmes, which contributed to evaluate and improve the levels of sustainability.

The main limitation of the study carried out comes mainly from the validation of the proposal through the application in a case study, and in the variables taken into account for the selection of this, so it is necessary in the future a generalisation of these results to other entities in the country, taking into account a representation of other organisations representative of the various sectors of the country's economy.

## References

1. WCED (1987) Our common future. United Nations
2. Fritz MMC, Schöggl J-P, Baumgartner RJ (2017) Selected sustainability aspects for supply chain data exchange: towards a supply chain-wide sustainability assessment. J Clean Prod 141:587–607. https://doi.org/10.1016/j.jclepro.2016.09.080
3. Morais D, Silvestre BS (2017) Sustainable supply chain management: the missing link of social sustainability. Sao Paulo, Brazil
4. Morioka SN, Carvalho MM (2016) Measuring sustainability in practice: exploring the inclusion of sustainability into corporate performance systems in Brazilian case studies. J Clean Prod 136:123–133. https://doi.org/10.1016/j.jclepro.2016.01.103
5. Nawaz W, Koç M (2017) Development of a systematic framework for sustainability management of organizations. J Clean Prod. https://doi.org/10.1016/j.jclepro.2017.10.011

6. Soler JAP, Delgado FM, García MN (2023) Metodología para contribuir a la sostenibilidad desde el proceso de dirección estratégica. Ciencias Administrativas 112–112. https://doi.org/10.24215/23143738e112.
7. Cook N, Sarver E, Krometis L-A (2015) Putting corporate social responsibility to work in mining communities: exploring community needs for central appalachian wastewater treatment. Resources 4:185–202. https://doi.org/10.3390/resources4020185
8. Khor K-S, Thurasamy R, Ahmad NH, Halim HA, May-Chiun L (2015) Bridging the gap of green IT/IS and sustainable consumption. Glob Bus Rev 16:571–593. https://doi.org/10.1177/0972150915581101
9. CEPAL (2012) Memoria del seminario realizado en la CEPAL. Santiago, Chile
10. Conesa JAB, de Nieto CN, Peñalver AJB (2016) Aplicación de un modelo de ecuaciones estructurales para analizar los sistemas de gestión en la integración de la RSC y su influencia en la estrategia y el performance de las empresas tecnológicas. Revista de Métodos Cuantitativos para la Economía y la Empresa 21:77–102
11. Wang Y, Chen Y, Benitez-Amado J (2015) How information technology influences environmental performance: empirical evidence from China. Int J Inf Manag 35:160–170. https://doi.org/10.1016/j.ijinfomgt.2014.11.005
12. MINCOM (2017) Política Integral para el Perfeccionamiento de la Informatización de la Sociedad en Cuba. Ministerio de las Comunicaciones
13. Baumgartner RJ (2014) Managing corporate sustainability and CSR: a conceptual framework combining values, strategies and instruments contributing to sustainable development. Corp Soc Responsib Environ Manag 21:258–271. https://doi.org/10.1002/csr.1336
14. Engert S, Rauter R, Baumgartner RJ (2016) Exploring the integration of corporate sustainability into strategic management: a literature review. J Clean Prod 112:2833–2850. https://doi.org/10.1016/j.jclepro.2015.08.031
15. Stead JG, Stead WE (2016) Spiritual capabilities: keys to successful sustainable strategic management. In: Spirituality and sustainability. Springer, Cham, pp 89–103. https://doi.org/10.1007/978-3-319-34235-1_6
16. Ketprapakorn N, Kantabutra S (2022) Toward an organizational theory of sustainability culture. Sustain Prod Consum 32:638–654. https://doi.org/10.1016/j.spc.2022.05.020
17. Mitra N, Mukherjee D, Gaur AS (2020) Mandated CSR in India: opportunities, constraints, and the road ahead. In: CSR, sustainability, ethics & governance. Springer, pp 193–217
18. Chan C, Ananthram S (2020) A neo-institutional perspective on ethical decision-making. Asia Pac J Manag 37:227–262. https://doi.org/10.1007/s10490-018-9576-x
19. Bhaumik S, Driffield N, Gaur A, Mickiewicz T, Vaaler P (2019) Corporate governance and MNE strategies in emerging economies. J World Bus 54:234–243. https://doi.org/10.1016/j.jwb.2019.03.004
20. Kreipl C (2020) Compliance management. In: Verantwortungsvolle unternehmensführung. Springer Gabler, Wiesbaden, pp 129–217
21. ISO: ISO 31000:2018 (2018) Risk management—Guidelines, provides principles, framework and a process for managing risk. International Organization for Standardization, Ginebra, Suiza
22. Society of Corporate Compliance and Ethics: Compliance Dictionary. https://www.corporatecompliance.org/publications/compliance-dictionary. Accessed 16 May 2022
23. Dvorsky J, Belas J, Gavurova B, Brabenec T (2021) Business risk management in the context of small and medium-sized enterprises. Econ Res-Ekonomska Istraživanja 34:1690–1708. https://doi.org/10.1080/1331677X.2020.1844588
24. Sánchez YA, Jane LG, Soler JAP, Delgado FM (2022) Sustainability risk management for project-oriented organizations. In: Piñero Pérez PY, Bello Pérez RE, Kacprzyk J (eds) Artificial intelligence in project management and making decisions. Springer International Publishing, Cham, pp 155–169. https://doi.org/10.1007/978-3-030-97269-1_9
25. Naadia MP, Khairuddin AR (2021) Shariah compliance risk management in the provision of wakaf-zakat housings. Int J Islamic Thought 19:102–109. https://doi.org/10.24035/IJIT.19.2021.200

26. Brondolo JD, Chooi A, Schloss T, Siouclis A (2022) Compliance risk management: developing compliance improvement plans. Tech Notes Manuals 2022. https://doi.org/10.5089/9798400205910.005.A001
27. Krepysheva AM, Sergievskaya AA, Storchevoy MA (2020) ОПРЕДЕЛЕНИЕ И ИЗМЕРЕНИЕ РИСКА В КОМПЛАЕНС-МЕНЕДЖМЕНТЕ. Стратегические решения и риск-менеджмент 11:150–159
28. International Organization for Standardization: Compliance management systems—Requirements with guidance for use (2021). https://www.iso.org/obp/ui#iso:std:iso:37301:ed-1:v1:es
29. Seidel S, Recker J, Brocke Jvom (2012) Green business process management. In: Green business process management, pp 3–13
30. Teniente E, Weidlich M (eds) (2018) Business process management workshops: BPM 2017 international workshops, Barcelona, Spain, Revised Papers. Springer International Publishing
31. Guo D, DeFrancia K, Chen M, Filiatraut B, Zhang C (2015) Assessing sustainability: frameworks and indices. Columbia University, Research Program on Sustainability Policy and Management. Earth Institute
32. Glac K (2015) Triple bottom line. In: Wiley encyclopedia of management. Wiley
33. Idowu SO (2013) Encyclopedia of corporate social responsibility. Springer, Berlin Heidelberg
34. Andrady AL (2015) Plastics and environmental sustainability: fact and fiction. Wiley, New Jersey, USA
35. Laasch O, Conaway R (2015) Principles of responsible management: glocal sustainability, responsibility. Cengage Learning, Stanford, USA
36. Robinson H, Symonds B, Gilbertson B, Ilozor B (2015) Design economics for the built environment: impact of sustainability on project evaluation. Wiley, West Sussex, UK
37. United-Nations: Indicators of Sustainable Development: Guidelines and Methodologies (2007). United Nations publication, New York, USA
38. Salvado MF, Azevedo SG, Matias JCO, Ferreira LM (2015) Proposal of a sustainability index for the automotive industry. Sustainability 7:2113–2144. https://doi.org/10.3390/su7022113
39. Nawaz W, Koç M (2019) Exploring organizational sustainability: themes, functional areas, and best practices. Sustainability 11:4307. https://doi.org/10.3390/su11164307
40. Evans S, Vladimirova D, Holgado M, Van Fossen K, Yang M, Silva EA, Barlow CY (2017) Business model innovation for sustainability: towards a unified perspective for creation of sustainable business models. Bus Stratege Environ 26:597–608. https://doi.org/10.1002/bse.1939
41. Frangopoulos CA, Keramioti DE (2010) Multi-criteria evaluation of energy systems with sustainability considerations. Entropy 12:1006–1020. https://doi.org/10.3390/e12051006
42. Hill RC, Bowen PA (1997) Sustainable construction: principles and a framework for attainment. Constr Manag Econ 15:223–239. https://doi.org/10.1080/014461997372971
43. Plasencia Soler JA, Marrero Delgado F, Nicado García M (2017) Metodología para evaluar el nivel ético en las organizaciones. Ingeniare, Revista Chilena de Ingeniería 25:170–179
44. Plasencia-Soler JA (2018) Modelo para contribuir a la sostenibilidad de entidades de las tecnologías de la información y las comunicaciones desde la gestión estratégica [Tesis de doctorado]. Universidad de La Habana, La Habana, Cuba
45. Medel-González F (2017) Evaluación y mejora de la gestión de la sostenibilidad empresarial con enfoque sistémico y multicriterio en la generación distribuida de Villa Clara
46. Mustapha MA, Manan ZA, Wan Alwi SR (2017) Sustainable green management system (SGMS)—An integrated approach towards organisational sustainability. J Clean Prod 146:158–172. https://doi.org/10.1016/j.jclepro.2016.06.033
47. Souza JPE, Alves JM (2018) Lean-integrated management system: a model for sustainability improvement. J Clean Prod 172:2667–2682. https://doi.org/10.1016/j.jclepro.2017.11.144
48. Stefano NM, Laux RO (2017) Sustentabilidade empresarial por meio do Sistema de Gestão Integrada (SGI). SBIJOURNAL. 0
49. Plasencia-Soler JA, Marrero-Delgado F, Nicado-García M, Aguilera-Sánchez Y (2017) Procedimiento para la priorización de Factores Críticos de Éxito. DYNA 84:26–34. https://doi.org/10.15446/dyna.v84n202.62084

50. Hernández-Oro RM (2015) Contribución a la evaluación de la fiabilidad de proyectos de instalaciones y obras hidráulicas en empresas cubanas de investigaciones y proyectos de ingeniería
51. Saaty TL, Sagir M (2015) Ranking countries more reliably in the summer olympics. Int J Anal Hierarchy Process 7:589–610. https://doi.org/10.13033/ijahp.v7i3.341
52. Oppio A, Corsi S, Torrieri F, Mattia S (2017) Infrastructure development and territorial vulnerability. the role of composite indicators for addressing siting decisions. In: Appraisal: from theory to practice. Springer, Cham, pp 277–290. https://doi.org/10.1007/978-3-319-49676-4_21
53. Zuo X, Hua H, Dong Z, Hao C (2017) Environmental performance index at the provincial level for China 2006–2011. Ecol Ind 75:48–56. https://doi.org/10.1016/j.ecolind.2016.12.016

# A Formal Representation of Standards for Project Management: Case PMBOK

**Nemury Silega, Vyacheslav S. Lapshin, Yury I. Rogozov, Gilberto F. Castro Aguilar, Inelda Martillo Alcívar, and Katya M. Faggioni**

**Abstract** In the last years, plenty of proposals have been created to guarantee the success of projects. However, the rate of failure projects is still high. In that sense, it has been demonstrated that adopting a standard that guide the development of all the stages of a project is a factor that considerably contribute to its success. However, the fact that the information about standards is provided only in text documents (expressed in natural language) hinders its automatic analysis. For example, the task of choosing a standard taking into account the features of the project might become a complex task. Moreover, it is not easy of detect inconsistencies and ambiguities using natural language. On the other hand, ontologies represented in Ontology Web Language (OWL) have demonstrated good results to represent knowledge. OWL ontologies can be analyzed by a reasoner in order to automatically validate and check the consistency of the specifications. Considering the significant advantages of ontologies, the main aim of this paper is to present an ontological model to represent and analyze standards or guides for project management. This ontology can be used

N. Silega (✉) · V. S. Lapshin · Y. I. Rogozov
Department of System Analysis and Telecommunications, Southern Federal University, Taganrog, Russia
e-mail: silega@sfedu.ru

V. S. Lapshin
e-mail: lapshin@sfedu.ru

Y. I. Rogozov
e-mail: rogozov@sfedu.ru

N. Silega · G. F. C. Aguilar
Facultad de Ingeniería, Universidad Católica de Santiago de Guayaquil, Guayaquil, Ecuador
e-mail: gilberto.castroa@ug.edu.ec; gilberto.castro@cu.ucsg.edu.ec

N. Silega · G. F. C. Aguilar · I. M. Alcívar · K. M. Faggioni
Facultad de Ciencias Matemáticas y Físicas, Universidad de Guayaquil, Guayaquil, Ecuador
e-mail: inelda.martilloa@ug.edu.ec

K. M. Faggioni
e-mail: katya.faggionic@ug.edu.ec

 
P. Y. Piñero Pérez et al. (eds.), *Computational Intelligence in Engineering and Project Management*, Studies in Computational Intelligence 1134,
https://doi.org/10.1007/978-3-031-50495-2_13

to represent the knowledge of several project management standards as well information about their implementation in particular projects. In general, this ontology can support the organizations to make decisions. In general, this ontology might reduce the ambiguity of guides or standards descriptions for project management and enhance its comprehensibility.

**Keywords** Project management standards · Knowledge representation · Ontologies · Software projects · PMBOK

## 1 Introduction

Though in the last years, plenty of proposals have been created to guarantee the success of projects, a rate of projects that still fail is high [1]. To successfully develop a project is critical to follow good practices. For example, the adoption of international standards that guide the entire project lifecycle is a suitable practice. Standards are models that are defined for specific authorities or organizations [2]. In the area of project management, *The Guide to the Project Management Body of Knowledge* (PMBOK) is one of the main references. It includes a group of specifications that cover the entire project lifecycle.

The adoption of standards provides important benefits to organizations. However, it can be a complex task. For example, to analyze the information about each version of PMBOK might be a tough task. In part, because the information is presented in large text documents. Since, this information is represented only in natural language, it can be analyzed only manually. For example, to analyze different versions of PMBOK or to compare with another standard usually is a time-consuming and complex task. In this type of specifications, usually there are different problems with the definition of concepts and their respective terms. These problems affect the appropriate adoption of standards in general. For example, to select the best standard for a project based its particular features can be a difficult and time-consuming task.

To use ontologies to describe the knowledge of standards might alleviate some of the aforementioned problems. Specifically, ontologies in Ontology Web Language (OWL). Several works have documented the important advantages of ontologies to represent and analyze knowledge [3–13]. By means of ontologies is possible to describe the knowledge of a domain in a machine readable form [14]. They can support the knowledge sharing and the communication between people or software agents [15].

To take advantage of the benefits of ontologies and overcome some of the aforementioned problems, this paper introduces a formal model based on ontologies to manage the knowledge about project managements standards or guides. Specially we will focus the ontology on representing the specifications described in PMBOK. A well-known methodology was adopted to guide the development of the ontology. The execution of the seven stages allowed to develop our ontology. In this ontological process, several standards were analyzed in order to identify general concepts. Once, the ontology is populated with information of different standards, new knowledge can be infered by a reasoner. In this case, we used the ontology to represent the information about PMBOK, a prominent guide in the field of project management. This ontology is a powerful means to support the description of project management standards and consequently contribute to its successful adoption.

By adopting this approach, the general advantages of ontologies [16] can be leveraged to: Standardize the activities and artifacts produced in different phases of the project life-cycle; Reduce problems of inconsistencies between different artefacts generated during the project life cycle that usually rise due to the multi-perspective of involved stakeholders; Facilitate the enforcement of traceability of these artefacts; Augment sharing and interoperation of artefacts created by stakeholders; Help identifying and reusing existing components.

Once the general structure of the ontology is created based on the specifications of PMBOK, it can be instanced to describe results of specific projects. Therefore, this ontology can be used to retrieve information depending on the needs of users. For example, to know the specifications that are common in successful projects or those that are common in failed projects. In general, the project manager might use the insights that this ontology provides to make decisions.

The paper is structure as follows. In Sect. 2, we briefly analyzed different elements related to standards. In Sect. 3, our ontology is described. In Sect. 4, some examples where to show the applicability of our proposal are described. Finally, the conclusions are presented.

## 2 Background

To follow a guide for project management might contribute to guarantee the success of a project. Specifically for projects of software development, they include specifications to correctly apply the software engineering [17]. In general, standards focus either on the process or the product [18]. The Guide to the Project Management Body of Knowledge (PMBOK) defines a set of specifications that can help to manage a project correctly. PMBOK includes several stages and processes that cover the entire lifecycle of projects lifecycle and their processes [2]. PMBOK has been used as a guide to develop projects in a wide variety of domains. PMBOK is made up of a group of knowledge area which cover the entire project lifecycle. In additions, for each area a group of processes is defined. A set of activities, tools, inputs and outputs is specified foreach process. The right application of the PMBOK specifications might be a valuable factor to successfully develop a project.

## *2.1 Ontologies*

An ontology allows information about a particular domain to be conceptualized. And it constitutes a formal and explicit representation of the processes and data in the domain in question [19]. The main elements of an ontology are concepts, axioms, or inference rules. Based on these elements can be inferred new knowledge. Ontologies have been extensively applied to detect ambiguities [20]. They can be used to represent, validate, share and analyze the knowledge of different domains. Hence, the understanding of the specifications might be improved and consequently new knowledge can be created.

In a process to develop an ontology, it is essential to define three elements: a mythology that guide the process, a powerful language to specify the ontology and a tool that support the implementation of the ontology in the selected language. We adopted the guide of Noy and McGuinness to develop our ontology [21]. This guide is one of the most used to support the development of ontologies [22]. It includes seven steps to obtain the main components of an ontology (classes, properties, restrictions and instances).

On the other hand, Web Ontology Language (OWL) was used to implement the ontology. With OWL is possible to represent concepts and a wide number of relationships [23]. The tool Protégé was used to construct the ontology. This tool is open-source and one of the most popular to create ontologies in OWL [24].

# 3 An Ontological Model to Manage the Knowledge of Project Management Standards

The main components (classes, properties, restrictions and instances) of the developed ontology are described in this section. Our ontology has the purpose of supporting the analysis of guides (or standards) for project management. It can improve the comprehension of a standard or guide. In addition, this ontology can be used to identify compatible concepts and terms that different standards have. A group of competence questions (CQs) were defined to evaluate the ontology.

*Classes and the class hierarchy*

After analyzing relevant terms in this domain, 48 classes were identified. Two of the main classes are **Standard** and **Part**. These two classes include other subclasses to characterize the individuals that compose it. The hierarchy of the class **Part** is shown in Fig. 1. A Standard includes **Main_Components** and **Sub_Components**. The specific components of a **Standard** depend on its features. For example, *PMBOK_Edition_6th* **Knowledge_Area** is made up of **Knowledge_Area**. Furthermore, A **Knowledge_Area** is made up of several **Processes**. On the other hand, there are a group of classes to classify a **Standard** according to its goal.

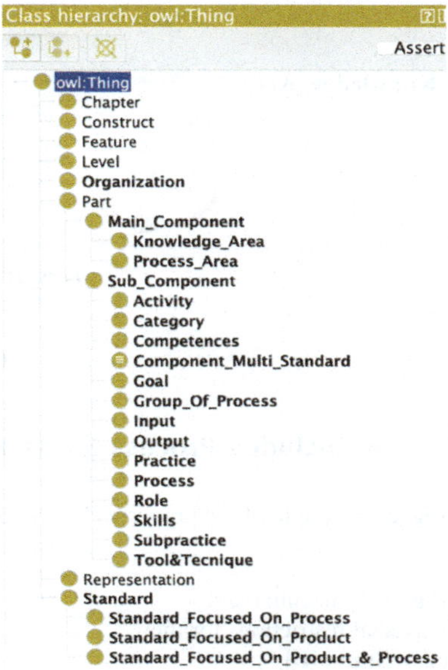


**Fig. 1** Excerpt of the class hierarchy

*Properties*

Our ontology includes 67 object properties. Each property has a domain and range to explicitly constraint the classes that can be part of a property. For example, the class **Knowledge_Area** is the domain of the properties *Has_Work_Document* and *Includes_Process*. These two properties have as range the classes **Input** (and **Output**) and **Process** respectively. We also specified for each property its inverse. OWL allows to describe in details the properties. For example, the object property *Is_Compatible_With* was defined as symmetric. This specification is useful when the ontology is adopted to represent the knowledge of different standards which may share concepts.

The possibility to represent property chains is another feature of OWL. For example, it is useful to know all the documents that are used in a **Knowledge_Area**. However there not exist direct relation between Document (either **Input** or **Output**) with **Knowledge_Area**. Hence, to get this information we have to do it through a property chain. An example of a property chain (*Has_Work_Document*) is depicted in Fig. 2a. Figure 2b depicts the representation in Protégé of this property. The property chain *Has_WorkDocument* includes the properties *Includes_Process* and *Has_Input*. A similar property chain was defined but replacing the property *Has_Input* by *Has_Output*.

In addition to object properties, a set of data properties was defined. Some of the most relevant are *Has_Name*, *Has_Description*, *Has_Publication_Date* and *Has_Version*. Domain and Range were defined foreach data property.

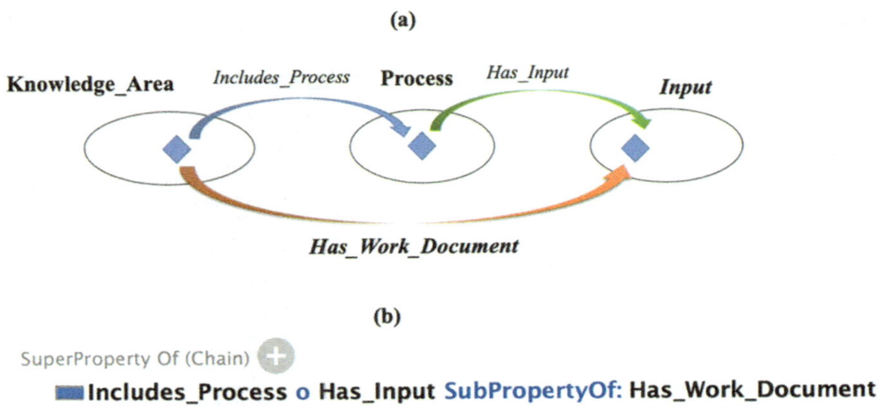


**Fig. 2** Example of the property chain *Has_Work_Document*

**Fig. 3** Existential and universal restrictions for the class **Standard**

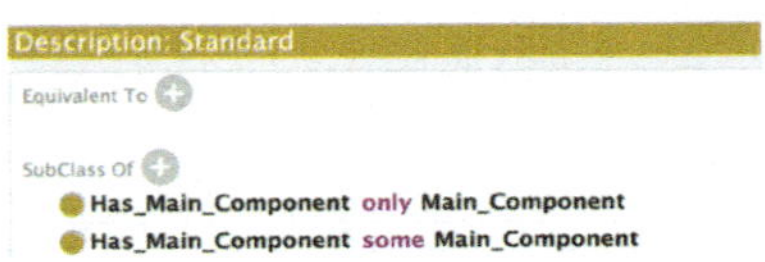


*Restrictions*

With OWL is possible to define universal (some) and existential (only) restrictions. Figure 3 depicts an existential restriction which states that the instances of the class **Standard** can be related to the instances of the class **Main_Component** through the property *Has_Main_Component*. Hence, it is mandatory for every **Standard** to have minimum one **Main_Component**. Figure 3 shows a universal restriction as well. This type of restrictions allows to classify an individual as a **Main_Component**. The next step in the ontological process is to create instances for each class and add different axioms according to the context.

## *3.1 Ontology Evaluation*

The ontology evaluation process considered the following elements:

- Its ability to satisfy requirements as a logical-formal system.
- In addition, satisfy the requirements of the stakeholders who requested its generation.
- Furthermore, the evaluation of the ontology was conceived following an iterative and gradual approach to validate each of its parts.

- The Pellet Reasoner was used to validate the consistency of the ontology. Throughout the entire life cycle, the reasoner was used to check if the ontology was consistent.

To demonstrate that the requirements of the ontology are satisfied, we checked that the ontology correctly answered all the competency questions that were defined. The results of this evaluation are described below. These examples also show that our ontology is useful and applicable.

## 4 Representing the Knowledge of PMBOK

By means the resulting ontology, it is possible to represent the knowledge of different standards. In this section, it is briefly described how the ontology was adopted to represent the knowledge of PMBOK. To evaluate the ontology, five competence questions were stated:

CQ 1. What are the main components of PMBOK?
CQ 2. How each component is made up?
CQ 3. What is the relation among different components?
CQ 4. What relations have a specific component of a standard?
CQ 5. What concepts of PMBOK have a correspondent concept in other standards?

We represented in our ontology the information about PMBOK 6th edition [25]. We created the individuals *PMBOK_6th_Edition* as instance of the class **Standard**. After representing the respective information *PMBOK_6th_Edition*, the ontology is able to answer the competence questions. For example, to answer CQ 1, the main parts of *PMBOK_6th_Edition* are depicted in Fig. 4. With these specifications, all the instances of **Knowledge_Area** will be classified as **Main_Component** by the reasoner.

The main elements in the analysis of a standard should be considered:

- Know its components and the other elements that make it up.
- When working with processes, it is convenient to clearly identify which are the main processes and which are the support processes.

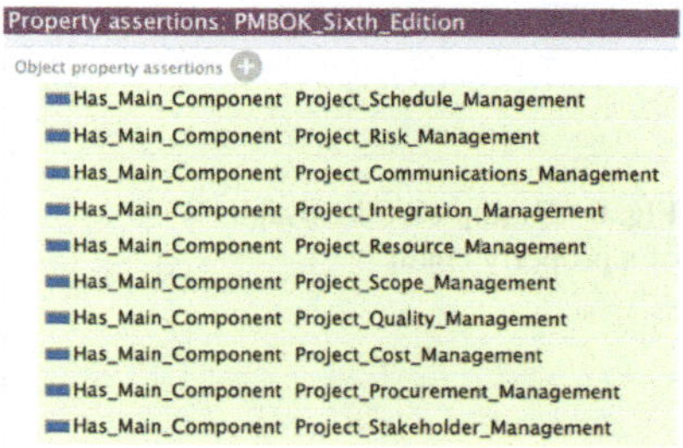


**Fig. 4** Inference of the main components of *PMBOK_6th_Edition*

- Identify the existence of common queries in the ontology and generalize them.
- Facilitate the generation of queries directly associated with the main elements and the generation of valor in the environment being worked on.

As an example of how the ontology can answer CQ.2, Fig. 5 depicts the processes that are part of the group of process *Executing* (individual ***Executing_Process_Group*** in the ontology).

Property chains allowed us to get information about individuals that are not directly related. For example, Fig. 6 shows the inputs/outputs of the **Knowledge_Area** *Poject_Resource_Management*. This is an example to demonstrate how the ontology is able to answer CQ.3.

In an ontology, the same individual may have different classifications based on their features. For example, a document may be an output document for a process and an input for another. To exemplify, the answer to CQ. 4, Fig. 7 depicts all the relations of the *Enterprise_Enviromental_Factors*. This figure shows the processes that use this document as an input and those processes that modify it.

In spite the main focus of the ontology is on PMBOK, it can be used to describe other standards. We defined the class **Component_Multi_Standard** to classify all

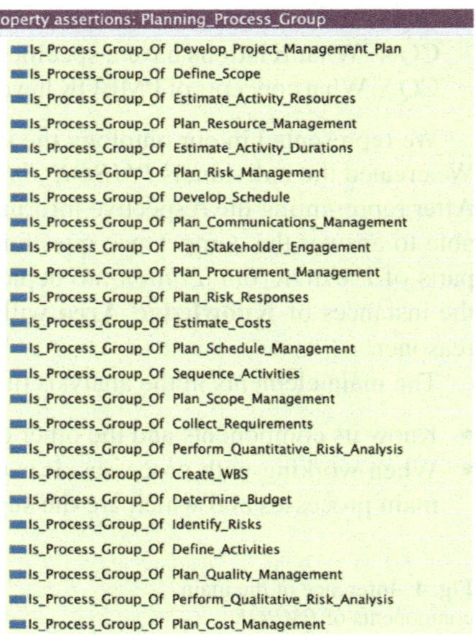


**Fig. 5** Processes that belong to the **Group_of_Processes** *Planning_Process_Group*

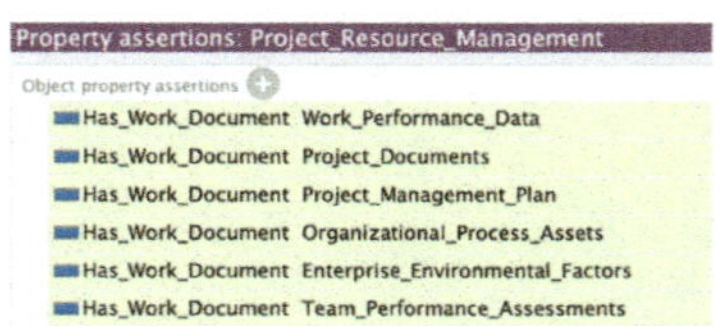


**Fig. 6** Example of the usage of a property chain

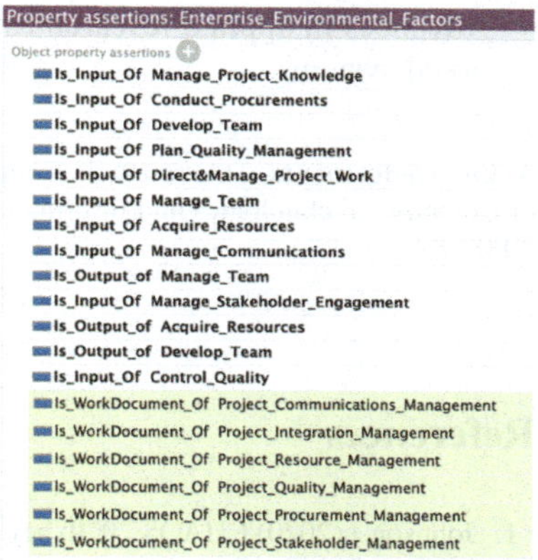


**Fig. 7** Relations of *Enterprise_Enviromental_ Factors*

the concepts that are included in different standards. Based on the set of necessary and conditions that we specified for this class, a reasoner can automatically identify its instances. These specifications allow the ontology to answer CQ 5. We included in the ontology information about the Capacity Maturity Model Integrated (CMMI) to specify the compatibility of some specifications of this standard with certain specifications of PMBOK. Using the specifications of the ontology we were able to define the compatible concepts between both standards.

## 5 Conclusions

The ontology presented allows us to represent and generalize knowledge about project management standards. In particular, it emphasizes the PMBok standard due to its high level of generalization.

This ontology might contribute to manage the knowledge about PMBOK as well to analyze the results of its adoption in different projects. The use of OWL as a formal language makes it easier to detect inconsistencies and ambiguities in the knowledge represented. In this way, contradictions and other negative effects are avoided in the decision-making processes where the constructed ontology intervenes. Currently we are working to extend the ontology with other concepts. In this sense, the identification of the common concepts that these standards have is an important task of our future work.

The authors advance applications of the proposed ontology in different real-world scenarios. The aim is to demonstrate the benefits of using ontologies in hybridization with other techniques and achieve the following objectives:

- Achieve greater agility in the ontology construction processes.
- Raise the quality of the analyses and ontologies built.

- Advances in applied research associated with the use of ontologies in conversational systems.

**Acknowledgements** This research is supported by a postdoc fellowship granted by the Institute of Computer Technologies and Information Security, Southern Federal University, project № PD/20-02-KT.

## References

1. Johnson J (2020) CHAOS 2020: beyond infinity Report. Standish Group
2. Institute PM (2017) A guide to the project management body of knowledge (PMBOK guide), 6th edn (PMBOK guide). Project Management Institute
3. Stevens R, Lord P (2009)Application of ontologies in bioinformatics. In: Handbook on ontologies. Springer, pp 735–756
4. Silega N, Rogozov YI, Lapshin VS (2022) Towards a software development process based on ontologies. In: XI All-Russian scientific conference "System Synthesis and Applied *Synergetics", 2022*
5. Happel HJ, Seedorf S (2006) Applications of ontologies in software engineering. In: Proceedings of the workshop on sematic web enabled software engineering (SWESE) on the ISWC, pp 5–9
6. Tapia-Leon M, Rivera AC, Chicaiza J, Luján-Mora S (2018) Application of ontologies in higher education: a systematic mapping study. In: 2018 IEEE global engineering education conference (EDUCON). IEEE, pp 1344–1353
7. Ma Z, Schultz MJ, Christensen K, Værbak M, Demazeau Y, Jørgensen BN (2019) The application of ontologies in multi-agent systems in the energy sector: a scoping review. Energies 12(16):3200
8. Antunes AL, Cardoso E, Barateiro J (2022) Incorporation of ontologies in data warehouse/business intelligence systems-a systematic literature review. Int J Inf Manag Data Insights 2(2):100131
9. de Lusignan S et al (2020) COVID-19 surveillance in a primary care sentinel network: in-pandemic development of an application ontology. JMIR Public Health Surveill 6(4):e21434
10. Cao Q, Zanni-Merk C, Reich C (2018) Ontologies for manufacturing process modeling: a survey. International conference on sustainable design and manufacturing. Springer, pp 61–70
11. Nikulina Y, Shulga T, Sytnik A, Frolova N, Toropova O (2019) Ontologies of the fire safety domain. International conference on information technologies. Springer, pp 457–467
12. Sinha PK, Gajbe SB, Debnath S, Sahoo S, Chakraborty K, Mahato SS (2022) A review of data mining ontologies. Data Technol Appl 56(2):172–204
13. Ahmad A, Bandara M, Fahmideh M, Proper HA, Guizzardi G, Soar J (2021) An overview of ontologies and tool support for COVID-19 analytics. In: 2021 IEEE 25th international enterprise distributed object computing workshop (EDOCW). IEEE, pp 1–8
14. Iqbal MW, Naqvi MR, Khan MA, Khan F, Whangbo T (2022) Mobile devices interface adaptivity using ontologies. Comput Mater Continua 71(3)
15. Osman MA, Noah SAM, Saad S (2022) Ontology-based knowledge management tools for knowledge sharing in organization—A review. IEEE Access 10:43267–43283
16. W3C EPT, Pan JZ, Oberle D, Wallace E, Uschold M, Boeing, Kendall E. Ontology driven architectures and potential uses of the semantic web in systems and software engineering. http://www.w3.org/2001/sw/BestPractices/SE/ODA/. Accessed 6 de Febrero 2011
17. Roger SP, Bruce RM (2015) Software engineering: a practitioner's approach. McGraw-Hill Education

18. Sommerville I (2011) Software engineering, 9th edn. p. cm (ISBN-13). Addison-Wesley, pp 978-0
19. Delgoshaei P, Heidarinejad M, Austin MA (2022) A semantic approach for building system operations: Knowledge representation and reasoning. Sustainability 14(10):5810
20. Amna AR, Poels G (2022) Ambiguity in user stories: a systematic literature review. Inf Softw Technol 145:106824
21. Suárez-Figueroa MC, Gómez-Pérez A, Motta E, Gangemi A (2012) Ontology engineering in a networked world (Ontology Engineering in a Networked World), pp 1–435
22. Silega N, Varén E, Varén A, Rogozov YI, Lapshin VS, Alekseevich SA (2022) Exploiting an ontological model to study COVID-19 contagion chains in sustainable smart cities. Information 13(1):40
23. Keet M (2018) An introduction to ontology engineering. Maria Keet Cape Town
24. Horridge M (2009) A practical guide to building OWL ontologies using Protégé 4 and CO-ODE tools edition 1.2. The University of Manchester, Manchester
25. Čabarkapa J (2019) Analysis and comparison of ISO 21500-Guidance on project management and PMBOK 6th guide. In: 5th IPMA SENET project management conference (SENET 2019). Atlantis Press, pp 266–271

# Ranking of Success Forecasts for Computer Engineering Students Based on Computing with Words

**Julio A. Telot González, Maylí Estopiñán Lantigua, and Lirianne Gutiérrez Sánchez**

**Abstract** In the field of the teaching-learning process, multiple attempts are made to obtain an integral evaluation of the student, which includes his academic situation, as well as other factors that contribute to have a more complete idea of the student's performance and his prospects for professional success. In the present work, a way of measuring performance and prospects of success is approached, by means of the 2-tuple linguistic model for computing with words, which integrates the academic performance in a given period with their subjects and can build a ranking of the students of a given group, taking into account the importance of each subject in an initial period of the training of the specific computing professional, as well as a criterion that includes the contribution of these subjects to the development of the students' creative thinking. For this purpose, surveys were used, the results of which were used to construct a parallel weighting of subjects by means of the aforementioned indicators. With the results of the evaluation given to each of the students in a group, and the integrated weights of the subjects, an integral evaluation of each student and a ranking of performance and prospects for professional success was obtained, which was processed with the help of the FLINTSTONES software.

**Keywords** 2-tuple linguistic model · Creativity · Students' performance · Students' ranking · Weighting

## 1 Introduction

For a long time, attempts have been made to objectively measure the contribution of the different contents of disciplines and subjects in the training of specialists in different branches of knowledge. On the other hand, although the importance of these contents in the preparation of higher education graduates is vital for the conformation of the curriculum, its implementation and, within it, the possible contribution of these

J. A. Telot González (✉) · M. E. Lantigua · L. G. Sánchez
Universidad de Matanzas, Matanzas, Cuba
e-mail: juliotelot@gmail.com



P. Y. Piñero Pérez et al. (eds.), *Computational Intelligence in Engineering and Project Management*, Studies in Computational Intelligence 1134,
https://doi.org/10.1007/978-3-031-50495-2_14

disciplines and subjects in the development of creativity in the student, constitute factors that must be taken into account to assess the training of that student.

Measuring creativity and importance of subjects in professional development can be a complex task, as both concepts are subjective and depend on various factors such as individual perspective, cultural context, and industry norms. However, there are a few methods that can be used to assess these factors.

Overall, it is important to note that measuring creativity and importance of subjects can be challenging and subjective. Therefore, a combination of objective assessments and subjective evaluations may be necessary to obtain a comprehensive understanding of these factors in professional development.

There are various experiences for measuring the importance of different subjects in professional training [1–3]. In some cases, they group subjects according to a certain level of importance and in others subjects are selected according to the criteria of researchers and experts. In these cases, a bias is observed that corresponds, in most cases, to the interests of the researchers. There are other authors [4] who use elements such as aspects of students' backgrounds, commitment and motivations to determine which of these affect students' success in their training as programmers and in general in aspects of computer science, especially the need for creativity to achieve success. There is also information about experiments carried out with students preparing for competitions and Olympiads [5], in which they train with special instruments for these purposes.

Regarding the development of creativity, different points of view and criteria have been developed about the ways of measuring it. The development of creativity in human beings is closely related to creative learning and many authors [6–8] consider this type of learning as the basis of creativity.

This paper deals with the experience of developing a parallel weighting of the importance of the subjects, together with the contribution to creativity that each of these adds to the comprehensive evaluation of the student, which allows projecting the perspective of success in their studies and as a senior specialist. A group of first-year Computer Engineering students from the University of Matanzas, Cuba, subject to Study Plan E for this degree, established by the Cuban Ministry of Higher Education [9], and adjusted to the characteristics of the higher level institution where the experiment is carried out.

In order to propose the importance of each subject in the student's training, what was stated in the aforementioned study plan was considered, where the roles to be played by the graduate of the degree are established and the level to be reached in each one of them is indicated these roles. This information is presented in Table 1.

To establish the weight of the different subjects in terms of their contribution to creativity, metrics proposed by authors who investigate this topic were used [10–14]. Among them, some aspects were selected that would contribute to defining an objective weight for each subject according to its contribution to creativity. Table 2 shows the dimensions and indicators that were taken into account to define the weights of the subjects in terms of their contribution to the creativity of the computer professional.

**Table 1** Roles of the profession and their levels of development in the career

| Role | Level | | |
|---|---|---|---|
| | High | Half | Basic |
| Programmer | X | | |
| Tester | | X | |
| Configuration and change manager | | | X |
| Business analyst | X | | |
| System analyst | X | | |
| Data analyst | | X | |
| Architect | | X | |
| Software designer | X | | |
| Database designer | X | | |
| Human-machine interface designer | | | X |
| Security specialist | | | X |
| Writer-exposer of technical works | X | | |
| Project manager | | X | |
| Information technologies infrastructure enabler | | | X |
| Decision facilitator | | X | |

*Source* Educación Superior [9]

## 2 Materials and Methods

To collect the corresponding information, were used surveys in order to decide the weights of subjects, according to importance and contribution to creativity.

The values of these weights were defined with the objective of processing students' achievement data and what these values can represent in their training. To obtain the comprehensive use of students, taking into account the importance of the subjects and the contribution to creativity, the computation method with linguistic 2.tuple words was used [15]. The study of the use of these weights in the proposed model has been studied by important authors in this field [16, 17].

For the processing of the information corresponding to the evaluation and prognosis of the students, the FLINTSTONES tool [18] was used.

## 3 Results

The subjects that first-year students of the Computer Engineering degree at the University of Matanzas receive are:

1. Mathematics 1.
2. Introduction to programming.

**Table 2** Dimensions and indicators that characterise creative learning

| Dimensions | Indicators |
|---|---|
| 1. Customisation of IT processes | 1.1. Prepares a personalised synthesis of concepts, procedures, models, systems and/or computer algorithms to be determined within the framework of the project<br>1.2. Develops personalised forms for the recording of information received on concepts, procedures, models, systems and/or computer algorithms determined within the framework of the project<br>1.3. Discriminates the relevant information from the knowledge the learner possesses about the information associated with the project obtained in the different learning spaces where the learner is involved in developing the project<br>1.4. Individualises the new concepts, procedures, models, systems and/or computer algorithms obtained in the different learning spaces where the learner is involved in order to develop the project<br>1.5. Applies the concepts, procedures, models, systems and/or computer algorithms obtained in the different learning spaces where the learner is involved in order to develop the project, adjusting them to the characteristics of the project |
| 2. Confrontation with the already given computer processes that make the emergence of subjective meanings possible | 2.1. Questions the concepts, procedures, models, systems and/or computer algorithms obtained in the different learning spaces where the learner is involved in order to develop the project<br>2.2. Argues the project development processes considered most efficient and effective on the basis of the concepts, procedures, models, systems and/or computer algorithms obtained in the different learning spaces where the learner is involved<br>2.3. Argues the tensions detected during the execution of the project development processes based on the concepts, procedures, models, systems and/or computer algorithms obtained in the different learning spaces where the learner is involved<br>2.4. Selects the people with the greatest potential to make up the project development teams considered on the basis of selection criteria obtained in the different learning spaces in which the learner is involved<br>2.5. Modifies opinions on the basis of valid opposing criteria during the execution of the project development processes on the basis of the concepts, procedures, models, systems and/or computer algorithms obtained in the different learning spaces in which the learner is involved |

(continued)

**Table 2** (continued)

| Dimensions | Indicators |
|---|---|
| 3. Production, generation of own and "new" ideas during the execution of an IT project | 3.1. Proposes new hypotheses during the execution of the project development processes based on the concepts, procedures, models, systems and/or computer algorithms obtained in the different learning spaces where the learner is involved<br>3.2. Selects the most efficient and effective route during the execution of the project development processes based on the concepts, procedures, models, systems and/or computer algorithms obtained in the different learning spaces where the learner is involved<br>3.3. Develops new projects that provoke satisfaction for what has been achieved and the generation of new ideas linked to his/her training as a computer engineer<br>3.4 Collaborates with the people involved during the execution of the project development processes on the basis of the concepts, procedures, models, systems and/or computer algorithms obtained in the different learning spaces where the learner is involved<br>3.5. Produces computer concepts, procedures, models, systems and/or algorithms needed in the framework of the project |

*Source* González Hernández et al. [13]

**Table 3** Weighting of the subjects based on the teachers' criteria

| Subject | WI | WC | W |
|---|---|---|---|
| Mathematics 1 | 0.21 | 0.21 | 0.21 |
| Introduction to programming | 0.25 | 0.26 | 0.25 |
| Discrete mathematics | 0.24 | 0.24 | 0.24 |
| Introduction to computing | 0.20 | 0.18 | 0.20 |
| Philosophy | 0.11 | 0.11 | 0.11 |

*Source* Authors' elaboration

3. Discrete mathematics.
4. Introduction to computing.
5. Philosophy.

In order to weigh the criteria about the importance of each subject, as well as its contribution to the development of creativity in the training of the computer engineer, criteria were collected from the professors who teach classes that year, as well as from the professors who work in the Computer Science Department of the University of Matanzas (24 in total), who were considered experts for the weighting of these criteria. An anonymous survey was prepared in which each teacher, for each subject, had to issue a value between 0 and 10 for the importance of the subject in the career and another similar one to weigh the contribution to creativity of the subject in question.

For example, the Fig. 1 shows the weights given by the experts to the importance of Mathematics 1.

In this case, there are five subjects, for each of which the expert must provide two weightings between 0 and 10: one for the importance of the subject (PI) and another for the assessment of what that subject contributes to creativity (PC). In this way, two parallel weightings PIij and PCij are established, where PI is the weighting

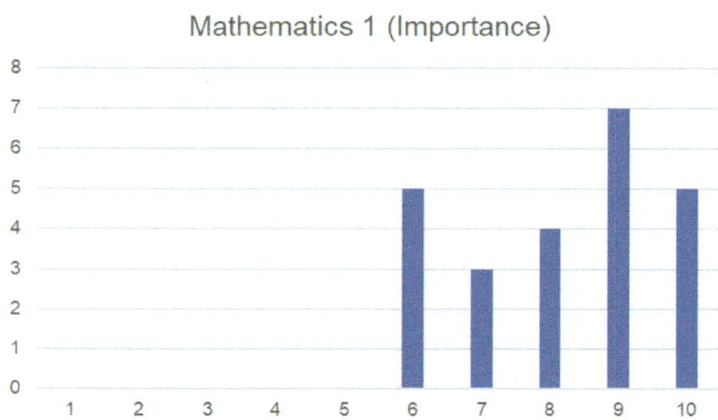


**Fig. 1** Experts' criteria about the importance of Mathematics 1. *Source* Authors' elaboration

established by expert i for subject j in terms of importance and PC is the weighting established by that expert for the contribution to creativity in that subject.

In this way, the weight in importance of a subject j (WIj) is calculated as shown in Eq. 1.

$$WI_j = \frac{\sum_{i=1}^{n} PI_{ij}}{\sum_{i=1}^{n} \sum_{j=1}^{m} PI_{ij}} \tag{1}$$

where n is the number of experts and m the number of subjects. Thus, in the case of Fig. 1, if it is considered that subject Mathematics 1 is represented as j = 1, its weight in importance would be, from the fact that $\sum_{i=1}^{n} PI_{i1} = 196$, and $\sum_{i=1}^{n} \sum_{j=1}^{m} PI_{ij} = 951$, the weight $WI_j = 0.21$, which is interpreted as that, regarding the importance of the subjects of the first period of the first year of Computer Engineering, the subject Mathematics 1 would have a weight of 0.21, according to its importance within the training of the computer engineer.

In a similar way, the weights of the subjects are calculated according to the contribution to creativity WCj of each of them (see Eq. 2).

$$WC_j = \frac{\sum_{i=1}^{n} PC_{ij}}{\sum_{i=1}^{n} \sum_{j=1}^{m} PC_{ij}} \tag{2}$$

From the collected data, the weights of the different subjects were obtained from the point of view of their importance within the curriculum (WIi), the contribution to creativity that this subject represents (WCi) and an integrated weight value of the subject (Wi), which results from considering that the importance of the subject within the curriculum is weighted at 0.65 and the contribution to creativity at 0.35, a distribution that was obtained in the form of individual interviews that achieved a consensus among the teachers interviewed, although this value needs further refinement from the psycho-pedagogical point of view, since opposing criteria, and a considerable dispersion among some experts, are observed.

Thus, the integrated weight value Wi is calculated as shown in Eq. 3.

$$W_i = 0.65WI_i + 0.35WC_i \tag{3}$$

From the data obtained in the surveys, the weights of the subjects from the point of view of their importance within the curriculum (WIi), the contribution to creativity that this subject represents (WCi) and the integrated weight of the subject (Wi) are shown in Table 3, where it can be seen that, in this case, the values of WI and WC do not differ substantially and, due to the weight assigned to the importance of the subject within the curriculum (WI) in relation to the weight of the contribution to creativity (WC), the values of the integrated weight (W) are closer to WI.

In the 2023 school year, there were 30 students in this group, who are systematically evaluated in each subject and, on a monthly basis, a criterion is issued by each teacher about each student's achievement in the subject in question.

To obtain the ranking of the students' performance, and to predict their chances of success in their training, the 2-tuple linguistic model is used.

*Stage 1. Definition of the framework*

Set of alternatives: in this case each student is considered as an alternative. For this paper, it has been considered as Student = {St1, St2, St3, …, St30}, where the students are not identified by their real names, in order to maintain the confidentiality of the information.

Set of Experts: the experts are the subjects through their respective teachers. Thus, there are five experts, who evaluate each student. Subject = $\{e_i | i \in (1,\ldots,p)\}$, such as: E = {Mathematics 1, Introduction to Programming, Discrete mathematics, Introduction to computing, Philosophy}}.

Set of Attributes: There is only one attribute considered in this case, it is the student performance in the considered stage.

Domain of expression: These student evaluations are provided in the form of linguistic values taken from the domain of possible evaluations S = {Very bad, Bad, Regular, Good, Excellent}, so in this case multigranularity is not used to carry out the aggregations, since all the subjects use the same domain of evaluation.

Each subject (considered as expert) has different relevance, as it was explained before. So, a vector of weights W = {0.21, 0.25, 0.24, 0.20, 0.11}, according to Table 4.

*Stage 2. Compilation of the experts' preferences*

Teachers' criteria on the students' achievement in an evaluation stage are shown in Table 4, and have been used to establish a ranking that aims to project the preparation that students are reaching in said stage.

Subsequently, the criteria issued by the subject teachers are transformed into the 2-tuple linguistic model, which is shown in Table 5, only for the first five students, with a view to simplifying the situation in this article.

*Stage 3. Analysis (aggregation)*

In this case, the selected method is the linguistic model based on 2-tuples, without the use of multigranularity, with multiple experts. After having the evaluations of the experts, and the transformation to 2-tuples, the collective preferences for each alternative (student) criterion are calculated using the weighted arithmetic average aggregation operator, considering the weight of each expert, since there is only one criterion.

*Stage 4. Selection (exploitation)*

After the preferences aggregation, according to the criteria, the value that reaches the Evaluation for each student is obtained, according to the teachers' evaluations. This value permits to identify the best or the worst alternative. In this case, it is obtained the best value (Good, 0.0), for alternative St4. It means that student St4

**Table 4** Evaluation received by the students in the stage analyzed

| Student | Mathematics 1 | Introduction to programming | Discrete mathematics | Introduction to computing | Philosophy |
|---|---|---|---|---|---|
| St1 | Good | Bad | Regular | Good | Excellent |
| St2 | Good | Good | Good | Good | Regular |
| St3 | Good | Regular | Good | Regular | Good |
| St4 | Excellent | Regular | Good | Regular | Excellent |
| St5 | Good | Regular | Bad | Excellent | Regular |
| St6 | Good | Bad | Regular | Good | Excellent |
| St7 | Good | Regular | Good | Good | Good |
| St8 | Bad | Bad | Regular | Bad | Regular |
| St9 | Bad | Bad | Bad | Bad | Regular |
| St10 | Good | Regular | Regular | Good | Good |
| St11 | Excellent | Good | Good | Good | Excellent |
| St12 | Good | Regular | Good | Good | Excellent |
| St13 | Good | Bad | Good | Good | Excellent |
| St14 | Good | Good | Good | Good | Good |
| St15 | Good | Good | Good | Good | Good |
| St16 | Good | Bad | Good | Good | Bad |
| St17 | Good | Regular | Good | Good | Good |
| St18 | Excellent | Good | Good | Regular | Good |
| St19 | Good | Regular | Good | Good | Good |
| St20 | Excellent | Regular | Good | Good | Good |
| St21 | Excellent | Excellent | Good | Good | Excellent |
| St22 | Regular | Regular | Bad | Good | Bad |
| St23 | Excellent | Excellent | Good | Good | Excellent |
| St24 | Excellent | Excellent | Good | Good | Good |
| St25 | Good | Regular | Good | Good | Regular |
| St26 | Good | Good | Good | Regular | Regular |
| St27 | Good | Bad | Good | Good | Excellent |
| St28 | Good | Bad | Good | Good | Regular |
| St29 | Good | Bad | Good | Good | Good |
| St30 | Good | Bad | Good | Regular | Good |

*Source* Authors' elaboration

obtained an exact Good evaluation. The worst evaluation corresponded to alternative St5 (Regular, 0.4), it means that it is better than Regular, and slightly near to Good value zone. In the real experiment, with 30 students, the best evaluation corresponded to St23 (Excellent, −0.4), and the worst one was St9 (Bad, 0.2).

**Table 5** Preferences transformed into 2-tuple linguistic

| Student | Mathematics 1 | Introduction to programming | Discrete mathematics | Introduction to computing | Philosophy |
|---|---|---|---|---|---|
| St1 | (Good, 0) | (Bad, 0) | (Regular, 0) | (Good, 0) | (Excellent, 0) |
| St2 | (Good, 0) | (Good, 0) | (Good, 0) | (Good, 0) | (Regular, 0) |
| St3 | (Good, 0) | (Regular, 0) | (Good, 0) | (Regular, 0) | (Good, 0) |
| St4 | (Excellent, 0) | (Regular, 0) | (Good, 0) | (Regular, 0) | (Excellent, 0) |
| St5 | (Good, 0) | (Regular, 0) | (Bad, 0) | (Excellent, 0) | (Regular, 0) |

*Source* Authors' elaboration

## *3.1 Solution of the Problem by Means of FLINSTONES*

The example described above was developed with the help of the FLINSTONES software (Fuzzy Linguistic Decision Tools Enhancement Suite) [18].

After structuring the framework, as explained above, where only one domain is used for the criteria, and the only criterion is evaluation, the information is collected.

Figure 2 shows the result obtained in domain Eval where there were five alternatives (students), and five experts (subjects). The best result corresponds to alternative St4 (Good, 0.0), and the worst one to St5 (Regular, 0.4), clearly better than Regular. Results for St1 and St3 are overlapped, as can be observed in upper side of figure.

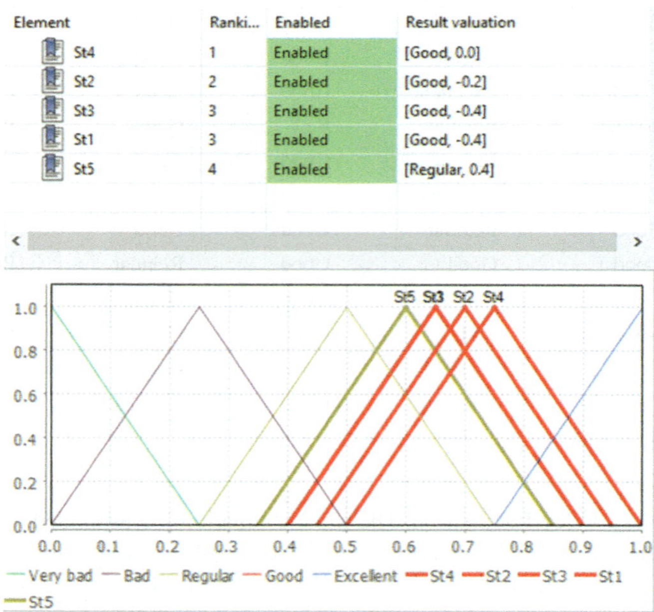


**Fig. 2** Ranking results for students from St1 to St5

### 3.2 Results After the Analyzed Period

From the results obtained for one stage of the analyzed period, it was possible to compare the forecast ranking of the group of students with the final results of the period. These results can be seen in Table 6, where it can be observed that the student St21, who had occupied the first place in the forecast for the total number of students, was first in the final results, both in the arithmetic mean of his grades and in the weighted mean (according to the weights previously mentioned) and in the ranking by means of the 2-tuple linguistic model. This student was the only one who was close to the Excellent value. In the remaining cases the behavior was similar, although there are students who do not meet the prediction. At the end of the list, also the predictions were close to reality.

It is important to note that these final results only refer to academic grades; they do not include skills in the performance of professional tasks, an aspect that will be considered in further research.

This information can be observed for the first five students in Table 5 and in the solution of the problem using FLINTSTONES in Table 7, where it is shown that the students (except for student St5, who, after a first negative stage, made a positive jump out of the predictions), maintained a ranking similar to the predictions, both in the 2-tuple linguistic model and in the arithmetic mean and weighted mean.

**Table 6** Preferences student academic results at the end of the period

| Student | Arithmetic mean | Weighted mean | Results of 2-tuple linguistic model |
|---|---|---|---|
| St21 | 4.60 | 4.69 | (Excellent, −0.4) |
| St14 | 4.40 | 4.52 | (Good, 0.0) |
| St24 | 4.20 | 4.27 | (Good, 0.4) |
| St17 | 4.00 | 4.08 | (Good, −0.2) |
| … | … | … | … |
| St3 | 2.60 | 2.57 | (Bad, 0.4) |
| St8 | 2.20 | 2.13 | (Bad, 0.2) |

*Source* Authors' elaboration

**Table 7** Academic results of students from St1 to St5

| Student | Arithmetic mean | Weighted mean | Results of 2-tuple linguistic model |
|---|---|---|---|
| St2 | 3.40 | 3.52 | (Regular, 0.4) |
| St5 | 3.40 | 3.47 | (Regular, 0.4) |
| St4 | 3.40 | 3.34 | (Regular, 0.4) |
| St1 | 2.80 | 2.68 | (Regular, −0.2) |
| St3 | 2.60 | 2.57 | (Regular, −0.4) |

*Source* Authors' elaboration

These results will continue to be studied for subsequent periods and refined in the selection of weights and parameters to be considered. Students will also be informed in order to stimulate their potential in terms of creativity and achievement.

## 4 Conclusions

Obtaining an accurate and fair criterion of the performance status of students in a given period of their studies and projecting their chances of success in a certain professional field is a multifactorial and generally subjective process, but it is convenient to have tools that help to lead the process and guide the development of students and the educational process itself. In this case, the weight that the subjects have for a given profession and what each subject can contribute to the development of creativity in the future graduate are considered. These weights were built based on the criteria of the professors of this subject and a group of experts from related fields, with which a weight was obtained for each subject of the first period of the first year of Computer Engineering, which is intends to extend to the entire career.

With these weights and the criterion of an evaluative stage made by the teachers to a group of students, a ranking and a projection of possible success of the students was obtained, for which the 2-tuple linguistic model was used.

With the collection of these data and an extension of the sample, both in students and in subjects and years, it is intended to carry out a more complete analysis of the contents and structuring of the Computer Engineering career, in order to motivate, to a greater extent, the development of creativity and obtaining a more competent and motivated professional.

## References

1. Alonso Álvarez Y, Pazos Couto JM (2020) Importancia percibida de la motricidad en Educación Infantil en los centros educativos de Vigo (España). Educação e Pesquisa 46.
2. Pérez-Cañaveras RM, Juan Herrero JD (1994) Importancia de las asignaturas y materias del curriculum en la formación de los Profesionales de Enfermería. Enfermería Científica (142–143):69–72
3. Estopiñán Lantigua M, Telot González JA (2017) Contribución de la matemática discreta a la formación del ingeniero informático. Atenas, Revista Científico Pedagógica 3(39):18–30
4. Rõõm M, Luik P, Lepp M (2023) Learner success and the factors influencing it in computer programming MOOC. Educ Inf Technol 28(7):8645–8663
5. Weigend M et al (2019) Computational thinking education through creative unplugged activities. Olymp Inform 13:171–192
6. de Almeida P, Mitjáns Martínez A (2020) La emergencia del aprendizaje creativo. Alternativas cubanas en psicología 8(23):95–111
7. Torres Oliveira C, Mitjáns Martínez A (2020) Expresiones de la subjetividad social del aula y creatividad en el aprendizaje: un estudio de caso. Alternativas cubanas en Psicología 8(23):126–144

8. Telot González JA, Estopiñán Lantigua M (2012) Creatividad y regularidad en informática. Monografías UM, 2022
9. Educación Superior MD (2017) Plan de Estudio E. Carrera de Ingeniería Informática. Habana, Cuba, p 90
10. Hershkovitz A et al (2019) Creativity in the acquisition of computational thinking 27(5–6):628–644
11. Mukhamadovna TM, Sharipovna HA, Supkhonovna HN (2020) The system of development of professional competence in future primary school teachers 7(13):2020
12. Rasulova ZD (2020) Conditions and opportunities of organizing independent creative works of students of the direction technology in higher education. Int J Sci Technol Res 9(3):2552–2155
13. González Hernández W, Petersson Roldán M, Moreno García M (2022) Aprendizaje creativo de la informática: conceptualización. In: Convención Internacional de la Universidad de Matanzas. Matanzas, Cuba
14. Araya-Fernández E, Garita-González G (2020) Habilidades blandas: elementos para una visión holística en la formación de profesionales en informática (Soft Skills: Elements for a Holistic View in the Training of IT Professionals). Trilogía Ciencia Tecnología Sociedad 12(23)
15. Martínez L, Rodriguez RM, Herrera F (2015) The 2-tuple linguistic model computing with words in decision making. Springer International Publishing, Switzerland, p 168
16. Shu Z et al (2023) Clustering using ordered weighted averaging operator and 2-tuple linguistic model for hotel segmentation: the case of tripadvisor. Expert Syst Appl 213
17. Tao Z et al (2019) Rank aggregation based multi-attribute decision making with hybrid Z-information and its application. J Intell Fuzzy Syst 37(3):4231–4239
18. Labella Á, Martínez L (2020) FLINTSTONES 2.0 an open and comprehensive fuzzy tool for multi-criteria decision analysis. Springer International Publishing, Cham

# Computing with Words to Assess the Perceived Quality of IT Products and Projects

**Marieta Peña Abreu, Juan Carlos Mejias Cruz, Cynthia López Valerio, and Pedro Yobanis Piñero Pérez**

**Abstract** This paper proposes a method based on computing with words paradigms to assess the perceived quality of IT services that contribute to decision making. The proposal is divided into four phases. In the first one, the experts and criteria for the evaluation of the perceived quality of the IT services are selected. In the second one, the preferences of the users are collected. In the third one, the evaluation is carried out using the 2-tuple model and in the fourth one, the achieved result is integrated for a more profound analysis during decision making. The experts are selected through the Delphi method. The evaluation criteria are defined using the Focal Group technique and weighted with values between (0, 1) according to their importance. Valuations are expressed in the linguistic domain and aggregation operators are used that avoid the loss of information. As a result, the pair (level of perceived quality, precision) is obtained for each evaluated service. The method was validated with three IT services, in which 24 users evaluated 12 criteria selected by 10 experts. The perceived quality level of service $S_1$ was excellent with $-0.33$ of precision, service $S_2$ was very high with $-0.16$ of precision and $S_3$ reached an excellent value with $-0.45$ of precision. The proposed method is a favorable solution to assess the perceived quality of IT services in uncertain environments. Its use contributes to the decision making and to the improvement of the IT services evaluated.

**Keywords** Computing with words · Evaluation of perceived quality · Decision making

M. Peña Abreu (✉)
Facultad de Tecnologías Educativas, Universidad de las Ciencias Informáticas, Havana, Cuba
e-mail: mpabreupc@gmail.com; mpabreu@uci.cu

J. C. Mejias Cruz
Gerente - Project Manager en Departamento IT Advisory, KPMG Uruguay, Montevideo, Uruguay

C. López Valerio
Universidad Instituto Tecnológico de Costa Rica, Alajuela, Costa Rica
e-mail: cylopez@itcr.ac.cr

P. Y. Piñero Pérez
Artificial Intelligence for Sustainable Development Group, IADES, La Habana, Cuba
e-mail: iades@businessredmine.com



P. Y. Piñero Pérez et al. (eds.), *Computational Intelligence in Engineering and Project Management*, Studies in Computational Intelligence 1134,
https://doi.org/10.1007/978-3-031-50495-2_15

## 1 Introduction

Offering products or services with quality and satisfying the user has become one of the most important objectives for organizations today. For this reason, the scientific community has taken great interest in the study and evaluation of perceived quality and user satisfaction, as well as their antecedents and consequences. Thus, it is postulated that both offering products or services with quality and satisfying the user will provide the organization with competitive advantages and, most importantly, with satisfied users who will continue to purchase its products or services and who will recommend their use and consumption to others [1, 2].

However, there is some controversy over the conceptualization of perceived quality, its dimensionality, and its influence on user satisfaction. Likewise, user satisfaction has been confused on numerous occasions with perceived quality, and, therefore, the relationship between the two concepts is not very consensual in the literature. Thus, conducting research on this subject will make it possible to successfully reconcile the objectives of organizations with those of the users to whom they offer their products or services.

Knowing users' perceptions about the quality of a product or service provided is considered a relevant factor in explaining the competitiveness and viability of organizations [3, 4]. Similarly, knowing the origin or cause of positive experiences is a fundamental aspect of the implementation of any project, particularly if the aim is to improve the efficiency and effectiveness of the processes and achieve user loyalty [5–7].

Since the 1980s, numerous contributions have been made to evaluate perceived quality [2, 8, 9]. Mainly focused on the area of services. These proposals can be grouped under two schools: the Nordic school and the North American school [10] and have been debated and questioned at the theoretical, methodological and empirical levels by the international scientific community.

In general, they use traditional decision-making methods [11, 12] among which are those of multi-criteria evaluation [13], with numerical outputs, which makes it difficult to analyze the results in decision-making. In addition, when the criteria are evaluated independently, different scoring scales are used, making the process of unifying the results and their interpretability cumbersome.

In the specific case of IT services, there is no precise definition of criteria from several dimensions that affect the evaluation of perceived quality since software is an intangible asset, making its evaluation more complex. IT services have become essential for the development of society. On the other hand, the evaluation of perceived quality does not take into account the uncertainty inherent in this evaluation environment [14, 15]. Uncertainty in the problem-solving process can be present both in the data of the problem instance being solved and in the knowledge being used to solve it [16]. In this case, it is more associated with knowledge.

This implies that it is necessary to modify these traditional methods and use new techniques to obtain more accurate results in decision-making [17].

To manage uncertainty, researchers have used different soft computing techniques depending on the context [18–21]. Among the paradigms that stand out the most is fuzzy mathematics [22] and Computing with Words, successfully to solve similar problems in other areas [23, 24]. Despite their usefulness in previous research, they have very domain-specific characteristics.

In contrast to these investigations, this paper proposes a method to evaluate the perceived quality of IT services in uncertain environments, which allows working with multiple experts and multiple criteria and facilitates decision-making. In this sense, the method defines the inputs, outputs, and the flow of activities to be performed in each of the proposed phases. It also proposes a set of candidate criteria for the evaluation of perceived quality in the case of IT services.

The document is structured as follows: in the next section, the activities of the proposed method are explained. Then, a synthesis of its experimental application to three computer services is presented. Finally, conclusions and bibliographical references are presented.

## 2 Methodology

As shown in the introduction to this research, currently existing proposals for assessing perceived quality have been mostly oriented toward the quality of services and do not take into account the existing uncertainty. To address this problem, it is advisable to use novel and internationally accepted techniques for decision-making. In this regard, it has been shown that the adoption of proposals based on CWW is an appropriate option.

The linguistic model based on 2-tuples is selected [25] because it is easy to understand and maintains a fuzzy representation of the linguistic information where its results have assigned syntax and semantics as defined by the ELD, making it feasible for the treatment of the uncertainty associated with the evaluation of the perceived quality in the present research. On the other hand, the virtual provides a numerical result that cannot be represented linguistically, while the 2-tuple proportional uses two consecutive linguistic labels.

Another aspect to take into account is that the 2-tuple model has been successfully used to treat multi-granular linguistic information [25], unbalanced linguistics [26], heterogeneity and uncertainty [27–29]. In addition, it is used in several areas of knowledge, as well as in the areas of project management and software engineering.

This method will be organized into four phases, each describing its inputs, outputs, and activities to be carried out. Among the defined elements that support the method are its principles and premises.

Principles:

- Adaptive capacity to be integrated with new evaluation criteria that make it possible to obtain more detailed and accurate results in the evaluation of perceived quality.

- Strategic nature for organizations, since it enables a better evaluation of perceived quality, contributing to effective decision-making.
- Flexibility for generalization in software development organizations.

Premises:

- Institutional support is needed to adopt the practices defined in the proposed method lines.
- Existence of qualified personnel for its application.

The method can be used to evaluate one or several IT services during the same evaluation. The result is an ordered list of various services according to their overall perceived quality. The method consists of the following phases: "Definition of evaluation criteria",

## *2.1 Phase 1 "Definition of Evaluation Criteria"*

Description: In this phase, it is proposed to define the criteria that will be used in the evaluation of the perceived quality. Therefore, a group of criteria is proposed, which will be used as a basis for the selection by the experts. In order to carry out the proposal, combinations of traditional techniques used in other research [29, 30] and the 2-tuple linguistic modeling approach [27] were used to reduce uncertainty. For the development of this phase, we modified and adapted algorithms proposed by Peña Abreu [29].

This phase's inputs are a list of experts and a list of candidate criteria for the evaluation of perceived quality. While its output is a list of evaluation criteria to be applied in the context in question, conceptualized and weighted.

In this phase the following activities must be executed:

Activity 1.1. Select experts who will determine the criteria to be used in the evaluation considering the characteristics of the product or service and the organization's objectives.
Activity 1.2. Define criteria for evaluating perceived quality.
Activity 1.3. Select criteria to be evaluated by experts.
Activity 1.4. Conceptualize criteria to be used in the evaluation process.

To implement these activities, 1.2 and 1.3 authors propose combining bibliographic analysis, the Delphi method [31] and the 2-tuple model. At this moment, the following tasks must be executed:

1. Identify a set of candidate criteria for analysis.
2. Apply round of the Delphi method, an open survey where experts can propose new criteria.
3. Evaluate the candidate criteria by the experts following Algorithm 1.
4. Eliminate criteria that have been evaluated with the no_consider label.
5. Obtain a list of basic criteria to be applied in the context in question.

**Algorithm 1: Evaluation of candidate criteria by experts using the 2-tuple model** *Inputs*:

*C—Candidate criteria for evaluating perceived quality* $c_q \in C : q \in [1..z]$
*V—Set of experts.* $v_l \in V : t \in [1..L]$

Step 1. *Define as a basic set of linguistic terms for the evaluation of the criteria*
Step 2. *(CBTL)* $S_T = \{NC, MPC, PC, MTC, C, MC, CS\}$.
Step 3. *For each criterion* $c_q \in C$ *make*
Step 4. *For each expert* $v_l \in V$ *make*
Step 5. *Issue participant preferences* $v_l$ *on the criterion* $c_q$ *using a seven-label linguistic scale The evaluation matrix is constructed, in which each element* $M_{ql}$ *represents the evaluation of the criterion q by the expert l.*
Step 6. *End of cycle step 3*
Step 7. *Calculate the q-th element* $C_q$ *evaluation criteria such that* $C_q$ *represents the aggregate expert assessment of the criterion* $r_q$*, once the 2-tuple method has been applied.*
Step 8. *End of cycle step 2*
Step 9. From the vector analysis $V_R$, sort the criteria according to their evaluation and select criteria $C_e$ that will ultimately be considered to assess perceived quality.
Step 10. Return the list of criteria $C_e$ *selected for the evaluation of perceived quality in this context.*

Activity 1.4: The objective of this activity is to achieve a complete conceptualization of each criterion, adding the domain and the scale of possible values. For this purpose, the Focus Group method is applied [32] in a single team with the following sequence of actions:

- Conduct first session to explain the objective of the work and present the basic criteria obtained in Activity 1.2.
- Present proposed domain and possible scale of values for each criterion.
- Conduct a final session where the focus group refines the proposal presented and makes recommendations for working with the criteria.
- Obtain a final list of base criteria to evaluate perceived quality in the context of conceptualized analyses.

Activity 1.5: In this activity the level of importance of the criteria is determined through the pairwise comparison matrix for elements of the same level of the Hierarchical Analysis Process method [33] To implement the activity, it is proposed to follow Algorithm 2 described below.

**Algorithm 2: Determine level of importance of criteria** *Inputs*:

C—list of criteria. $c_j \in C : j \in [1 \ldots m]$
E—Set of experts. $e_i \in E : i \in [1..n]$

Step 1. *For each expert* $e_i \in E$ *make*
Step 2. *For each criterion* $c_j \in C$ *make*

Step 3. *Define score for criterion $c_j$, comparing it with the rest, applying expert judgment.*
Step 4. *End of cycle step 2*
Step 5. *End of cycle step 1*
Step 6. *For each criterion $c_j \in C$ make*
Step 7. *Calculate the weight of criterion $c_j$, $w^{c_j}$ from calculating the geometric average of the scores given by the experts. It is satisfied that $\sum_{j=1}^{m} w^{c_j} = 1$*
*End of cycle step 7*
Step 8. *Return $W^c$—list of criteria weights. $w^{c_j} \in W^c : j \in [1..m]$*

In Step 4 an matrix is gradually constructed nxm, with the score of each criterion by the experts taking into account the scale of Saaty [34]. Then, this score matrix is normalized where each cell has values between 0 and 1.

## 2.2 Phase 2 "Collection of User Preferences"

In this phase, user preferences are collected to perform an evaluation of the perceived quality of IT services. Its phase inputs are a list of users and a list of criteria to be conceptualized and weighted, and its phase outputs are the user preferences. Finally, the following activities should be performed:

Activity 2.1. Collect user preferences.
Activity 2.2. In this activity, the preferences of the users are collected, and the users give their evaluation of each of the selected criteria, using the following basic set of linguistic terms (CBTL) $S_T = \{N; MB; B; M; A; MA; E\}$ whose semantics. The utility vector is used to express user preferences $X = \left(X_n^{pm}, \ldots, X_n^{pm}\right)$ where $X_n^{pm}$ represents the user's preference $u_m$ over the service $a_n$ according to the criterion $c_p$ [35].

## 2.3 Phase 3 "Evaluation of Perceived Quality"

In this phase, the perceived quality is evaluated using the 2-tuple linguistic modeling method. This method avoids the loss of information and facilitates communication with users using natural language. The input data for this phase are the user preferences, and the outputs are the aggregated preferences. To complete this phase, the following activities should be performed:

Activity 3.1. Transform user preferences into 2-language tuples.
Activity 3.2. The 2-tuple based linguistic representation model presented in Herrera [25], is based on the concept of symbolic translation and reduces the loss of information that occurs in classical fuzzy linguistic modeling.

Since in the proposed solution all criteria are issued on the same set of linguistic labels, the transformation of user preferences is performed directly, assuming that the symbolic translation of the given value is equal to zero. As a result of this activity, all user preferences should be transformed into a 2-tuple linguistic term, thus leaving the information ready for the next phase.

## 2.4 Phase 4 "Interpretation of Results"

In this phase, the results obtained from the evaluation of the criteria are integrated independently using 2-tuples to transform the data set into a single representative element of each service. The results allow decision-makers to interpret them and make decisions accordingly. The inputs to the phase are the user preferences transformed to a 2-tuple, while the outputs are the collective evaluation value of each criterion for each evaluated service, the overall quality value perceived per service, and a list of services ordered according to their envelope—all perceived quality. To develop this phase, the following activities must be carried out:

Activity 4.1. Calculate the collective value of each criterion for each service. The collective value of a criterion for a service is calculated by applying the aggregation operator with which it is decided to work according to the case. It is obtained as a result of the aggregation of all the preferences issued by the users of the service with respect to the criterion in question

Activity 4.2. Aggregate the value of the criteria for each service. The evaluation of the overall perceived quality of the service is calculated from the aggregation of the collective values of the service criteria.

Activity 4.3. Interpret results and make decisions. The objective of this activity is to analyze the results of evaluating all the services as a whole and obtain an ordered list of them according to their overall perceived quality. For this purpose, the 2-tuple comparison operator must be applied. The comparison of linguistic information represented by means of 2-tuples is performed according to a lexicographical order.

With the results of the ordered criteria, pertinent analyses can be carried out on the elements that could be affecting the perceived quality of the services that have been the object of the evaluation.

The collective values of the criteria of a service will allow decision-makers to identify which criteria have the greatest impact (positively and negatively) on the perceived quality of services. The value of the overall perceived quality will make it possible to obtain a criterion for measuring the level of quality perceived by the user and to compare different services with each other or the same services in different evaluations.

## 3 Analysis Results

In order to demonstrate the applicability of the proposed method, a case study was carried out, considering the requirements proposed by Vermaas [36, 37] regarding the fact that it should not be artificial or specific and that it should show the usefulness of the proposal.

The case study was conducted at the E-Government Center (CEGEL), software developer with experience in the country's computerization process. CEGEL's mission is to satisfy the needs of government clients through the development of products, services and integral solutions of high reliability, quality, competitiveness and efficiency. Three IT services offered by this development center were selected for the implementation of the proposed method S = $\{S_1, S_2, S_3\}$.

$S_1$: Development Service of the Management System for the Attention to the Population for the National Assembly of the People's Power of the Republic of Cuba.

$S_2$: Price Indexes data collection service for the National Bureau of Statistics and Information.

$S_3$: Digital Documentary Heritage Management and Preservation Service (GEPADD) for the Digital Library of the Office of the Attorney General of the Republic of Cuba.

In phase 1 of applying the method, the following dimensions of criteria were proposed to experts: technical, interactive, corporate, process and result. Associated with these dimensions, the following criteria were established:

Technical dimension refers to the technical elements of the organization offering the IT service that are perceived by the user.

- T1 Facilities: This refers to the condition of the facilities, whether they are visually attractive, well-maintained, and suitable for providing the service.
- T2 Equipment: refers to the condition of the equipment on which the service is performed, or the product is used (equipment with a modern appearance in good condition).
- T3 Materials and Media: The necessary materials and means (computers, internet, user manuals, etc.) are available for the development of the service or use of the product.
- T4 Reliability: It is the ability to perform the promised service or product with formality and accuracy, correctly, reliably, carefully and smoothly.
- T5 Employees: Sufficient personnel are available to develop the product or service.
- T6 Price of the product or service to the user.

The interactive dimension groups the subjective characteristics of the interaction between the organization and the users during the execution of the IT service.

- I1 Trust: It refers to the trust that can be transmitted by the members of the organization.
- I2 Empathy: the provision of friendly and individualized attention to users.

- I3 Professionalism: The organization's knowledge, skills and experience to develop the product or service.
- I4 Courtesy: The attention, friendliness, respect and kindness of the organization's personnel in contact with users.

The corporate dimension refers to the aspects associated with the image of the organization offering the IT service.

- C1 Comprehension refers to the company's effort to establish good communication with its users and to know their needs and desires through personalized attention.
- C2 Credibility: The organization's ability to create an environment of trust with users through truthfulness, belief and honesty in the development of the product or service.
- C3 Feedback: The ability of the organization to respond to user's questions, concerns, complaints, or claims in a satisfactory manner.

The process dimension refers to the importance given by users to the IT service during its execution.

- P1 Need It refers to the sensation of lack of something and the desire to satisfy it on the part of the users, and influences the perception and the sense of increasing or decreasing the stimuli according to whether or not it helps to solve this lacking state.
- P2 Motivation: It refers to the degree of motivation of the user and can be appreciated.
- P3 Previous experiences: The user has previous experience with the organization and the product or service. This makes it possible to determine the influence on perceptions, since it can be justified according to the experience accumulated by the users.

The result dimension describes the user's perception of the quality characteristics of the IT service result.

- R1 Security refers to the non-existence of hazards, risks or doubts related to the provision of a service or the use of a given product in a specific context of use.
- R2 efficiency refers to the ability to enable users to achieve specific goals with accuracy and completeness in a specific context of use.
- R3 functionality refers to the ability of a product or service to provide functions that meet stated needs under specified conditions.
- R4 suitability refers to the ability of a product or service to maintain an appropriate set of functions to meet specified user tasks and objectives.
- R5 satisfaction refers to the ability of a product or service to satisfy users in a specific context of use.
- R6 Ease of Learning: The ability of a product to allow the user to learn about its use
- R7 Attractiveness: the ability of a product or service to be attractive to the user

- R8 Comprehensibility: the ability of a product to enable the user to understand when it is suitable and how it can be used for a particular task under certain conditions of use.
- R9 Accuracy of Information: It focuses on the intrinsic nature of information quality. Information accuracy is the degree to which information is correct, unambiguous, reliable, objective, and verifiable.
- R10 Adequacy of Information focuses on the contextual nature of information quality. Emphasizes the importance of transmitting appropriate information according to user-oriented tasks and goals.
- R11 Accessibility: In the case of a product, it highlights the importance of technical aspects in terms of making content more accessible to different users.
- R12 Contents The contents present in the final result are of interest to the user.
- R13 technology refers to the perceived degree of technological updating (hardware and software).
- R14 Adequacy of expectations refers to the adaptation of a product or service to the cognitive and linguistic level of the user.

Considering the characteristics of these services, the Delphi method was applied as described in Activity 1.1 of Phase 1 to evaluate the group of candidate experts. Initially, 18 peers were evaluated, and their scientific degree, years of experience in project management, and level of knowledge associated with perceived quality were measured.

For each candidate expert, the competence coefficient was calculated from the results of the survey. Finally, 10 experts with the following characteristics were selected for greater reliability in their considerations:

- 80% have a high proficiency rating, and the rest have a medium rating.
- 100% have worked for 6 years or more in project management.
- 60% are part of the faculty of the Master's Degree in IT Project Management at UCI, accredited for excellence in 2015, and include specialists from different areas (software production and quality, commercial department, technology transfer, and support).

From the pool of proposed criteria, the experts selected 12 criteria, 11 with the consider_always (CS) label, and another criterion given the characteristics of the services to be evaluated. For the criteria, a weight vector was defined according to the relevance given to them by the experts:

- T4 Reliability: 0.088
- I1Trust: 0.088
- I2 Empathy: 0.088
- C1 understanding: 0.088
- C3 Feedback: 0.088
- R3 Functionality: 0.088
- R4 suitability: 0.088
- R5 Satisfaction: 0.088
- R6 Ease of Learning: 0.088

- R9 Accuracy of Information: 0.032
- R10 Adequacy of Information: 0.088
- R14 Adequacy to Expectations: 0.088.

To collect user preferences, a survey was created and applied to a total of 24 users (10 users of Service 1, 8 users of Service 2, and 6 users of Service 3), in which they were asked to give their opinion on the computer service received, taking into account the selected criteria. The user preferences collected were unified and transformed into 2-language pairs for the evaluation of perceived quality.

Once the users' preferences had been unified, we proceeded to carry out the word computation processes. For this purpose, the following tools were used: FLINTSTONES, a name derived from Fuzzy Linguistic Decision Tools eNhacemEnt Suite, in its version 2.0.

The main objective of FLINTSTONES is to support the solution of decision-making problems based on the 2-tuples linguistic model and its extensions, facilitating the execution of the associated processes and the understanding of the results. The collective values achieved are shown in the column "collective values of the criteria" in Table 1. Finally, using the "weighted average for 2-tuples" operator, the overall perceived quality assessment of each service was calculated.

The data in Table 1 offers a wide range of options for analyzing the results and making decisions. Initially, the overall results of the evaluation of the perceived quality of individual IT services can be interpreted, and then, if necessary, the analysis can be refined by focusing on some of the criteria $\{S_1 > S_3 > S_2\}\{(E, -0.33) > (E, -0.45) > (MA, -0.16)\}$.

Criteria that show the best overall results are reliability, comprehension, adequacy of information, and adequacy to expectations. However, progress was achieved, but more discreetly in trust and empathy. The rest of the criteria evaluated had the same behavior in the same service.

**Table 1** Collective values of criteria and perceived quality

| Service $S_1$(E, −0.33) | Service $S_2$(MA, −0.16) | Service $S_3$(E, −0.45) |
|---|---|---|
| T4 (MA, 0.49) | T4 (A, 0.04) | T4 (MA, 0.21) |
| I1 (E, −0.33) | I1 (M, 0.1) | I1 (E, −0.45) |
| I2 (E, −0.33) | I2 (A, 0.04) | I2 (MA, 0.43) |
| C1 (E, −0.33) | C1 (MA, −0.16) | C1 (E, −0.45) |
| C3 (E, −0.33) | C3 (A, −0.06) | C3 (MA, 0.21) |
| R3 (E, −0.39) | R3 (A, 0.11) | R3 (MA, 0.23) |
| R4 (E, −0.39) | R4 (A, 0.41) | R4(MA, 0.23) |
| R5 (E, −0.39) | R5 (A, −0.4) | R5 (MA, 0.12) |
| R6 (E, −0.39) | R6 (A, −0.3) | R6 (MA, 0.43) |
| R9 (MA, 0.49) | R9 (A, 0.47) | R9 (MA, 0.23) |
| R10 (MA, 0.49) | R10 (MA, −0.4) | R10 (MA, 0.34) |
| R14 (E, −0.39) | R14 (A, −0.4) | R14 (MA, 0.43) |

The results obtained are in line with the perceived quality that can be appreciated when exchanged with the users participating in the evaluation. In the case of ONEI, it was recommended that the center's management work to improve the indicators of the interactive dimension.

In the particular case of the Area of Attention to the Population of the National Assembly of People's Power, the application of the method at the beginning of the development service made it possible to work on strengthening the elements identified as the weakest.

Thanks to these actions, the success of the service was guaranteed, not only in the closing of the project but also in the satisfaction of the end users. As a result, the ANPP has now become one of the center's potential clients, and because of its recommendations, we have already received several requests from government clients interested in our services. This has helped expand the market niche and increase profitability.

## *3.1 Qualitative Comparison with Other Models of Perceived Quality Assessment of Services*

In order to qualitatively compare the proposed method with the following five models for assessing the perceived quality of services: Grönroos, Model of Assessed Performance, Three-Component Model, Model IS SERVQUAL and Model IS SERVPERF.

In this sense, the six indicators that cover essential aspects of the evaluation process with multiple experts and multiple criteria under conditions of uncertainty were considered:

As shown in Table 2, it is possible to appreciate the superiority and quality of the proposed method, taking into account the management of uncertainty and foreseeing the loss of information. The use of computation with words provides the proposed solution with greater potential for this. The facilities for evaluating the perceived quality of computer services through linguistic variables provide greater flexibility for users to express their preferences. It also proposes specific criteria for IT services and makes an assessment of the overall perceived quality of each IT service evaluated, increasing the level of interpretability of the information for decision-making.

The combination of linguistic modeling with the treatment of information uncertainty and the ability to avoid loss of information during aggregation make the proposed method a more comprehensive option than the others.

# 4 Conclusions

At the end of the investigation, the following conclusions were reached:

**Table 2** Comparison of the method with existing models

| Comparisons criteria | Models |
|---|---|
| Tangible elements | Grönroos, three-component, SERVQUAL, SERV-PERF, and proposed model |
| Functional aspects | SERVQUAL, SERV-PERF, and proposed model |
| User evaluation | Grönroos, assessed performance, three-component, SERVQUAL, SERV-PERF and proposed model |
| Information uncertainty management | Only the proposed model |
| Information loss | Only the proposed model |
| Criteria for IT services | Only the proposed model |

- The evaluation of the perceived quality of IT services can be treated as a decision-making problem with heterogeneous evaluation criteria, evaluated by users with different levels of knowledge.
- The models proposed in the literature do not consider the uncertainty present in this process, an element that affects the evaluation results. In addition, there is no clear definition of the criteria to be used for IT services, so the proposed solution is novel in this regard.
- Among the most important elements included in the method are the definition of inputs and outputs and the flow of activities to be carried out in each of the defined phases. Its main contribution lies in the use of computing with words methods for information processing and managing uncertainty.
- The level of satisfaction of real users and the positive results of its implementation in the E-Government Center demonstrate its level of applicability and corroborate that its use contributes to improving decision-making in uncertain environments.
- The implementation costs of the proposal are very low, and its use can contribute to increasing the perceived quality of the evaluated IT services and building user loyalty with all the benefits that this provides.

## References


1. Henao Nieto DE, Giraldo Villa A, Yepes Delgado CE (2018) Instrumentos para evaluar la calidad percibida por los usuarios en los servicios de salud. Rev Gerenc y políticas salud 17(34):69–80
2. Izquierdo GRA, González YD, Martínez CCM (2020) Calidad percibida por los clientes de los hoteles de la provincia Los Ríos, Ecuador. Rev ciencias Soc 26(4):380–393
3. Pino Nicó Y, Castro Pimienta OD, Alvarez Alvarez A (2020) Evaluar la calidad de la comunicación visual para una mayor usabilidad del software de gestión empresarial. Rev Cuba. Ciencias Informáticas 14(3):1–17
4. Elzeki OM, Elfattah MA, Salem H, Hassanien AE, Shams M (2021) A novel perceptual two layer image fusion using deep learning for imbalanced COVID-19 dataset. PeerJ Comput Sci 7:1–35. https://doi.org/10.7717/PEERJ-CS.364/SUPP-2

5. Novita D, Husna N (2020) Peran ecolabel awareness dan green perceived quality pada purchase intention. J Manaj Maranatha 20(1):85–90
6. Özkan P, Süer S, Keser İK, Kocakoç İD (2020) The effect of service quality and customer satisfaction on customer loyalty: the mediation of perceived value of services, corporate image, and corporate reputation. Int J Bank Mark 38(2):384–405
7. Konuk FA (2019) The influence of perceived food quality, price fairness, perceived value and satisfaction on customers' revisit and word-of-mouth intentions towards organic food restaurants. J Retail Consum Serv 50:103–110
8. Hong S-J, Choi D, Chae J (2020) Exploring different airport users' service quality satisfaction between service providers and air travelers. J Retail Consum Serv 52:101917
9. Magasi C, Mashenene RG, Ndengenesa DM (2022) Service quality and students' satisfaction in tanzania's higher education: a re-examination of SERVQUAL model. Int Rev Manag Mark 12(3):18
10. Ros Gálvez A (2016) Calidad percibida y satisfacción del usuario en los servicios prestados a personas con discapacidad intelectual
11. Paul A, Shukla N, Paul SK, Trianni A (2021) Sustainable supply chain management and multi-criteria decision-making methods: a systematic review. Sustainability 13(13):7104
12. Cables E, García-Cascales MS, Lamata MT (2012a) The LTOPSIS: an alternative to TOPSIS decision-making approach for linguistic variables. Expert Syst Appl 39(2):2119–2126
13. Kabir G, Sadiq R, Tesfamariam S (2014) A review of multi-criteria decision-making methods for infrastructure management. Struct Infrastruct Eng 10(9):1176–1210
14. Begoli E, Bhattacharya T, Kusnezov D (2019) The need for uncertainty quantification in machine-assisted medical decision making. Nat Mach Intell 1(1):20–23
15. Sadeghi B, Cohen DR (2023) Decision-making within geochemical exploration data based on spatial uncertainty—A new insight and a futuristic review. Ore Geol Rev 161:105660
16. Bello R, Verdegay JL (2010) Los conjuntos aproximados en el contexto de la soft computing. Rev Cuba. Ciencias Informáticas 4(1–2):5–24
17. Srivastava M, Rai AK (2013) Investigating the mediating effect of customer satisfaction in the service quality-customer loyalty relationship. J Consum Satisf Dissatisfaction Complain Behav 26:95–109
18. Fong SJ, Li G, Dey N, Crespo RG, Herrera-Viedma E (2020) Composite Monte Carlo decision making under high uncertainty of novel coronavirus epidemic using hybridized deep learning and fuzzy rule induction. Appl Soft Comput 93:106282
19. Falcone R, Lima C, Martinelli E (2020) Soft computing techniques in structural and earthquake engineering: a literature review. Eng Struct 207:110269
20. Chen Z, Ming X, Zhou T, Chang Y (2020) Sustainable supplier selection for smart supply chain considering internal and external uncertainty: an integrated rough-fuzzy approach. Appl Soft Comput 87:106004
21. Das NN, Kumar N, Kaur M, Kumar V, Singh D (2022) Automated deep transfer learning-based approach for detection of COVID-19 infection in chest X-rays. IRBM 43(2):114–119. https://doi.org/10.1016/J.IRBM.2020.07.001
22. Kaufmann A, Gil Aluja J (1992) Técnicas de gestión de empresa: previsiones, Decisiones y Estrategias, no 658, KAUt
23. Kali A, Shamoi P, Zhangbyrbayev Y, Zhandaulet A (2022) Computing with words for industrial applications. In: Proceedings of SAI intelligent systems conference. Springer, pp 257–271
24. Tamir D, Neumann S, Rishe N, Kandel A, Zadeh L (2019) Computing with words—A framework for human-computer interaction. In: Augmented cognition: 13th international conference, AC 2019, held as part of the 21st HCI international conference, HCII 2019, Orlando, FL, USA, Proceedings 21. Springer, pp 356–372
25. Herrera F, Martínez L (2000) A 2-tuple fuzzy linguistic representation model for computing with words. IEEE Trans Fuzzy Syst 8(6):746–752. https://doi.org/10.1109/91.890332
26. Liu D, Chen X, Peng D (2018) Distance measures for hesitant fuzzy linguistic sets and their applications in multiple criteria decision making. Int J Fuzzy Syst 20:2111–2121

27. Verdecia Vicet P, Piñero Pérez PY, Pérez Pupo I, García Vacacela R, Villavicencio Bermúdez N (2022) Combining artificial intelligence and project management techniques in ecosystem for training and innovation. In: Piñero Pérez PY, Bello Pérez RE, Kacprzyk J (eds) Artificial intelligence in project management and making decisions. Springer International Publishing, Cham, pp 259–275
28. Wen TC, Chang KH, Lai HH (2020) Integrating the 2-tuple linguistic representation and soft set to solve supplier selection problems with incomplete information. Eng Appl Artif Intell 87:103248. https://doi.org/10.1016/J.ENGAPPAI.2019.103248
29. Peña Abreu M, Rodríguez Rodríguez CR, Piñero Pérez PY, García García Y (2022) New methods for feasibility analysis of investment projects in uncertain environments. In: Studies in computational intelligence. Springer Science and Business Media Deutschland GmbH, pp 143–154. https://doi.org/10.1007/978-3-030-97269-1_8/COVER
30. Peña Abreu M, Rodríguez Rodríguez CR, Piñero Pérez PY (2016) Computación con palabras para el análisis de factibilidad de proyectos de software. Rev Tecnura 20(50):69–84
31. Nasa P, Jain R, Juneja D (2021) Delphi methodology in healthcare research: how to decide its appropriateness. World J Methodol 11(4):116
32. Sim J, Waterfield J (2019) Focus group methodology: some ethical challenges. Qual Quant 53(6):3003–3022
33. Lázaro-Palacio L, Aranda-Camacho Y, Lázaro-Palacio L, Aranda-Camacho Y (2019) An analytical model to evaluate the performance of associative seed producer organizations in the framework of Plan Semilla-Agrosavia (Colombia). Agron Colomb 37(2):173–188. https://doi.org/10.15446/AGRON.COLOMB.V37N2.76948
34. Saaty TL (2008) Decision making with the analytic hierarchy process. Int J Serv Sci 1(1):83–98
35. Al-Subhi SHS, Mahdi GSS, Alava MV, Pérez MPYP, Vázquez ML (2017) Operador media potencia pesada lingüística y su aplicación en la toma de decisiones. Int J Innov Appl Stud 22(1):38–43
36. Vermaas PE (2014) Design theories, models and their testing: on the scientific status of design research. In: An anthology of theories and models of design: philosophy, approaches and empirical explorations. Springer, pp 47–66
37. Calle WAC, Betancourt ASG, Enríquez NJ (2019) Validation of the proof reversal on the inexistence of untimely dismissal by using neutrosophic IADOV technique. Neutrosophic Sets Syst 33(1):33–36

# GPSR Compliance

*The European Union's (EU) General Product Safety Regulation (GPSR) is a set of rules that requires consumer products to be safe and our obligations to ensure this.*

*If you have any concerns about our products, you can contact us on ProductSafety@springernature.com*

In case Publisher is established outside the EU, the EU authorized representative is:

Springer Nature Customer Service Center GmbH
Europaplatz 3
69115 Heidelberg, Germany